Groundwater Models for Resources Analysis and Management

Aly I. El-Kadi

LEWIS PUBLISHERS
Boca Raton   London   Tokyo

Library of Congress Cataloging-in-Publication Data

Groundwater models for resources analysis and management / Aly I. El-Kadi, editor.
   p. cm.
   Includes bibliographical references and index.
   ISBN 1-56670-100-7
   1. Groundwater flow—Mathematical models. I. El-Kadi, Aly I.
GB1197.7.G765 1995
551.49'01'5118—dc20   94-31123
                                                                                         CIP

This book contains information obtained from authentic and highly regarded sources. Reprinted material is quoted with permission, and sources are indicated. A wide variety of references are listed. Reasonable efforts have been made to publish reliable data and information, but the author and the publisher cannot assume responsibility for the validity of all materials or for the consequences of their use.

Neither this book nor any part may be reproduced or transmitted in any form or by any means, electronic or mechanical, including photocopying, microfilming, and recording, or by any information storage or retrieval system, without prior permission in writing from the publisher.

All rights reserved. Authorization to photocopy items for internal or personal use, or the personal or internal use of specific clients, may be granted by CRC Press, Inc., provided that $.50 per page photocopied is paid directly to Copyright Clearance Center, 27 Congress Street, Salem, MA 01970 USA. The fee code for users of the Transactional Reporting Service is ISBN 0-56670-100-7/95/$0.00 + $.50. The fee is subject to change without notice. For organizations that have been granted a photocopy license by the CCC, a separate system of payment has been arranged.

CRC Press, Inc.'s consent does not extend to copying for general distribution, for promotion, for creating new works, or for resale. Specific permission must be obtained in writing from CRC Press for such copying.

Direct all inquiries to CRC Press, Inc., 2000 Corporate Blvd., N.W., Boca Raton, Florida 33431.

© 1995 by CRC Press, Inc.
Lewis Publishers is an imprint of CRC Press

No claim to original U.S. Government works
International Standard Book Number 0-56670-100-7
Library of Congress Card Number 94-31123
Printed in the United States of America 1 2 3 4 5 6 7 8 9 0
Printed on acid-free paper

To Faten, Shereen, Aladdin, and Enjy

Preface

Groundwater modeling has advanced greatly over the past two decades. Models are useful for understanding the physicochemical processes involved, for analyzing field results, and for making predictions for water resources management. No doubt models have greatly benefitted our efforts regarding the first two uses. There is a need, however, for model assessment to study the successes and failures of our efforts in making predictions. In general, our efforts in this regard are hindered by the lack of data and the constraints of both software and hardware capabilities. For example, efficient three-dimensional models that are suitable for large aquifers hardly exist. Therefore, a decision is always needed regarding overlooking variability in the third dimension.

This book includes invited and selected papers from those presented at the 1994 Pacific Northwest/Oceania Conference, Assessment of Models for Groundwater Resources Analysis and Management, which was held in Hawaii, March 21–23, 1994. This conference was aimed at bringing together water, land, and environmental managers and groundwater scientists and modelers to assess: (1) the efficiency of current generation of models in addressing environmental problems, (2) the success of models in advancing understanding of groundwater systems, and (3) the needs for the next generation of models.

To achieve its objectives, the conference program covered state of the art in both theory and application. Papers also addressed the role of models in decision making and the problems facing decision makers regarding potential conflict with nontechnical factors, e.g., social or political.

The conference was sponsored by the University of Hawaii Water Resources Center and the Department of Geology and Geophysics, the U.S. Environmental Protection Agency, the National Science Foundation, and the International Ground Water Modeling Center. Without such support, the success of the meeting, and the subsequent publication of this book, would have been impossible. We greatly appreciate the continued support of Dr. Roger S. Fujioka, Director of the Water Resources Research Center. The help of the publication and secretarial staff of WRRC has surely eased our effort, and we thank them. We also thank Evelyn Norris of the Department of Geology and Geophysics for her help in various stages of preparation of the conference.

The Conference Committee consisted of Aly I. El-Kadi (Chair), L. Stephen Lau, James E. T. Moncur, Philip Moravcik (all from WRRC), and William Souza from the U.S. Geological Survey, Honolulu.

Aly I. El-Kadi

DISCLAIMER

The conference on which this book is based was funded in part by the United States Environmental Protection Agency under cooperative agreement CR821935010 to the University of Hawaii Water Resources Research Center. The research reported here has not been subjected to the agency's peer and administrative review and therefore may not necessarily reflect the views of the agency, and no official endorsement should be inferred.

Partial support for the meeting was also provided by the National Science Foundation under grant number BES94–03359. Any opinions, findings, and conclusions or recommendations expressed in this publication are those of the authors and do not necessarily reflect the views of the National Science Foundation.

The Editor

Aly I. El-Kadi obtained his undergraduate and masters degree in civil engineering from Ain Shams University, Cairo, Egypt, and his doctoral degree in civil and environmental engineering from Cornell University. The author's area of expertise is in model development and application. His research activities are related to application and assessment of various types of groundwater models, helium gas as a tracer in groundwater aquifers, multiphase flow and transport of hydrocarbons, numerical modeling, flow and transport in field soils, geothermal modeling, agricultural contamination of soils and groundwater, and databases and geographic information systems. He is jointly appointed as an Associate Professor at the Department of Geology and Geophysics, and a Researcher with the Water Resources Research Center, University of Hawaii at Manoa, Honolulu. Before joining the University of Hawaii, he was employed at the International Ground Water Modeling Center (1983–1989) where he was involved in the technology transfer program of the Center, including workshop and conference organization.

Contributors

Mary P. Anderson
Department of Geology and
  Geophysics
University of Wisconsin–Madison
Madison, Wisconsin

Philip B. Bedient
Environmental Science and
  Engineering Department
Rice University
Houston, Texas

John D. Bredehoeft
U.S. Geological Survey
Menlo Park, California

T. J. Cheema
Department of Geology and
  Geological Engineering
South Dakota School of Mines and
  Technology
Rapid City, South Dakota

Joel P. Conte
Civil Engineering Department
Rice University
Houston, Texas

M. Yavuz Corapcioglu
Department of Civil Engineering
Texas A&M University
College Station, Texas

Aly I. El-Kadi
Department of Geology and
  Geophysics
Water Resources Research Center
University of Hawaii at Manoa
Honolulu, Hawaii

Stephen B. Gingerich
Department of Geology and
  Geophysics
Water Resources Research Center
University of Hawaii at Manoa
Honolulu, Hawaii

Steven M. Gorelick
Stanford University
Palo Alto, California

Mark T. Hagley
Barr Engineering
Minneapolis, Minnesota

Maged M. Hamed
Environmental Science and
  Engineering Department
Rice University
Houston, Texas

Wade E. Hathhorn
Department of Civil and
  Environmental Engineering
Washington State University
Pullman, Washington

Peter W. Huntoon
Department of Geology and
  Geophysics
University of Wyoming
Laramie, Wyoming

J. L. Hutson
Department of Soil, Crop and
  Atmospheric Sciences
Cornell University
Ithaca, New York

M. R. Islam
Department of Geology and
 Geological Engineering
South Dakota School of Mines and
 Technology
Rapid City, South Dakota

Kiran K. R. Kambham
Department of Civil Engineering
Texas A&M University
College Station, Texas

Leonard F. Konikow
U.S. Geological Survey
Reston, Virginia

L. Stephen Lau
University of Hawaii
Water Resources Research Center
Honolulu, Hawaii

Rajasekhar Lingam
Hydro Geologic Inc.
Herndon, Virginia

Joel W. Massmann
Department of Civil Engineering
University of Washington
Seattle, Washington

John F. Mink
Mink and Yuen, Inc.
Honolulu, Hawaii

Frank L. Peterson
Department of Geology and
 Geophysics
Water Resources Research Center
University of Hawaii at Manoa
Honolulu, Hawaii

Wayne A. Pettyjohn
Oklahoma State University
Stillwater, Oklahoma

Thomas A. Prickett
Consulting Water Resources Engineers
Urbana, Illinois

Eric G. Reichard
U.S. Geological Survey
San Diego, California

H. S. Rifai
Department of Environmental Science
 and Engineering
Rice University
Houston, Texas

Paul K. M. van der Heijde
International Ground Water Modeling
 Center
Colorado School of Mines
Golden, Colorado

R. J. Wagenet
Department of Soil, Crop and
 Atmospheric Sciences
Cornell University
Ithaca, New York

Marylynn V. Yates
Department of Soil and Environmental
 Sciences
University of California
Riverside, California

T.-C. Jim Yeh
Department of Hydrology and Water
 Resources
The University of Arizona
Tucson, Arizona

Contents

SECTION 1
GENERAL MODEL ASSESSMENT

1. A Comparison of Model and Parameter Uncertainties in Groundwater Flow and Solute Transport Predictions, *J. W. Massmann and M. T. Hagley* 3
2. A Statistical Discussion of Model Error in the Use of the Advection-Dispersion Equation, *W. E. Hathhorn* 25
3. Model Testing: A Functionality Analysis, Performance Evaluation, and Applicability Assessment Protocol, *P. K. M. van der Heijde* 39
4. The Value of Postaudits in Groundwater Model Applications, *L. F. Konikow* 59
5. Groundwater Modeling in the 21st Century, *M. P. Anderson* 81
6. Needs for the Next Generation of Models 97

SECTION 2
ON MODELS AS MANAGEMENT TOOLS

7. If It Works, Don't Fix It: Benefits From Regional Groundwater Management, *J. D. Bredehoeft, E. G. Reichard, S. M. Gorelick* 103
8. Uncertainty Analysis of Subsurface Transport of Reactive Solute Using Reliability Methods, *M. M. Hamed, J. P. Conte, and P. B. Bedient* 125
9. Groundwater Modeling and Litigation, *T. A. Prickett and W. A. Pettyjohn* 139

SECTION 3
ON UNSATURATED/MULTIPHASE FLOW AND TRANSPORT MODELING

10. On the Numerical Solutions of One-Dimensional Flow in the Unsaturated Zone, *A. I. El-Kadi* 151
11. Consequences of Scale-Dependency on Application of Chemical Leaching Models: A Review of Approaches, *R. J. Wagenet and J. L. Hutson* 169
12. Stochastic Modeling of Water Flow and Solute Transport in the Vadose Zone, *T.-C. J. Yeh* 185
13. Modeling Multiphase Contaminant Flow in Groundwater Aquifers, *M. Y. Corapcioglu, K. K. R. Kambham, and R. Lingam* 231

SECTION 4
ON ISLAND MODELING

14. Groundwater Modeling in Hawaii: A Historical Perspective,
    L. S. Lau and J. F. Mink 253
15. Modeling Atoll Groundwater Systems, F. L. Peterson and
    S. B. Gingerich 275

SECTION 5
ON BIODEGRADATION/VIRUS TRANSPORT MODELING

16. A Review of Biodegradation Models: Theory and Applications,
    H. S. Rifai and P. B. Bedient 295
17. Evaluation of the Groundwater Disinfection Rule "Natural Disinfection"
    Criteria Using Field Data, M. V. Yates 313

SECTION 6
ON FRACTURE FLOW MODELING

18. A New Modeling Approach for Predicting Flow in Fractured Formations,
    T. J. Chema and M. R. Islam 327
19. Is It Appropriate to Apply Porous Media Groundwater Circulation
    Models to Karstic Aquifers, P. W. Huntoon 339

Index 361

SECTION 1

General Model Assessment

CHAPTER 1

A Comparison of Model and Parameter Uncertainties in Groundwater Flow and Solute Transport Predictions

Joel W. Massmann and Mark T. Hagley

INTRODUCTION

The purpose of this paper is to evaluate the relationship between the complexity of groundwater models and the overall uncertainty in model predictions. This study is based on the hypothesis that uncertainty and variability in model input parameters are often large enough that the efforts involved in applying sophisticated computer models may not be warranted.

Mathematical models are simply representations of reality. In these models, cause and effect relationships are translated into mathematical terms. Six general sets of information are required to develop and to implement mathematical models. These sets are: (1) equations describing the physical processes, (2) region of flow, (3) boundary conditions, (4) initial conditions, (5) material properties, and (6) a method of solution. The region of flow may involve one, two, or three dimensions. The flow and transport equations are usually continuous partial differential equations (e.g., Domenico and Schwartz, 1990). Boundary conditions are used to define the value of dependent variables on the boundaries of the region to be modeled. Initial conditions describe the state of the system at the start of the simulation. Material properties such as hydraulic conductivities and dispersivities must be defined over the region of interest (Franke and Reilly, 1987). Finally, the mathematical equations describing the system must be solved. There are two general methods for solving these equations: analytical methods and numerical methods.

Analytical methods generally involve the direct solution of the continuous partial differential equations. These analytical models have advantages over other methods. They are relatively simple to use, often requiring only pencil, paper, and a calculator. They do have one major drawback, however. These models generally require the assumption of homogeneous and isotropic aquifer properties as well as very simple aquifer geometry. These situations rarely exist in nature.

The need for flexible models which could simulate complex groundwater systems brought about the development of numerical modeling methods. Numerical models are today the most widely used groundwater models. In fact, in some situations, such as licensing a high-level radioactive waste repository, their used is required (Domenico and Schwartz, 1990). They are most appropriate for aquifers having irregular boundaries, heterogeneous materials, or highly variable stresses in space and time. Most field situations are in this category.

Because complex field geometries and heterogeneities can be incorporated into numerical models, the quality of the results of numerical methods does not change appreciably with changing system complexity. Analytical solutions generally require the assumption of simple geometries and homogeneous materials, however. When dealing with a relatively simple field situation which meets many of the assumptions necessary for using analytical modeling methods, the difference between the analytical and numerical methods will be small. As the complexity of the system increases, the assumptions necessary for analytical solutions will no longer be met. The quality of the analytical results will therefore decrease, causing a greater difference between analytical and numerical solutions.

If numerical methods required no greater effort than analytical methods, had no greater potential for misuse, and there were no uncertainties associated with parameters such as hydraulic conductivity and system geometry, then the improved accuracy of numerical methods would justify their use in all but the simplest of situations. However, there is a great amount of parameter uncertainty. In simple field situations that approximately meet the assumptions necessary for analytical methods, these parameter uncertainties almost certainly outweigh the model uncertainties, which can be defined as the difference between the results of the different modeling methods. As the complexity of the field situations increases, a point must be reached where the results of the analytical solutions are sufficiently inaccurate that the model uncertainty becomes greater than the parameter uncertainty. At this point the use of the simpler analytical methods is no longer justified.

The purpose of this paper is to define this point for selected flow and transport systems. Several hypothetical systems will be simulated using models of varying complexity in an effort to find how complex of a system can be adequately simulated with each of the model types. The effects of parameter uncertainty on model predictions will be taken into consideration during this comparison. The errors due to the misuse of models were not considered in this study. These errors may exist in field studies, however, and should not be overlooked.

UNCERTAINTIES IN GROUNDWATER FLOW PREDICTIONS

The first system used to evaluate the effects of model complexity and parameter uncertainty was unconfined groundwater flow with discharge to a river. The flow field consists of unconsolidated sand of glacial outwash origin, overlying flat lying, low-permeability bedrock. This field was modeled in cross-section, parallel to the groundwater flow direction. The output variables from the models are the total discharge and the travel time to the river. Figure 1a shows the dimensions of the cross section. This general hypothetical field situation was considered to exist in five levels of complexity, from a very simple situation to a relatively complex situation. The degree of complexity of each system depended on variables such as the number of aquifer materials, whether these materials were homogeneous or isotropic, and whether any infiltration was present. For example, the most complex situation that was modeled

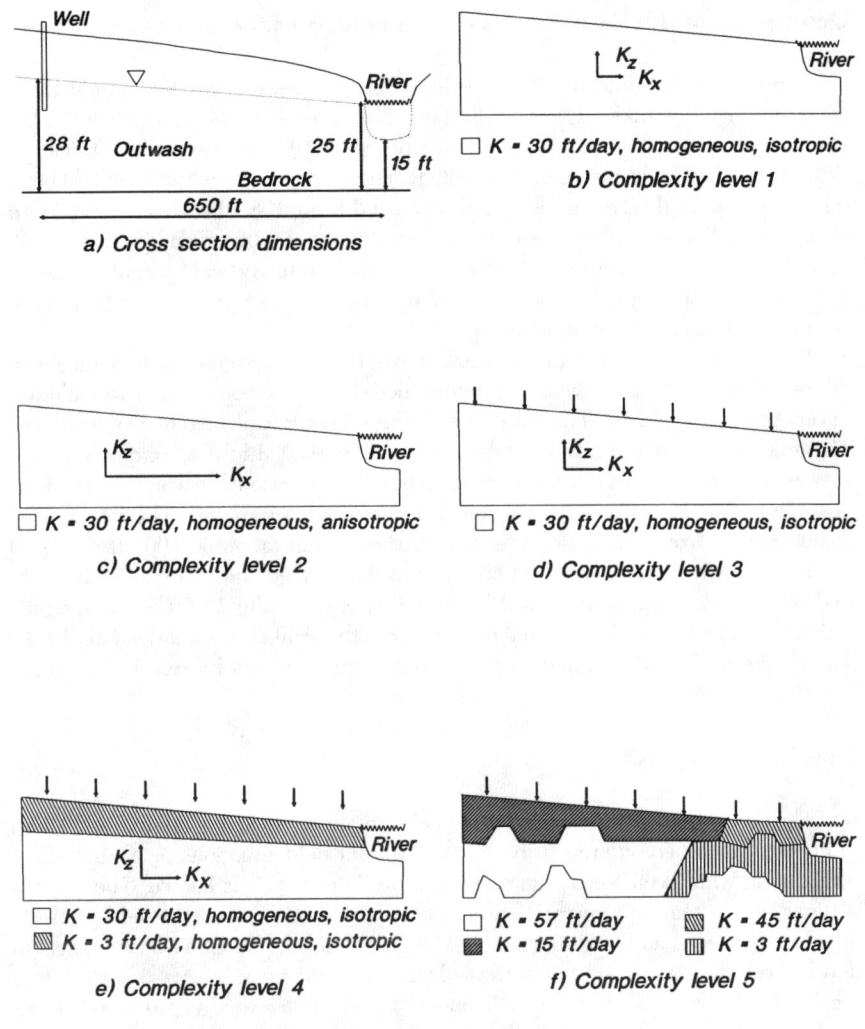

Figure 1. Cross-sections showing simulated field situations.

included infiltration and four different aquifer materials, each of which was heterogeneous and anisotropic. Each of these levels of system complexity was modeled using four distinct modeling methods, from Darcy's law to a relatively sophisticated finite element numerical model.

Models were also run with variations in input parameters to determine what effect these have on the model solutions. The most sophisticated numerical model was assumed to give the most accurate results. For each level of complexity, the results of this model, using the best estimate for each of the parameters, were considered to be "reality". The results of subsequent runs using the other models and varying the input parameters were then compared to this reality. In this manner, error resulting from the type of model could be compared in a relative sense to the error due to parameter uncertainty.

Description of the Base Case and Parameter Variations for Flow

A base case was first developed using typical values of hydraulic conductivity, groundwater table gradient, and infiltration. The base hydraulic conductivity of the sandy outwash was 30 ft/day (Freeze and Cherry, 1979) and the saturated thickness above the bedrock was between 28 and 25 ft. The gradient of the water table (0.005), is fairly typical for this type of flow field (Ryan and Kipp, 1985; Freyberg, 1986; Ward et al., 1987). The rate of infiltration was arbitrarily set at 0.002 ft/day (8.76 in./year). All anisotropic materials in the study have a horizontal to vertical hydraulic conductivity ratio of 10:1. These parameter values were assumed to represent reality, and will be referred to as the base parameters.

Because there is uncertainty associated with each of these parameters, ranges of values for each parameter were also determined. Hydraulic conductivity was allowed to vary by a factor of 2.07. The reasoning for this choice is explained in a later section. The gradient of the groundwater table ranged from 0.0033 to 0.0066, or a variation in water level of one foot per 600 horizontal ft, which is reasonable when dealing with possible errors caused by surveying and water level measurements. The rate of infiltration ranged from zero infiltration to a maximum infiltration of 0.005 ft/day (21.9 in./year), which was considered to be equal to the average annual rainfall. Depth to bedrock varied by approximately 15%, which is close to the 10% error commonly estimated to be involved in seismic refraction depth calculations (Zohdy et al., 1974). These ranges of possible values for the input parameters will be referred to as parameter uncertainty.

Levels of Complexity

The cross section was modeled using field situations of varying complexity. These situations can be considered to be either different field situations or a single field situation in which complexity increases as more information is known. The five situations can be seen in Figure 1. The most simple situation, termed level 1, consists of a homogeneous and isotropic flow field with no infiltration. Level 2 is identical to level 1 with the exception of anisotropic hydraulic conductivity. Level 3 is also identical to level 1, except infiltration is included. Level 4 consists of a nonhomogeneous and isotropic flow system with infiltration. The aquifer consists of two materials with different hydraulic conductivities. Level 5, the most complex, is heterogeneous, with four different aquifer materials. Each of these materials is anisotropic. In addition, the bedrock underlying the aquifer is undulating and infiltration is included. These five variations in the flow system will be referred to as the level of system complexity.

Description of Discharge Models

Several models of varying complexity were used to simulate flow in each of the five flow systems described above. The simplest model used is Darcy's law:

$$Q = K A \, dh/dl \qquad (1)$$

where

Q = discharge to river (L^3/t)
K = hydraulic conductivity (L/t)
A = cross sectional area of the flow field (perpendicular to flow direction) (L^2)
dh/dl = groundwater gradient

Darcy's law assumes a homogeneous, isotropic, and confined aquifer of constant thickness. Although the situation modeled was not confined and the thickness varied, Darcy's law was applied as a first approximation. Solution time for Darcy's law averaged less than one minute using a hand calculator.

The second model that was applied is the Dupuit method. This method is somewhat more complex and better fits the situation to be modeled because it assumes unconfined flow and includes infiltration. Further assumptions necessary for the Dupuit method include a homogeneous and isotropic aquifer, hydraulic gradient equal to the slope of the water table, and horizontal flow (Fetter, 1988). The equation has the following form:

$$Q = K(h_1^2 - h_2^2)/2L - w(L/2 - x) \qquad (2)$$

where

Q = discharge per unit width (L^3/t)
K = hydraulic conductivity (L/t)
h_1 = upgradient hydraulic head (L)
h_2 = downgradient hydraulic head (L)
L = horizontal length of cross section (L)
w = infiltration rate (L/t)
x = horizontal location of point of interest (in all cases this was equal to L)

The solution time again averaged less than one minute using a hand calculator.

Two numerical models were used, and the same computer code was used for both cases. This code is a finite element model based on the general equation for transient flow through saturated porous media. Iterations and a deforming mesh are used to identify the location of the water table. The two numerical models differed only in the refinement of the finite element mesh used to discretize the flow field. The simpler numerical model, which will be referred to as the coarse model, consisted of a finite element mesh of 51 elements. The development and debugging of this mesh took 2.5 h and the average run time was less than one minute on a 496 PC. Changes to the input parameters for subsequent runs took from one to ten minutes. The more complex, or refined, model consisted of a mesh of 454 elements. The development and debugging of the dataset for this refined model set took 10 h, while the average computer run time was approximately one minute on a 486 PC. Changes to the input data took up to 30 min. It should be noted that the development time does not include the several days that were spent learning to use the model.

Methods for Evaluating Results

Each of the five levels of system complexity was modeled using the base parameters and the most refined finite element model. These solutions were considered to be "reality". Once this "real" solution was obtained for each level of complexity, the solutions of all subsequent modeling efforts were compared to it. In this way, it was possible to determine the error stemming either from the modeling method used or from parameter uncertainty.

The solutions for the different modeling methods were obtained by simply applying these models using the base parameters. The difference between these solutions and reality is termed the model uncertainty. To determine the uncertainty due to the various parameters, the refined finite element model was run using the maximum and minimum values for hydraulic conductivity, water table gradient, and infiltration. Aquifer thickness was run only as a minimum. The difference between these solutions and reality is termed the error due to parameter uncertainty.

The situations modeled do not meet all of the assumptions necessary for the application of the two analytical models. Some generalizations were made to use the models. For example, because Darcy's law assumes constant aquifer thickness, it does not provide for the sloping water table in the field problem. The aquifer thickness used was the average thickness over the horizontal range of the aquifer. Neither of the analytical methods are able to deal with heterogeneous or layered materials. This problem was handled in two ways. In the case of level 4, with one thin and one thick layer of different hydraulic conductivities, a weighted-average hydraulic conductivity was used for the analytical methods. In the case of level 5, with four material types of approximately equal areas, an arithmetic mean of the four hydraulic conductivities was used for the analytical methods. Darcy's law also does not account for infiltration. This was included by simply adding the total input due to infiltration to the computed discharge.

Results of Discharge Predictions

The results for predictions of discharge to the river are presented in Table 1. To allow comparison among the different levels of system complexity, all discharge values were converted to percentages relative to their respective realities using the following equation:

$$R = D_m / D_r \cdot 100 \tag{3}$$

where

R = percentage of reality
D_m = discharge predicted using model or parameter in question (L^3/t)
D_r = discharge for corresponding reality (L^3/t)

The results for each level of system complexity are graphically demonstrated in Figure 2. These graphs compare the error associated with each of the models with

Table 1. Results of Discharge Predictions

"Reality"	"Reality"	Complexity Level				
		1	2	3	4	5
		3.92[a]	3.83	4.49	3.60	3.18
Effects of Models	Coarse Golder	101%	100%	99%	98%	83%
	Dupuit	102%	104%	102%	102%	125%
	Darcy	102%	104%	115%	119%	144%
Effects of Parameters	Large K	207%	207%	193%	190%	191%
	Small K	48%	48%	54%	56%	55%
	Large Gradient	136%	136%	131%	128%	131%
	Small Gradient	65%	66%	70%	72%	70%
	Thinner Aquifer	85%	86%	87%	84%	89%
	Larger Infilt.	NA	NA	119%	124%	122%
	No Infilt.	NA	NA	87%	84%	86%

[a] $ft^3/day/ft$.

the parameter uncertainties. Each bar shows the range of error and uncertainty for each model type and parameter. The center line at 100% represents reality. It is important to note that the axis showing percentage of uncertainty is on a logarithmic scale.

The uncertainty associated with the type of model is negligible when compared to the uncertainty associated with the input parameters in the simpler groundwater systems, levels 1 and 2. The graph for complexity level 3 shows that there is some uncertainty associated with using Darcy's law, even though it is still small when compared to the hydraulic conductivity and gradient uncertainties. Only with the most complex system, level 5, is there discernible uncertainty associated with the coarse finite element and Dupuit methods, although they are still negligible. Even when dealing with this complex field situation, the uncertainty associated with using the two analytical methods is still overshadowed by the uncertainty associated with hydraulic conductivity.

The results in Table 1 indicate that the analytical models worked quite well for estimating discharge to the river, even for the most complex situation. Predictions for travel times to the river were also made. The travel time reflects how long it would take a water molecule to get from the upper left surface of the field area to the river. It was assumed that the more tortuous travel path of the water in the more complex flow fields might cause a greater variation in model results. Travel time analysis was done only for levels 1 and 5. The results of the different models were compared only to the results of the hydraulic conductivity variability, since these had already been shown to have the greatest uncertainty.

For the analytical models, groundwater velocity was estimated by simply dividing the discharge in each case by the cross-sectional area perpendicular to flow direction and multiplying by a porosity of 0.35. The distance to the river was then divided by this linear velocity to estimate the travel time to the river. Particle tracking routines were used to estimate travel times for the numerical models.

The results for the travel time analysis are given in Table 2 and are shown in Figure 3. As expected, level 1 showed almost no variation between the modeling methods. Level 5 showed variation between the modeling methods, although it was still less than that caused by the hydraulic conductivity uncertainty. These results demonstrate that for travel times, the error due to parameter uncertainty is still larger than the differences between modeling methods.

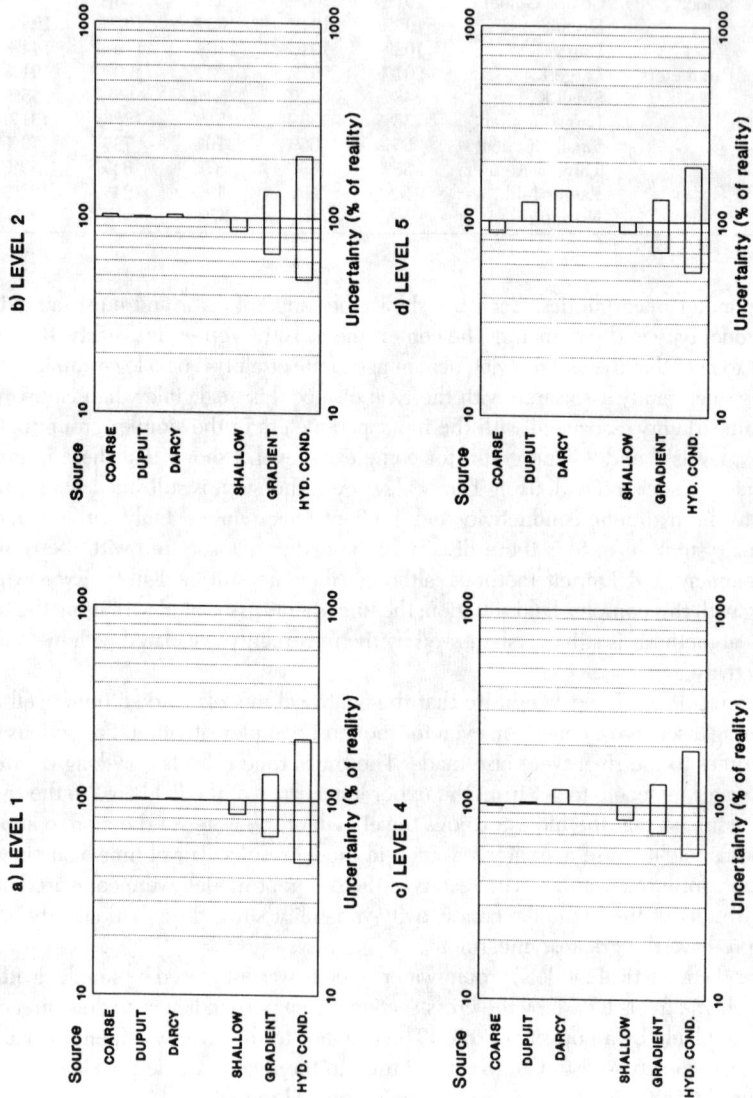

Figure 2. Comparison of model uncertainty and parameter for groundwater flow predictions.

Table 2. Results of Travel Time Predictions

"Reality"	"Reality"	Complexity Level	
		1	5
		3.78	4.63
Effects of Models	Coarse Golder	101%	138%
	Dupuit	101%	70%
	Darcy	101%	61%
Effects of Parameters	Large K	48%	49%
	Small K	207%	205%

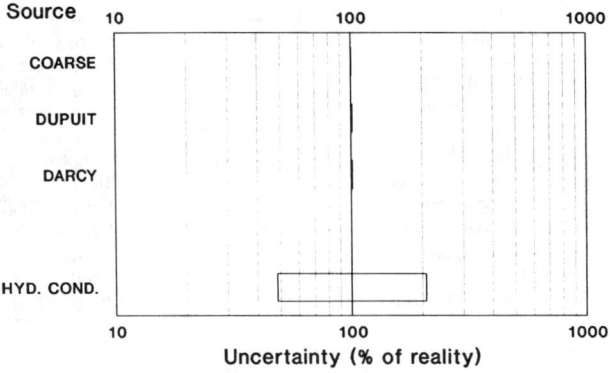

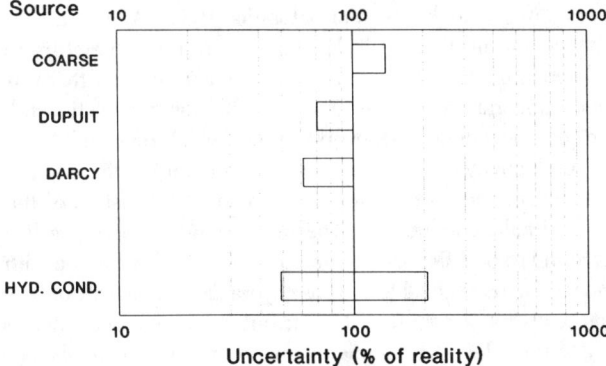

Figure 3. Comparison of model uncertainty and parameter for groundwater travel time predictions.

Table 3. Summary of Hydraulic Conductivity Data

Std. Dev. of ln K	Material	No.	Source
0.62	Sand	90	Russo and Bresler (1981)
0.85	Coarse sand	330	Willardson and Hurst (1965)
0.94	Fine sand	287	Willardson and Hurst (1965)
0.93	Coarse sand	352	Willardson and Hurst (1965)
1.24	Sandy loam	64	Gumaa (1978)
1.19	Sandy loam	5	Babalola (1978)
0.44	Sand and gravel	NA	McMillan (1966)
0.72	Sand and gravel	42	McMillan (1966)
0.54	Sand and gravel	16	McMillan (1966)
0.86	Loamy sand	121	Willardson and Hurst (1965)
0.80	Outwash sand	NA	Smith (1978)
1.90	Sand and gravel	NA	Hufschmied (1985)
0.60	Outwash sand	NA	Sudicky (1986)
1.80	Sediment	46	Cearlock et al. (1975)
0.39	Outwash sand	28	Walker et al. (1965)
0.50	Glacial till	38	Burris et al. (1981)
0.99	Glacial drift	60	Kempton et al. (1982)
1.08	Glacial drift	23	Kempton et al. (1982)
1.17	Glacial drift	71	Kempton et al. (1982)
1.05	Glacial drift	60	Kempton et al. (1982)
0.89	Glacial drift	58	Kempton et al. (1982)
0.42	Sand and gravel	19	Broom and Lyford (1981)
0.89	Alluvial deposits	58	Arteaga (1980)
1.35	Alluvial deposits	64	Arteaga (1980)
1.95	Sandy gravel	30	Metzger et al. (1973)
2.14	Sand	15	Woodman et al. (1978)
1.86	Sand and gravel	35	Prych (1983)

Discussion of the Hydraulic Conductivity Range Used

It was assumed in this study that the hydraulic conductivity of each material type was measured at one randomly selected point. Point measurements of hydraulic conductivity vary erratically in space due natural geologic variability and due to measurement errors (e.g., Hoeksema and Kitanidis, 1985). Although the measurement error may be significant, there is little information in the literature to quantify this error for field conditions. The range of hydraulic conductivity that was used to evaluate parameter uncertainty was based on the geologic variability within an aquifer.

A literature review identified several studies which reported the spatial variation of hydraulic conductivity in sandy materials. A summary of these findings is included in Table 3. The variability is reported as the standard deviation of the natural logarithm of the hydraulic conductivity. The mean standard deviation from all of these sources was found to be 1.04. A hydraulic conductivity distribution with this standard deviation and with a mean equal to the base hydraulic conductivity of 30 ft/day was used to represent the hydraulic conductivity distribution in the hypothetical field situations.

To evaluate the effects of parameter uncertainty, the hydraulic conductivity was varied by a factor of 2.0 about the mean value. These hydraulic conductivity values are located at plus or minus 0.7 standard deviations from the base hydraulic conductivity. The probability of the measured hydraulic conductivity being between these two values is approximately 50%. Considering the probable presence of additional uncertainty due to measurement error, the choice of this relatively narrow hydraulic conductivity range gives a somewhat conservative estimate of the impact of uncertainty in hydraulic conductivity.

Summary of Discharge Results

A groundwater flow system, existing in five levels of complexity, was modeled using both analytical and numerical models of different levels of sophistication. The variation caused by using different models was compared to the variation caused by uncertainties in input parameters. In all cases, the variation between modeling methods was less than that due to possible parameter uncertainty. Hydraulic conductivity was found to be the parameter with the largest associated output uncertainty. Finally, the accuracy of the less sophisticated models did diminish as the situation being modeled became more complex, although their associated error was still less than that due to parameter uncertainty.

UNCERTAINTIES IN GROUNDWATER TRANSPORT PREDICTIONS

The second situation that was used to evaluate the effects of model complexity and parameter uncertainty was contaminant transport. The cross section that was used was the same as for the flow problems described in the previous section. A source of contamination has been added, as shown in Figure 4a. This source extends a short distance into the saturated zone. The concentration at this source was set at one. The output variable is the relative concentration a point near the river 5 years after the source is emplaced. This point (shown in Figure 4a), is 550 ft from the source and one foot below the surface of the groundwater table. All other characteristics of the flow field are the same as described in the previous section.

The flow system is again thought to exist in five levels of complexity. These five levels are identical to those used for the groundwater flow portion of the study as shown in Figure 2. In effect, this study of contaminant transport is simply a continuation of the flow study, similar to the steps commonly taken by modelers studying a real-world problem. As was the case with the study of flow, each of the levels of system complexity was modeled using four distinct modeling methods. Two of the methods were analytical and two were numerical. Once again, the most sophisticated numerical model was assumed to give the most accurate results. For each level of system complexity, the results of this model using the base parameters were considered to be reality. The results of subsequent runs using the other models and varying the input parameters were then compared to this reality to see how closely they matched. In this manner, errors resulting from the type of transport model used could be compared in a relative sense to the error due to parameter uncertainty.

Description of the Base Case and Parameter Variations for Transport

There are five general mechanisms involved in solute transport in saturated groundwater flow systems (Domenico and Schwartz, 1990). These mechanisms are

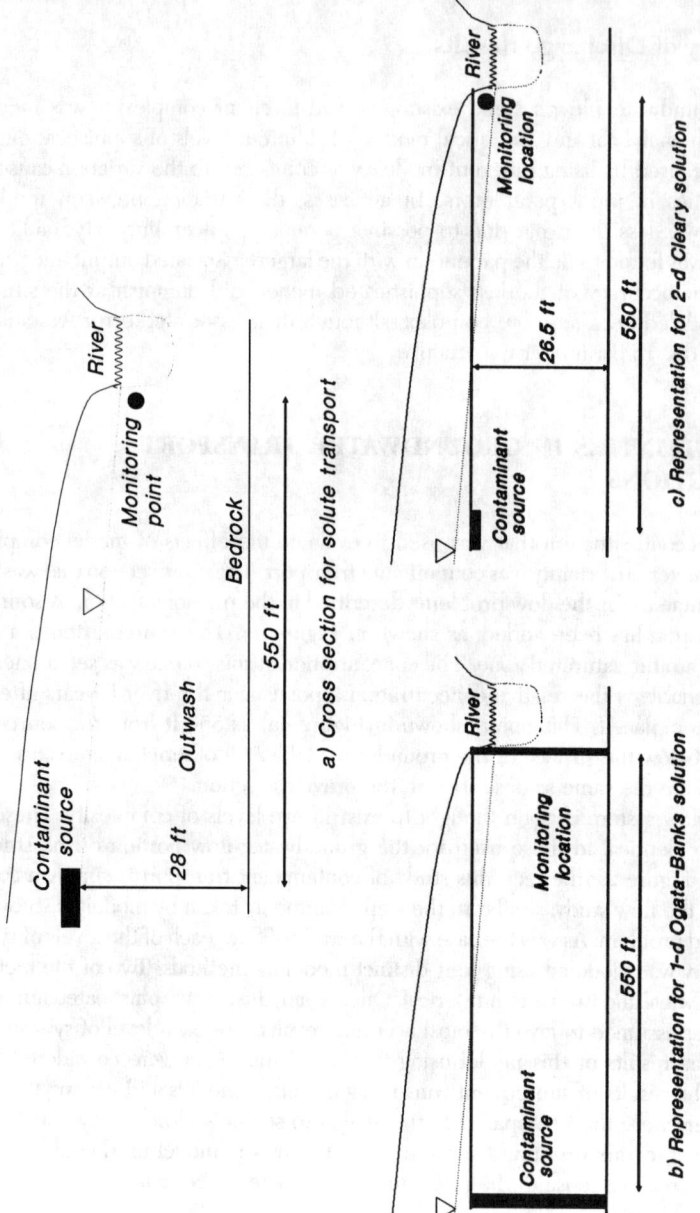

Figure 4. Cross-sections showing field situations simulated in solute transport analysis.

advection, diffusion, dispersion, retardation, and decay. Advection, diffusion, and dispersion are principally physical mechanisms while retardation and decay are principally chemical and biological mechanisms. The relative importance of each mechanism depends on geology, hydraulic gradients, groundwater chemistry, and the scale of the particular problem that is being assessed.

As described previously, the transport systems consist of a river flowing through an area of unconsolidated outwash overlying bedrock. Values of dispersivity, porosity, and diffusion were chosen to be representative of field situations. The base dispersivity was 20 ft longitudinally, with a longitudinal to transverse dispersivity ratio of 100:1. This ratio was held constant throughout the study. These dispersivities are representative of values used in field studies described in the literature (Anderson and Woessner, 1990; Gelhar, 1986). The porosity, 0.35, and the diffusion coefficient, 4.6×10^{-4} ft^2/day, are typical of sandy materials (Freeze and Cherry, 1979). The groundwater flow velocities were taken directly from the output of the numerical flow models from the previous section. In the case of the analytical models, these velocities were computed using the discharge that was calculated using the set of base parameters for each complexity level. The size of the containment source projecting into the aquifer was arbitrarily chosen as 10 ft wide and 1.875 ft deep. Retardation was ignored, since the aquifer material, sand, is assumed to be relatively free of organic materials which would cause retardation. The effects of decay were also neglected.

Because there is uncertainty associated with each of these parameters, ranges of values for each parameter were also determined. The diffusion coefficient had no measurable effect on the predictions due to the relatively high groundwater flow velocities assumed in the examples. For this reason, diffusion was ignored for the remainder of the study. The depth of the contaminant source ranged from 2 to 6 ft. Finally, because hydraulic conductivity variations caused the major uncertainty in the flow output, these variations were also included in the study of transport uncertainty. This was accomplished by using the velocity output for the maximum and minimum hydraulic conductivity for each level of system complexity as input into the solute transport model. Porosity varied from 0.32 to 0.38, which is within one standard deviation of the normal spatial variability found for sand in the field (EPRI, 1985).

Dispersivity can be used as a means of compensating at field scale for heterogeneities that are at a scale too small to be realistically measured. Dispersivities used in field-scale studies typically range from 1 to 1000 ft (Gelhar, 1986; Anderson, 1979). The dispersivity increases as the scale of the problem increases due to an increase in the number of heterogeneities encountered in the larger field problems (Gelhar et al., 1992). In this study, longitudinal dispersivity varied from 10 to 40 ft, which can be considered a conservative range. In an actual field situation, the value used may vary by much more than this, especially since dispersivity values used are commonly little more than educated guesses, following "conventional dispersion theories" (Ward et al., 1987).

Description of Transport Models

Four different models were used: two analytical models and two numerical models. The simplest model was the solution developed by Ogata and Banks to the one-dimensional, advection-dispersion equation (Domenico and Schwartz, 1990).

The analytical solution to this equation was developed assuming a homogeneous, isotropic, one-dimensional flow field with constant and uniform groundwater velocity. The situation modeled can be seen in Figure 4b. The solution time for this model averaged less than one minute using a hand calculator.

The second analytical modeling method used was a three-dimensional analytical model developed by Cleary (1978). This model also assumes a homogeneous, isotropic aquifer. Although it can incorporate three dimensions, for this application it was used to model only two. The situation modeled can be seen in Figure 4c. This model was also run on a PC using a FORTRAN program. Solution time for this method averaged about 10 sec. Initial set-up time was approximately 10 min, while changes to the data set again took about 1 min.

Two numerical models were used. Each model used the Galerkin finite element method to solve the transport equation for a conservative solute. The only difference between them was the refinement of the finite element mesh used. The same two meshes used in the flow modeling were also used to model transport. The total set-up time for each solute transport simulation was approximately 30 min for both the coarse mesh and the refined mesh. Changes to the input data took several minutes. Several days were spent becoming familiar with the solute transport of the numerical model.

One major difference between the analysis done for flow and that done for transport is that flow was considered to be steady state while the contaminant transport was transient. Solutions to transient problems introduce another variable into the problem: the number of time steps in the solution. For this analysis, the 5 years were broken down into 73 time steps, or one step every 25 days. This value was chosen by running the solute transport program four times with identical input parameters but with four different time steps. The concentrations at three different nodes in the flow field were compared for each of the four runs in an attempt to observe any changes in accuracy. The improvements to accuracy diminish rapidly below 25 days per step. For this reason, 25 days per time step was chosen for this study. Solution times for the coarse model were about 2 min while the refined model averaged approximately 10 min per run.

Methods for Evaluating Transport Results

The methods used to evaluate the models for transport were the same as those used to evaluate the flow models. The output of the refined numerical model using the base parameters was assumed to represent reality for each level of complexity. All subsequent results were compared to this reality to determine the error stemming either from the modeling method used or from parameter uncertainty. The solutions for the different modeling methods were obtained by applying these models using the base parameters. The difference between these solutions and reality is termed the model uncertainty. In order to determine the effects of parameter uncertainty, the refined finite element model was run using the maximum and minimum values for dispersivity, porosity, and hydraulic conductivity, and the maximum for source thickness. The differences between these solutions and reality is the error due to parameter uncertainty.

PARAMETER UNCERTAINTIES IN GROUNDWATER FLOW

Table 4. Results of Transport Predictions

"Reality"	"Reality"	Complexity Level				
		1	2	3	4	5
		0.1087[a]	0.1141	0.0755	0.0041	0.0145
Effects of Models	Coarse Golder	112%	126%	145%	356%	146%
	Ogata	825%	772%	1273%	20217%	5934%
	Cleary	93%	88%	142%	2324%	697%
Effects of Parameters	Large Dispers.	121%	117%	130%	214%	197%
	Small Dispers.	93%	97%	87%	53%	49%
	Large Porosity	94%	94%	95%	77%	79%
	Small Porosity	104%	104%	104%	127%	124%
	Thick Source	255%	254%	245%	295%	280%
	Large K	110%	109%	132%	266%	257%
	Small K	14%	14%	10%	2%	3%

[a] Relative concentration, C/C_o.

Results of Transport Predictions

The output variable used for this study of transport was the relative concentration at a point near the river after five years. The results are presented in Table 4. In order to compare the different levels of system complexity, all concentration values were converted to percentages relative to their respective realities using the following equation:

$$PR = Cmp / Cr \cdot 100 \qquad (4)$$

where

PR = percentage of reality
Cmp = concentration predicted using model or parameter in question (mass/L3)
Cr = concentration for corresponding reality (M/L3)

The results for each level of system complexity are shown graphically in Figure 5. These graphs compare the error associated with each of the models to the error associated with the parameter uncertainties. Each bar shows the range of error and uncertainty for each type of model and parameter. The center line at 100% represents reality. It should again be noted that the uncertainty is plotted on a logarithmic scale.

Discussion of Transport Results

The most obvious point regarding the results for transport is that the uncertainty associated with both the models and the parameters is greater than it was when simulating groundwater flow. This result is especially noticeable in the more simple field situations. The one-dimensional Ogata-Banks model does a poor job of simulating these field situations; its solution is up to 200 times reality. This is because the one-dimensional model cannot account for lateral dispersion processes. However, it should

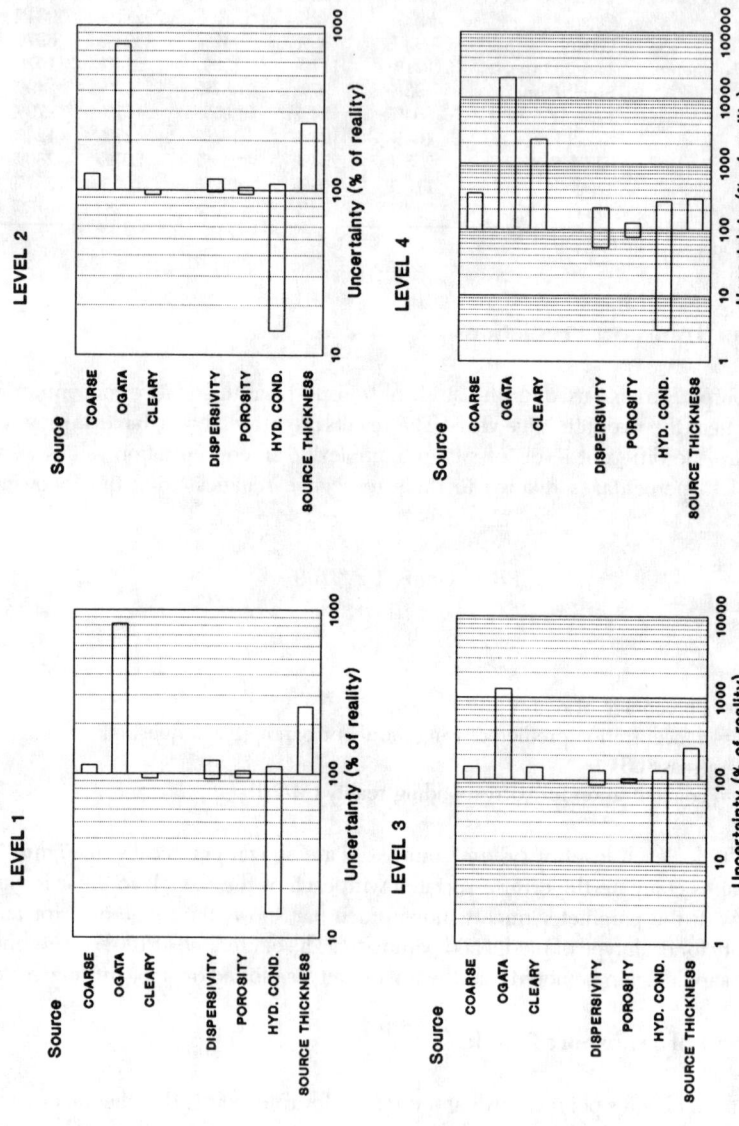

Figure 5. Comparison of model uncertainty and parameter for groundwater transport predictions.

be noted that the error of this model is approximately of the same order of magnitude as the error due to hydraulic conductivity uncertainty.

The analytical and coarse finite element models performed reasonably well when simulating the simpler solutions. However, the three-dimensional analytical model loses its accuracy when used to simulate complexity levels 4 and 5. The coarse finite element model does somewhat better, although its solution is consistently higher than reality. This situation can be partially explained by the fact that the coarse finite element grid was not discretized fine enough to model a source only 10 ft wide. Thus, its source was 50 ft wide, which contributed to the higher concentrations downstream. If the coarse mesh had been designed specifically to solve this problem, much of this inaccuracy could have been eliminated.

Once again, however, the uncertainty due to the models is overshadowed by uncertainty due to the parameters, especially hydraulic conductivity. The Cleary and coarse finite element models did not perform particularly well for complexity levels 4 and 5. Nevertheless, these model errors are still less than the error caused by the hydraulic conductivity uncertainty.

The uncertainty in the dispersivity values caused a relatively small error in output, smaller than may be expected. This result may be due to the relatively small range for this parameter. As mentioned earlier, dispersivity is a difficult parameter to estimate. In real-world applications, the error due to dispersivity may be much larger.

Retardation was not included in this study because it was assumed that the sandy outwash of the example flow field contained very little organic material which would cause adsorption of contaminants. In many situations, retardation would introduce another source of parameter uncertainty.

Analysis of Complexity Level 6

The original hypothesis of this study was that at some point of increasing system complexity the error related to analytical models would surpass the error due to parameter uncertainty. However, this did not occur with any of the five levels of system complexity, either when dealing with groundwater flow or contaminant transport. For this reason, a sixth level of complexity was introduced. This complexity level 6 can be seen in Figure 6. Although similar to level 5, level 6 contains a thin low permeability lens which covers the entire depth of the aquifer, essentially forming a "seal", slowing groundwater flow and advective transport. While this situation is admittedly contrived to lower the accuracy of the analytical models, it is certainly a reasonable field situation to expect. The average hydraulic conductivity of the flow field for this situation was raised to prevent the flow due to infiltration from overwhelming the flow due to groundwater gradient, which would cause flow out of the left side boundary of the flow field.

The flow and transport results for level 6 are presented in Tables 5 and 6 and graphically in Figure 6. As can be seen, the analytical solutions for flow are no longer within the error caused by parameter uncertainty. This result proves that there will be situations where the analytical solutions to flow problems are no longer sufficient. The analytical solution for transport did surprisingly well for level 6. As with flow, however, there are certainly situations in which the error associated with the analytical solution far outweighs the error due to parameter uncertainty.

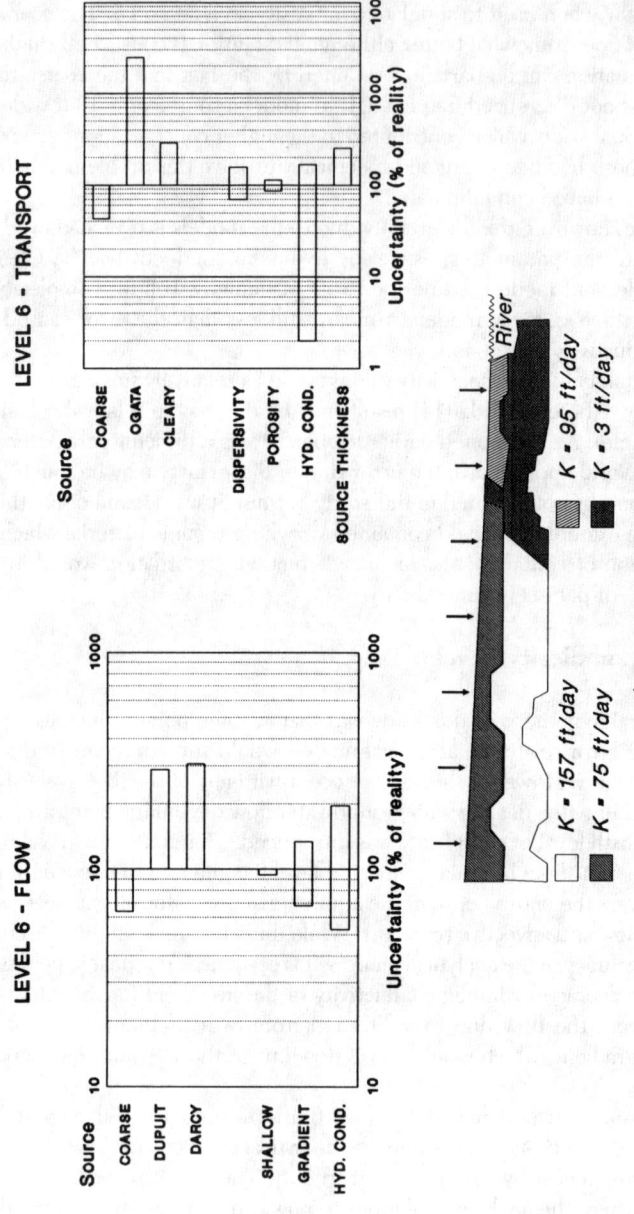

Figure 6. Comparison of model and parameter uncertainty for transport and flow predictions with level 6.

Table 5. Results of Level Six Flow Predictions

		Discharge	Percentage
"Reality"	"Reality"	3.44[a]	100%
Effects of Models	Coarse Golder	2.20	64%
	Dupuit	9.88	287%
	Darcy	10.48	305%
Effects of Parameters	Large K	6.80	198%
	Small K	1.81	53%
	Large Gradient	4.63	135%
	Small Gradient	2.33	68%
	Thinner Aquifer	3.26	95%
	Larger Infilt.	3.88	113%
	No Infilt.	3.14	91%

[a] ft^3/day/ft.

Table 6. Results of Level Six Transport Predictions

		Rel. Conc.	Percentage
"Reality"	"Reality"	0.0367	100%
Effects of Models	Coarse Golder	0.0158	43%
	Ogata	0.9170	2499%
	Cleary	0.1060	289%
Effects of Parameters	Large Dispers.	0.0577	157%
	Small Dispers.	0.0258	70%
	Large Porosity	0.0316	86%
	Small Porosity	0.0415	113%
	Thick Source	0.0939	256%
	Large K	0.0611	166%
	Small K	0.0006	2%

Summary of Transport Predictions

The groundwater flow situations which were studied in Section 2 were analyzed for solute transport, again using both analytical and numerical models of different sophistication levels. The variation caused by using different transport models was compared to the potential variation caused by input parameter uncertainty. It was found that the variation between modeling methods was less than the error due to possible parameter uncertainty. It was also found that the possible range of output error for solute transport was larger than that for groundwater flow. Hydraulic conductivity was again found to be the most important parameter in terms of its contribution to prediction errors.

SUMMARY AND CONCLUSIONS

This study compares the prediction error associated with using different models to the prediction error associated with parameter uncertainty. The analysis was done for both groundwater flow and solute transport models. The relationship between these parameter and model errors was examined for hypothetical field situations of varying levels of complexity. In many cases it was found that the error associated with parameter uncertainty outweighed the error due to the choice of model used.

The findings from this study illustrate that for simple field situations, or when little data are available, analytical models and simple numerical models are likely appropriate. As the complexity of the field situation increases, the accuracy of the analytical models decline. Each application or field situation will have a unique "break point" where the analytical solutions will no longer deliver the desired or required degree of accuracy. In light of the uncertainties associated with parameter errors, this break point may not be reached in many situations.

The amount and quality of information that is known about a field situation is related to what has been termed the complexity level. If there are little data concerning the geometry and material properties of a field area, it is reasonable to assume the simplest situation consistent with the data. This essentially makes it a simple situation to be modeled. Simple field situations described in this study could also refer to complex situations in which insufficient data are available to describe the complexities.

REFERENCES

Anderson, M. P., Using models to simulate the movement of contaminants through groundwater flow systems, *CRC Crit. Rev. Environ. Control,* 9(2), 97, 1979.

Anderson, M. P. and Woessner, W. W., *Applied Groundwater Modeling,* Academic Press, San Diego, 1992.

Arteaga, F. E., Mathematical model analysis of the eagle valley ground-water basin, west-central Nevada, U.S. Geol. Surv. Open File Rep., 80–1224, pp. 1–62, 1980.

Babalola, O., Spatial variability of soil water properties in tropical soils of Nigeria, *Soil Sci.,* 126, 269, 1978.

Broom, M. E. and Lyford, F. P., Alluvial aquifer of the cache and St. Francis River basins, northeastern Arkansas, U.S. Geol. Surv. Open File Rep., 81–476, pp. 1–55, 1981.

Burris, C. B., Morse, W. J., and Naymik, T. G., Assessment of a regional aquifer in central Illinois, Groundwater Rep. 6, Ill. Geol. Surv. and Ill. Water Surv. Coop., Champaign, IL, 1981.

Cearlock, D. B., Kipp, K. L., and Friedricks, D. R., *The Transmissivity Iterative Calculation Routine—Theory and Numerical Implementation,* Battelle Pacific Northwest Lab., Richland, WA, 1975.

Cleary, Robert W., Mathematical models and computer programs, Report No. 78-WR-15, Water Resources Program, Princeton University, 1978.

Domenico, P. and Schwartz, F. W., *Physical and Chemical Hydrogeology,* John Wiley, New York, 1990.

Electric Power Research Institute, Spatial variability of soil physical parameters in solute migration: a critical literature review, EPRI Report EA-4228, University of California at Riverside, 1985.

Fetter, C. W., Jr., *Applied Hydrogeology,* 2nd ed., Charles E. Merrill Publishing, Columbus, OH, 1988.

Franke, L. O. and Reilly, T. E., The effects of boundary conditions on the steady-state response of three hypothetical ground-water systems—results and implications of numerical experiments, U.S. Geological Survey Water-Supply Paper 2315, 1987.

Freeze, R. A. and Cherry, J. A., *Groundwater,* Prentice Hall, Englewood Cliffs, NJ, 1979.

Freyberg, D. L., A natural gradient experiment on solute transport in a sand aquifer 2. Spatial moments and the advection and dispersion of nonreactive tracers, *Water Resour. Res.,* 22(13), 2031, 1986.

Gelhar, L. W., Stochastic subsurface hydrology from theory to applications, *Water Resour. Res.,* 22(9), 135S, 1986.

Gelhar, L. W., Welty, C., and Rehfeldt, K. R., A critical review of data on field-scale dispersion in aquifers, *Water Resour. Res.,* 28(7), 1955, 1992.

Gumaa, G. A., Spatial Variability of In Situ Available Water, Ph.D. dissertation, University of Arizona, Tucson, 1978.

Hoeksema, R. J. and Kitanidis, P. K., Analysis of the spatial structure of properties of selected aquifers, *Water Resour. Res.,* 21(4), 563, 1985.

Hufschmied, P., Estimation of Three Dimensional Statistically Anisotropic Hydraulic Conductivity Field by Means of Single Well Pumping Tests Combined with Flowmeter Measurements, *Proc. Symp. Stochastic Approach of SubSurface Flow,* International Association for Hydraulic Research, Montvillargenne, France, 1985.

Kempton, J. P., Morse, W. J., and Visocky, A. P., Hydrogeologic evaluation of sand and gravel aquifers for municipal groundwater supplies in east-central Illinois, Rep. 8, Ill. Geol. Surv. and Ill. Water Surv. Coop., Champaign, IL, 1982.

McMillan, W. D., Theoretical analysis of groundwater basin operations, *Water Resour. Center Contrib.,* 114, 167, 1966.

Metzger, D. G., Loeltz, O. J., and Irelna, B., Geohydrology of the Parker-Blythe-Cibola area, Arizona and California, U.S. Geol. Surv. Prof. Paper 486-G, pp. G1–130, 1973.

Prych, E. A., Numerical simulation of groundwater flow in lower status creek basin, Yakima Indian Reservation, Washington, U.S. Geol. Surv. Water Resour. Inv. 82–4065, pp. 1–89, 1983.

Russo, D., and Bresler, E., Soil hydraulic properties as stochastic processes: 1. analysis of field spatial variability, *Soil Sci. Soc. Am. J.,* 45, 682, 1981.

Ryan, B. J. and Kipp, K. L., Jr., Low-Level Radioactive Ground-Water Contamination From a Cold-Scrap Recovery Operation, Wood River Junction, Rhode Island, U.S. Geol. Surv. Water-Supply Paper 2270, 1985.

Smith, L., A Stochastic Analysis of Steady State Groundwater Flow in a Bounded Domain, Ph.D. Dissertation, University of British Columbia, Vancouver, BC, 1978.

Sudicky, E. A., A natural gradient experiment on solute transport in a sand aquifer: spatial variability of hydraulic conductivity and its role in the dispersion process, *Water Resour. Res.,* 22(13), 2069, 1986.

Walker, W. H., Bergstrom, R. E., and Walton, W. C., Preliminary report on the groundwater resources of the Havana Region in west-central Illinois, Rep. 3, Ill. Geol. Surv. and Ill. Water Surv. Coop., Urbana, IL, 1965.

Ward, D. S., Buss, D. R., Mercer, J. W., and Hughes, S. S., Evaluation of a groundwater corrective action at the Chem-Dyne hazardous waste site using a telescopic mesh refinement modeling approach, *Water Resour. Res.,* 23(4), 603, 1987.

Willardson, L. S. and Hurst, R. L., Sample size estimates in permeability studies, *J. Irrig. Drain. Div.,* ASCE, 91, 9, 1965.

Woodman, J. T., Kier, R. S., and Bell, D. L., Hydrology of the Corpus Christi area, Texas, Res. Note 12, Bur. of Econ. Geol., University of Texas at Austin, 1978.

Zohdy, A. A. R., Eaton, G. P., and Mabey, D. R., Techniques of Water-Resources Investigations of the United States Geological Survey, Chapter D1, Application of Surface Geophysics to Ground-Water Investigations, U.S. Geological Survey, Washington, D.C., 1974.

CHAPTER 2

A Statistical Discussion of Model Error in the Use of the Advection-Dispersion Equation

Wade E. Hathhorn

INTRODUCTION

The adequacy of existing groundwater contaminant transport modeling continues to be a point of scientific concern. The desired goal is to produce models which will enable users to accurately predict plume evolution within complex natural geologic environments. In all cases, the formulation of transport is that founded on the classic advection-dispersion equation (ADE):

$$\frac{\partial C}{\partial t} + V(x) \frac{\partial C}{\partial x} - \frac{\partial}{\partial x}\left(D(x)\frac{\partial c}{\partial x}\right) = 0 \qquad (1)$$

where $C = C(x,t)$ is the solute concentration as a function of space and time, $V = V(x)$ is the (Eulerian) Darcian velocity field, and $D = D(x)$ is the coefficient of hydrodynamic dispersion. The principal feature of hydrogeologic research over the past two decades has been that of parameterizing and solving Equation (1) subject to varying degrees of natural lithologic and structural heterogeneity. The correctness of the model itself is adopted as being universally valid. The common belief is that the errors introduced by parametric uncertainty far outweigh those associated with the model itself. Yet, Dagan (1982) notes that ". . . there is is no a priori reason to believe that the diffusion type equation is valid at all. . . ." Nevertheless, the lack of recognition of this fact continues to define the norm rather than the exception.

At the heart of this issue are questions regarding the appropriateness of the two-term expansion of the ADE. With advection accounting for only the average translatory effects (or mean drift of the plume), the remaining phenomena must all be lumped into the the so-called hydrodynamic dispersion term. In order to better understand the ADE itself, we must then ask these questions. What is hydrodynamic dispersion? and What quantification does it provide? One of the most recognized discussions on this topic is that given by Bear (1972), who writes that, "Hydrodynamic dispersion is the macroscopic outcome of the actual movements of the individual tracer particles through the pores and the various physical and chemical phenomena

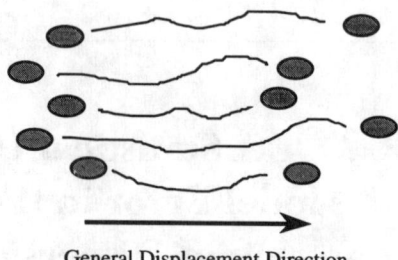

Figure 1. Illustration of random particle displacements.

that take place within the pores." Classically quantified as a combination of pure molecular diffusion and local scale variations in the fluid velocity field (i.e., mechanical mixing), hydrodynamic dispersion is used to model the spreading effects (or dispersive flux) within a plume under the traditional mathematical formalism of Fick's first law. In all but low Peclet number flow regimes (i.e., those typically produced under very low permeabilities), the spreading effects produced by mechanical mixing are thought to dominate the hydrodynamic dispersion. Thus, this paper will examine the adequacy of classic dispersion modeling in capturing the transport effects produced by heterogeneities within the fluid velocity field.

THE "TRUE" VISION OF MASS TRANSPORT

In developing the transport problem, the discussion will be limited to a conservative, nonretarded tracer which is being displaced within a heterogeneous, saturated groundwater formation. Using a continuum-based formulation, the transport of mass can be replicated by a collection of randomly moving particles subject to a random, Eulerian field-scale velocity field (see Figure 1). The source of this randomness lies within the natural heterogeneity of the medium itself, namely that of the hydraulic conductivity. In formulating the problem, it is important to point out that the particles described here should be interpreted as continuum packets of mass. The observation of interest for those particles is their relative position with respect to the ensemble of all such particles present.

Under these assumptions, the time-dependent position X_t of a particle may be described by:

$$X_t = x_o + \int_o^t V(x) \, ds \qquad (2)$$

where $V(x)$ is formally defined as a stationary, ergodic random space function (Dagan, 1987). In making this description, microscale spreading effects, such as those due to pure molecular diffusion, have been ignored. The overall shape of the plume, including both translation and spreading, are thought to be dictated solely by advection which in turn is controlled by a heterogeneous field of hydraulic conductivity.

Now, differentially, Equation (2) can be written as:

$$dX_t = X_{t+dt} - X_t = V(x)\, dt \tag{3}$$

Operationally, however, it may be convenient to express Equation (3) in finite terms:

$$\Delta = V(x)\, \tau \tag{4}$$

The random velocity term $V(x)$ may in turn be decomposed into two parts: (1) a deterministic component, U, representing the mean or average velocity and (2) a probabilistic term, $u(x)$, quantifying the associated randomness. Accordingly, Equation (3) may be written as:

$$\Delta = [U(x) + u(x)]\, \tau \tag{5}$$

where $U(x) = E[V(x)]$ and $u(x)$ is a random space function possessing zero mean and a probabilistic character identical to that of $V(x)$ (i.e., the same correlation structure and pointwise marginal pdf). More formally, Equation (5) is a forward difference representation of a continuous time random walk. In traditional form, Equation (5) would be written as:

$$dX_t = U(x)\, dt + \sigma(x)\, \xi(t) \tag{6}$$

where $\xi(t)$ is zero-meaned, unit variance random fluctuation (i.e., noise term), and $\sigma(x)$ is a variance scaling parameter, such that $\text{Var}[u(x)] = \sigma(x)$. Note also that, other than the change to a unit variance, the marginal probability density function and correlation structure for $\xi(t)$ must remain equal to that of $u(x)$ and $V(x)$.

Recognizing that $u(x)$ is controlled by the randomness in field-scale hydraulic conductivity, the probabilistic character of $\xi(t)$ is likely non-Gaussian and correlated over significant integral scales (Dagan, 1986; Gelhar, 1986). It is that same physical description which is assumed herein to form the "true" hydrogeologic vision of transport. For further details regarding this topic, the reader is referred to Hathhorn and Charbeneau (1994).

THE STATISTICAL ORIGINS OF THE ADE

If Equation (5), as described, is in fact an adequate representation of the transport process, the question then becomes one of addressing the origins of the ADE and its connection to similar stochastic descriptions of transport. Certainly, most readers are familiar with the classic advective-diffusive theories of Einstein (1905) and Taylor (1921). In developing an alternative approach to the same theory, Langevin (1908: see citation in Risken, 1987) developed the stochastic differential equation which now bears his name:

$$\frac{dX_t}{dt} = \alpha(x,t) + \beta(x,t)\,\eta(t) \tag{7}$$

Note that Equations (4) and (7) are identical in form; the two differ only with respect to the their noise terms, $\xi(t)$ and $\eta(t)$. Here, it will be assumed that $\eta(t)$ is distinctly a Gaussian, dirac-delta (white) noise, while that for $\xi(t)$ in Equation (6) is both non-Gaussian and significantly correlated.

In order to proceed, one must recognize that the ADE is nothing more than a solution to Equation (7) under stochastic diffusion criteria (Gihman and Skorohod, 1972; Arnold, 1974; Gardiner, 1985). A diffusion, by formal definition, is nothing more than a Markov process whose incremental transition probability density function (pdf) possesses (a) a definable mean and covariance:

$$\lim_{\tau \to 0} \frac{1}{\tau} \int_{|\Delta|<\epsilon} \Delta\, p(\Delta \mid \tau)\, d\Delta = \alpha(x,t) \tag{8}$$

$$\lim_{\tau \to 0} \frac{1}{\tau} \int_{|\Delta|<\epsilon} \Delta\,\Delta'\, p(\Delta \mid \tau)\, d\Delta = \beta^2(x,t)$$

and (b) satisfies a condition of probabilistic continuity:

$$\lim_{\tau \to 0} \frac{1}{\tau} \int_{|\Delta|>\epsilon} p(\Delta \mid \tau)\, d\Delta = 0 \tag{9}$$

Here, $p(\Delta \mid \tau)$ is the pdf of the displacements of size Δ which are occurring within the small observed time interval τ.

Recognizing that Equation (8) is simply the mean and covariance of Δ, our interest then is drawn to Equation (9). In theoretical terms, Equation (9) represents a condition on the stochastic character of the realizations for X_t, namely that large incremental displacements (Δ) become improbable as $\tau \to 0$. More importantly, this condition acts to limit the type of Δ pdf and correlation structure applicable as a diffusion. Although these concepts are only briefly introduced here, Equation (9) will become the focal point of discussion in subsequent sections.

In returning to the question at hand, the unusual property of a diffusion is that under certain regularity conditions the solution to Equation (8) can be rendered via the total transition pdf $p(x,t)$ obtained from the well-known Fokker-Plank equation:

$$\frac{\partial p(x,t)}{\partial t} + \alpha \frac{\partial p(x,t)}{\partial x} - \frac{\beta^2}{2}\frac{\partial^2 p(x,t)}{\partial x^2} = 0 \tag{10}$$

where $p(x,t) = p(x,t \mid x_o, t_o)$ is the pdf of displacing to a point x in time t given the starting point of x_o at time t_o (Risken, 1987). In relating Equation (10) to the ADE, the law of large numbers (i.e., the presence of a sufficiently large number of particles), can be employed to equate $p(x,t)$ to a plume concentration $c(x,t)$ using:

$$C(x,t) = \int C_o(x_o)p(x,t)\, dx_o \tag{11}$$

where $C_o(x_o)$ is the initial solute concentration at the release point, x_o (Bhattacharya and Gupta, 1983; Sposito et al., 1986). Through Equation (11), one can state that same partial differential equation which yields a solution to $p(x,t)$ also determines $C(x,t)$. That is to say, the diffusion model of Equation (10) is in fact identical to that of the ADE, both in terms of mathematical and physical interpretation. It follows then that Equation (6), subject to the diffusion criteria of Equations (8) and (9), form the stochastic model giving rise to the ADE.

With this in mind, it is of interest to identify exact quantifications for the type of Δ pdf and correlation structure applicable to diffusions and compare those processes to the "true" hydrogeologic model developed earlier.

THE LIMITING CONDITIONS FOR THE ADE

In establishing a point of comparison, it is important to recognize that the Δ process being discussed here is that of "steps" in a random walk. At issue are: (a) the qualifications for that walk which will lead to an exact representation under the ADE, and (b) the quantification of errors, if any, in such representation as a model of our original hydrogeologic construct. We know, however, from the discussion above that in order for the ADE to be applicable, the Δ process must satisfy: (1) a Markovian construct and (2) the continuity principle identified in Equation (9). For all practical purposes, the Markovian criteria limits the autocorrelation of the observed sequences to a dirac-delta function, that is:

$$<\Delta_{t+\tau}, \Delta_t> = \delta(\tau) \qquad (12)$$

The crucial feature of this statement is the size of τ. First, τ must necessarily be much smaller than the overall elapsed time of observation (i.e., the relative size of τ must approach zero), and second, τ must be at least as large as the average travel time across any physical integral scale for the true domain (i.e., the individual steps of the walk must be sequentially uncorrelated).

Nevertheless, it is often argued that the ADE will eventually become valid provided sufficient time elapses for the central limit theorem to act. The adequacy of such arguments are, however, directly influenced by the presence and strength of correlation within the underlying Δ process. Such presence of correlation may in fact tend to negate the viability of the ADE, even if that correlation is weak (Fox, 1986). When comparing these ideas to those of natural hydrogeologies, significant correlation lengths may exist, particularly for such physical properties as the hydraulic conductivity and corresponding fluid velocities. In fact, it is common to observe structural correlations of several hundred meters in the horizontal plane (Hoeksema and Kitanidis, 1985). If this is true, then on an incremental time scale (τ) for systems in which the hydraulics are those described under Darcy's law, the corresponding sequential displacement of particles (either fluid or solute) cannot be independent. Therefore, the Markovian requirement for the displacement process Δ is almost surely violated under the normal application of continuum-based hydrogeologic theories.

Secondly, and probably most significant, one must also consider the requirement of path continuity for Δ. Hathhorn (1990) has shown that for all practical purposes the only Δ pdf which yields an exact representation under the ADE is that of a Gaussian distribution, namely:

$$p(\Delta \mid \tau) = \frac{1}{\sqrt{2\pi\beta^2\tau}} \exp\left\{-\frac{(\Delta-\alpha\tau)^2}{2\beta^2\tau}\right\} \qquad (13)$$

Moreover, for a fixed value of τ, the above statement is equivalent to limiting the random character of $V(x)$ in Equation (4) to that of a Gaussian distribution. This fact would place Equation (13) in direct conflict with our original non-Gaussian physical based model. To explain, recall that $V(x)$ is directly dependent on the hydraulic conductivity (K), as would be indicated by Darcy's law. In turn, K is lognormally distributed. Thus, the true $V(x)$ cannot be Gaussian. This fact and the presence of significant correlation in the physical realm give rise to serious questions regarding the applicability of the ADE as an appropriate model of transport in highly heterogeneous systems.

As such, the question of attempting to quantify those errors for a broad class of hydrogeologic settings becomes the issue. To do this, comparisons can be made between the numerical simulation of Equation (4) and an analytic solution of the ADE. These ideas are discussed in further detail in the following section.

A RUDIMENTARY EXAMINATION OF ERRORS

In addressing one of the most important questions in hydrogeology today, Sposito et al. (1986) asked; For what broad class of spatially varying functions $V(x)$ can the solution of the transport problem be approximated by the ADE? Recall, the basic description of transport is one dominated by random advection. Following the traditional theory of Darcy's law, that advection can be quantified as:

$$V(x) = KI/n \qquad (14)$$

where K is a second rank tensor of spatially varying hydraulic conductivity, I is the vector of hydraulic gradient, and n the porosity scalar. In a formal sense, K, I, and n may each be considered as random fields. Here, however, it is argued that the variability in K dominates both that of I and n and, in turn, dominates the character of $V(x)$. Accordingly, both $V(x)$ and Δ are assumed to be log-normally distributed and possess a similar field-scale correlation structure to that of K (Dagan and Nguyen, 1989; Cvetkovic et al., 1991; Dagan et al., 1992). If Equations (4) and (14) are then adequate descriptors of the true hydrogeologic transport process, the real question is not the character of $V(x)$ itself but rather the adequacy of the ADE to accurately predict that transport for lognormally distributed and correlated random fields of advection.

In order to address that question, a means must be established for making comparisons between solutions of the ADE and those obtained by simulation of Equations

(4) and (14). In establishing a comparative basis, consider the case of a point release of a conservative, nonretarded mass tracer into a one-dimensional, infinite domain. Resolution of Equation (1) subject to:

$$C(x,0) = \frac{C_o V_o}{n} \delta(x) \text{ and } C(-\infty,t) = C(\infty,t) = 0 \qquad (15)$$

yields

$$\frac{C(x,t)}{C_o} = \frac{V_o}{2n\sqrt{\pi D t}} \exp\left[-\frac{(x-Ut)^2}{4Dt}\right] \qquad (16)$$

where V_o is the volume of mass released at a concentration of C_o, n is the deterministic porosity, and U and D are the corresponding seepage velocity and hydrodynamic dispersion coefficient for the ADE. The reader should recognize Equation (16) as the well-known Gaussian plume model. According to Equation (16), the spatial distribution of mass at any instant in time is Gaussian in shape, regardless of the particular values of U or D selected and/or their spatial heterogeneity. Moreover, it is important to recognize that the choice of Equation (16) as a comparative model is simply a matter of convenience. The inclusion of other (finite) boundary and/or initial conditions would not affect the discussion being presented herein.

Now, in order to simulate the so-called "true" hydrogeologic process, a means must be established for numerically analyzing Equations (4) and (14). To do this, a random walk is constructed based on a log-normally distributed and correlated sequence of steps. The main algorithm is one derived from a modification of the procedures outlined by Clifton and Neuman (1982). The highlights of that procedure are outlined below:

Let Y be the (natural) logrithmic transform of Δ. If Δ is log-normally distributed, then Y must be normally distributed. To generate an N-sized sequence of Δ for a given particle, one may use:

$$Y = <Y> + C\Phi \qquad (17)$$

where $<Y>$ is the mean of the log-transformed displacements, Φ is an n-sized vector of standard (0,1) Gaussian deviates, and C is an $N \times N$ lower triangular matrix formed from the Cholesky decomposition of the covariance matrix for Y. The Y-vector can be converted to produce an N-sized vector of Δ's using the transform:

$$\Delta = \exp(Y) \qquad (18)$$

This process can be repeated in Monte Carlo fashion to simulate the desired number of realizations for a finite number of corresponding particles, each released at a known starting location, x_o.

From this, one can numerically synthesize the transport of mass subject to a log-normally distributed and correlated advection field. The results of the simulated spatial distribution of mass, for a given heterogeneity and domain length, can be compared to that of Equation (16) and its corresponding Gaussian distribution. Differences, in terms of higher order moments and residuals, can be computed and used as means of quantifying the errors between the "true" hydrogeologic transport model and that of the ADE.

THE SIMULATION EXERCISES AND THEIR RESULTS

The first task in conducting the simulation exercises is to parameterize the random walk. The items of information needed include the mean (μ_Y), variance (σ_Y^2), and correlation structure for the log-transformed displacements, Y. Short of simply guessing values, one can rely on the dominant character of the hydraulic conductivity (K), to establish bounds for these parameters. Here, an exponential correlation structure (R_Y) was used, where:

$$R_Y = \exp\left(-\frac{\tau}{\lambda}\right) \tag{19}$$

along with a mean displacement size of unity, that is:

$$<\Delta> = 1 = \exp(\mu_Y + 0.5\,\sigma_Y^2) \tag{20}$$

The latter choice was made as a matter of convenience. In this way, the average total displacement of a realizations of particles at any time will simply be equal the N number of steps taken. The reader should also recognize that the right-hand side of Equation (20) is derived from an assumed log-normally distributed Δ. The equation shown is part of the standard relations for equating the moments of log-transformed data with those of the original (log-normal) data (see Haan, 1977). Moreover, the term λ in Equation (19) is that of the integral scale for Y. An exact value for this parameter is not important. Rather, it's the total number of steps taken by each particle in relation to the size of λ (i.e., the number of integral scales traversed N/λ) which tends to characterize a particular simulation execution.

In conducting the simulations themselves, the displacement generation procedures outlined earlier were used to produce numerical realizations of a non-Gaussian, correlated random walk. For each realization, N number of random "steps" were generated, each covering an assumed incremental time interval of $\tau = 1$. (N representing the effective domain length.) The final position for each particle was determined using $X_t = \Sigma\Delta$ at time $t = N\tau$. The procedure was then repeated in Monte Carlo fashion for 10,000 particles. The use of the large number of particles permitted a sufficient, stable resolution of the higher order spatial moments. Once complete, the 10,000 final positions were ranked (in ascending order) and their normalized

MODEL ERROR IN THE ADVECTION-DISPERSION EQUATION

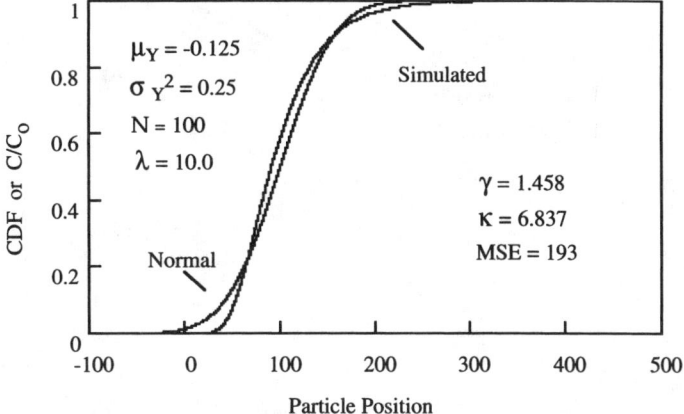

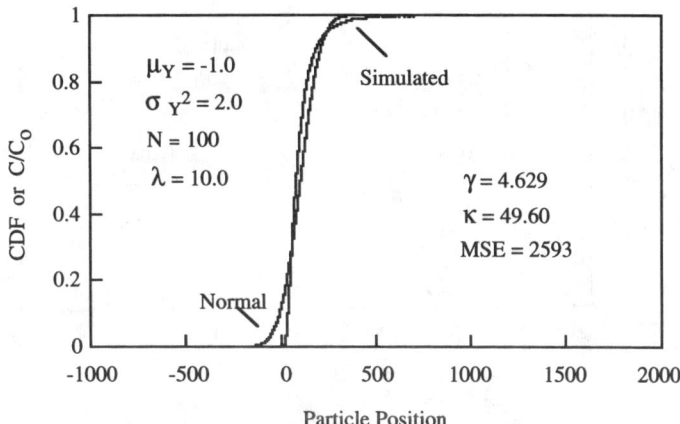

Figure 2. Simulated results for $N = 100$ and $\lambda = 10.0$.

cumulative mass distribution determined according to Cunnane's (1978) unbiased plotting position. In addition, the mean, variance, skew, and kurtosis of the final positions were also computed.

In displaying typical results, the outcomes for a select set of parameterizations is provided in Figures 2 and 3. In these figures, two separate plots are shown: (1) the "simulated" outcomes from the numerical random walk and (2) the "normal" result produced from the Gaussian solution of Equation (16) using $U = <\Delta>/\tau = 1$ and $D = \text{Var}[X_t]/2t$, where $\text{Var}[X_t]$ is the variance of the final particle positions determined from the numerical walks. Here, since $U = 1$, $<X_t> = N$, wherein providing a simple means of normalizing the outcomes. As measures of "fit", values for the higher order spatial moments (i.e., skewness-γ and kurtosis-κ) for the simulated outcomes are displayed, along with the mean-squared error computed between the two. Note, if the

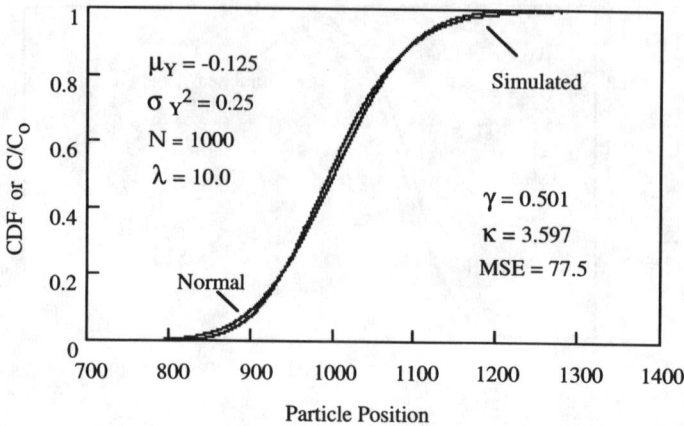

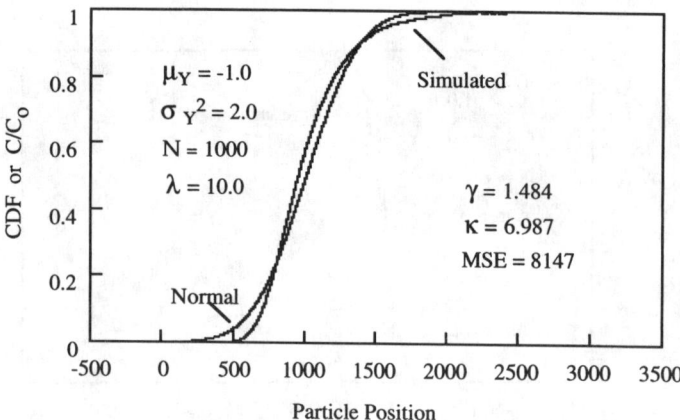

Figure 3. Simulated results for $N = 1000$ and $\lambda = 10.0$.

simulations were to match the Gaussian ADE model, γ and κ should approach 0.0 and 3.0, respectively, and the MSE should be near 0.0. Any deviation from these standards would be indicative of the relative error of the ADE in predicting the so-called "true" simulated result.

Although only a small sample of the results run are portrayed here, a review of their collection indicate some very important trends:

1. For a particular value of N/λ, the non-Gaussian character of the plume profile increases with increasing σ_Y^2;
2. For a particular N and σ_Y^2, the non-Gaussian character of the plume profile increases with increasing λ; and
3. For a particular λ and σ_Y^2, the Gaussian character of the plume profile improves with increasing N (i.e., increased travel time).

Table 1. Summary of Selected Comparative Statistics for Various Trials Based on $\mu_Y = -0.125$, $\sigma_Y^2 = 0.5$ (on left) and $\mu_Y = -1.0$, $\sigma_Y^2 = 2.0$ (on right)

N	λ	γ	κ	MSE	N	λ	γ	κ	MSE
	0.5	0.61	3.94	3.84		0.5	1.7	12.65	60.3
	1	0.71	4.27	7.51		1	1.82	13.63	91.7
	2	0.94	5.12	22.2		2	2.13	15.56	206
100	5	1.44	7.58	109	100	5	3.09	25.06	840
	10	2.06	12.28	353		10	4.63	49.6	2,590
	20	2.89	21.77	992		20	7.31	125.9	7,145
	50	3.55	26.44	2,580		50	7.84	117.5	16,310
	0.5	0.35	3.32	14.2		0.5	0.61	4.08	129
	1	0.44	3.47	32.8		1	0.63	3.85	196
	2	0.54	3.66	91.8		2	0.80	4.24	546
1000	5	0.73	4.13	402	1000	5	1.16	5.68	2,560
	10	0.9	4.49	1,209		10	1.48	6.99	8,150
	20	1.18	5.48	3,936		20	2.46	17.3	31,970
	50	1.73	8.35	18,670		50	3.73	31.7	148,300
	100	2.28	12.0	56,340		100	4.63	42.4	408,170

These results are, however, somewhat obvious in that they portray similar trends to those already scientifically understood (see, e.g., Dagan, 1989). Nevertheless, a surprising fourth result was that simulated outcomes produced skewness-γ and kurtosis-κ values which were distinctly different from those of 0.0 and 3.0, even for large values of N/λ. (Typical results for this observation are shown in Table 1.) By contrast, under traditional theory, the asymptotic approach to Gaussianity for a displacing plume is argued via the central limit theorem in the large time approximation. Dagan (1989) and others have postulated validity for the ADE (through a Fickian approximation) for N/λ > about 60 to 80. The results indicated here, however, show that under a strict statistical interpretation (i.e., a skewness of 0.0 and a kurtosis of 3.0), the ADE within limited heterogeneous environments (i.e., for $\sigma_Y^2 = 1$) remains invalid even for N/λ > 200. Moreover, as σ_Y^2 increases, the initial validity of the ADE is in excess of N/λ > 2000. Granted, however, these results are generated for the limited numerical exercises described here. Nevertheless, they do indicate the errors produced in modeling contaminant transport within heterogeneous hydrogeologic environments.

In addressing the magnitude of those errors more directly, the results displayed in Figures 2 and 3 require clarification. On visual inspection, the absolute differences between the two outcomes shown may appear minimal. Hidden in the plot, however, is an elongated leading tail for the simulated outcomes, as shown in Figure 4. In cases where the first arrival time of mass is important, the errors generated by prediction under the ADE are most severe. In short, the ADE tends to delay such estimates of breakthrough, while consistently providing larger predictions of the arrival time for a given isochore (C/C_o). The source of this error is associated with the inability of the ADE to reconcile the extreme and persistent displacements of mass which occur under a log-normal and correlated velocity field. The complexity of such transport simply cannot be quantified through a two-term expansion whose inclusive information is limited to only the mean and variance of the velocity field.

Out of what seems a form of desperation, a great deal of effort has been put forth by other hydrogeologic researchers to find effective parameterizations for the ADE

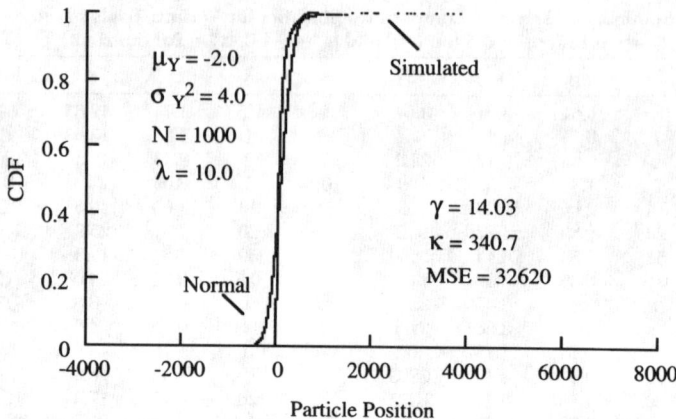

Figure 4. Illustration of extreme tail on "true" simulated outcomes.

which would "remedy its ails". One can show, however, that from a purely statistical viewpoint, such a remedy cannot be produced. The recognized complexities, at all scales, associated with natural hydrogeologic systems simply cannot be modeled under classic advective-diffusive based transport theories. In explanation, recall that the ADE is limited by the conditions outlined earlier for statistical diffusions, and as such, will always display significant modeling error for systems possessing strong non-Gaussian dynamical behavior and persistent internal correlations.

SUMMARY AND CONCLUSIONS

In summary, the numerical results from this study indicate that the model errors posed by the advection-dispersion equation (ADE) are genuine. For hydrogeologic systems possessing modest heterogeneity (i.e., $\sigma_Y^2 > 1$), those errors may be significant even for extended time frames. Unfortunately, the common scientific approach is to neglect these errors. The typical modeling exercise is based on an acceptance of the ADE as an exact representation of hydrogeologic transport. The rationale given in justifying that position is one of perceived overwhelming parametric uncertainty. The fact of the matter is that the errors produced by the use of the ADE itself may in many cases be as large as those associated with the incomplete knowledge of its parameters. Furthermore, no means of sophisticated parametric quantifications using current measurement or analytic techniques will ever reconcile this problem. Existing knowledge of the velocity fields, at all scales, remains an unknown and as such must be treated as a random variable. Within this framework, the ADE simply cannot capture the complexity of the natural heterogeniety which tends to dominate the transport in many hydrogeologic settings. The continued search for modified advection-dispersion terms under various forms of scaling principles, fractals, and other approaches is believed futile so long as the basis for that quantification remains founded on the ADE.

In closing, the author wishes to recognize that these concluding statements constitute a rather contentious position. Nevertheless, these remarks are provided with the hope that they will instill a renewed debate over alternative and pragmatic solutions to the problem of predicting mass transport within the subsurface. Examples of what the author believes are viable research avenues include the use of stochastic-convective methods and the development of improved downhole velocity measuring devices. In any case, the desire throughout this article has been to provoke stimulating discussion and provide a nontraditional examination of the problem of predicting subsurface contaminant transport.

REFERENCES

Arnold, L., *Stochastic Differential Equations: Theory and Applications,* Wiley-Interscience, New York, 1974, 228.
Bear, J., *Dynamics of Fluids in Porous Media,* Dover Publ., New York, 1972, 764.
Bhattacharya, R. N. and Gupta, V. K., A theoretical explanation of solute dispersion in saturated porous media at the Darcy scale, *Water Resour. Res.,* 19(4), 938, 1983.
Clifton, P. M. and Neuman, S. P., Effects of kriging and inverse modeling on conditional simulation of the Arva Valley in southern Arizona, *Water Resour. Res.,* 18(4), 1215, 1982.
Cox, D. R. and Miller, H. D., *The Theory of Stochastic Processes,* Chapman and Hall, New York, 1965, 398.
Cunnane, C., Unbiased plotting positions—a review, *J. Hydrol.,* 37, 205, 1978.
Cvetkovic, V., Dagan, G., and Shapiro, A., An exact solution of solute transport by one-dimensional random velocity fields, *Stochastic Hydrol. Hydraul.,* 5, 45, 1991.
Dagan, G., Stochastic modeling of groundwater flow by unconditional and conditional probabilities 2. The solute transport, *Water Resour. Res.,* 18(4), 835, 1982.
Dagan, G., Statistical theory of groundwater flow and transport: pore to laboratory, laboratory to formation, formation to regional scale, *Water Resour. Res.,* 22(9), 120S, 1986.
Dagan, G., Theory of solute transport by groundwater, *Annu. Rev. Fluid Mech.,* 19, 183, 1987.
Dagan, G., *Flow and Transport in Porous Formations,* Springer-Verlag, New York, 1989, 465.
Dagan, G., Cvetkovic, V., and Shapiro, A., A solute flux approach to transport in heterogeneous formations. 1. The general framework, *Water Resour. Res.,* 28(5), 1369, 1992.
Dagan, G. and Nguyen, V., A comparison of travel time and concentration approaches to modeling transport by groundwater, *J. Contam. Hydrol.,* 4(1), 79, 1989.
Einstein, A., *Investigations on the Theory of the Brownian Movement,* Dover Publ., New York, 1956, 122.
Fox, R., Uniform convergence to an effective Fokker-Plank equation for weakly colored noise, *Phys. Rev. A,* 34(5), 4525, 1986.
Gardiner, C. W., *Handbook of Stochastic Methods for Physics, Chemistry, and the Natural Sciences,* 2nd ed., Springer-Verlag, New York, 1985, 442.

Gelhar, L. W., Stochastic subsurface hydrology from theory to applications, *Water Resour. Res.,* 22(9), 135S, 1986.

Gihman, I. I. and Skorohod, A. V., *Stochastic Differential Equations,* Springer-Verlag, New York, 1972, 354.

Haan, C. T., *Statistical Methods in Hydrology,* Iowa State Univ. Press, Ames, IA, 1977, 378.

Hathhorn, W. E., Diffusion Theory and the Passage Time Problem, Ph.D. dissertation, Univ. of Texas, Austin, UMI No. 9116869, Ann Arbor, MI, 1990, 128.

Hathhorn, W. E. and Charbeneau, R. J., Stochastic fluid travel times in heterogeneous porous media, *J. Hydr. Engr.,* ASCE, 120(2), 134, 1994.

Hoeksema, R. J. and Kitanidis, P. K., Analysis of the spatial structure of properties of selected aquifers, *Water Resour. Res.,* 21(4), 563, 1985.

Risken, H., *The Fokker-Plank Equation: Methods of Solution and Application,* Springer-Verlag, New York, 1987, 472.

Sposito, G., Jury, W. A., and Gupta, V. K., Fundamental problems in the stochastic convection-dispersion model of solute transport in aquifers and field soils, *Water Resour. Res.,* 22(1), 77, 1986.

Taylor, G. I., Diffusion by continuous movements, *Proc. London Math. Soc.,* ser. 2, Vol. XX, 196, 1921.

CHAPTER 3

Model Testing: A Functionality Analysis, Performance Evaluation, and Applicability Assessment Protocol

Paul K. M. van der Heijde

INTRODUCTION

Groundwater modeling has become an important methodology in support of the planning and decision-making processes involved in groundwater management. Groundwater models provide an analytical framework for obtaining an understanding of the mechanisms and controls of groundwater systems and the processes that influence their quality, especially those caused by human intervention in such systems. For managers of water resources, models may provide essential support for planning and screening of alternative policies, regulations, and engineering designs affecting groundwater. This is particularly evident with respect to groundwater resources development, groundwater protection, and aquifer restoration.

Assessment of the validity of modeling based projections is difficult and often controversial (e.g., van der Heijde and Park, 1986; Tsang, 1987, 1991; Konikow and Bredehoeft, 1992; Bredehoeft and Konikow, 1993). The three major components contributing to the success or failure of a modeling exercise are the availability of field information (i.e., quality and completeness of data), the type and quality of the analytical tools (e.g., geostatistical and hydrogeological software), and the competence of the team of experts involved in the preparation of the modeling based advise. This paper focuses on the assessment of the quality of the available tools, specifically computer codes for simulation of groundwater systems.

When discussing computer-based groundwater models distinction should be made between the development of a generalized, nonsite-specific, quantitative description of a groundwater system, the so-called *generalized groundwater model* or *(groundwater) simulation code*, and the application of such a generalized groundwater model to a specific site, often referred to as a *groundwater model application*. In this paper the terms *simulation code* and *groundwater model* indicate a generalized groundwater model. The development of such models or codes consists of three components: (1) research aimed at obtaining a quantitative understanding of the studied groundwater system; (2) software development; and (3) model testing and evaluation. In general, groundwater model application is part of a larger set of activities aimed at solving site- or problem-specific issues, and includes data collection, interpretation and storage, system conceptualization and model design, calibration, formulation of alternative problem-solving scenarios and engineering designs, and postsimulation analysis.

Model development is closely related to the scientific process of acquiring new, quantitative knowledge about nature through observation, hypothesizing, and verifying deduced relationships, which results in the establishment of a credible theoretical framework for the observed phenomena. The object of model research is a prototype system containing selected elements of a specific real world groundwater system, or group of groundwater systems (van der Heijde and Elnawawy, 1992). The conceptual model of the selected groundwater system forms the basis for determining the causal relationships among various components of the system and its environment. These relationships are defined mathematically, which results in a mathematical model. If the solution of the mathematical equations is complex, or when many repetitious calculations are required, the use of computers is essential. This requires the coding of the solution to the mathematical problem in a programming language, resulting in a computer code. The conceptual formulations, mathematical descriptions, and the computer coding constitute the prototype or generalized model.

Developing efficient and reliable software and applying it to groundwater management requires a number of steps, each of which should be taken conscientiously and reviewed carefully. Taking a systematic, well-defined and controlled approach to all steps of the model development and application process is essential for its successful utilization in management. Quality assurance (QA) provides the framework to ensure that decisions are based on the proper use of appropriate analysis techniques and adequate tools (van der Heijde and Elnawawy, 1992). An important part of quality assurance in code development is code testing and performance evaluation. This paper presents a comprehensive groundwater model testing and evaluation procedure.

MODEL EVALUATION AND TESTING PROCEDURE

Successful water management requires that decisions be based on the use of technically and scientifically sound data collection, information processing, and interpretation methods, and that these methods be properly integrated. As computer codes are essential building blocks of modeling supported management, it is crucial that before such codes are used as planning and decision-making tools, their credentials are established and their suitability determined through systematic evaluation of their correctness, performance characteristics, and applicability. Such a systematic approach, referred to as "model testing and evaluation protocol", consists of evaluation or review of the underlying concepts and mathematical formulations, a rather qualitative process, and extensive code testing, a more quantitative process. Often, a code's credentials are furthered by successful applications to a variety of site conditions and management problems.

An assessment of the effectiveness of a model testing and evaluation protocol should address the requirements of the different audiences interested in the application of such a protocol to a particular model: model researchers, code developers, code users, project and program managers, and regulators. In the context of this paper, model researchers are focussed on the model's underlying concepts and mathematical framework, code developers are concerned with the correct implementation of algorithms and other code design features in a computer program, code users deal with the correctness of model and code, as well as the applicability of the generalized model to the problem at hand, while the managers and regulators are primarily concerned with the reliability of the modeling based analysis. To be effective, the protocol should satisfy the needs of these different audiences and answers such questions as:

- Is the conceptual model valid for the prototype system for which it was developed?
- Does the mathematical model truly represent the original conceptual model, including the hydrogeologic framework present, and the processes and stresses present?
- Does the code correctly represent the model's mathematical framework?
- Will the model be able to represent the responses of the prototype system to the site-specific stress scenarios the user intends to investigate?

To evaluate groundwater modeling software in a systematic and consistent manner, the International Ground Water Modeling Center (IGWMC) has formulated a model development QA framework (van der Heijde and Elnawawy, 1992). The QA framework includes scientific and technical reviews, a functionality analysis, performance evaluation and applicability assessment protocol, and a three-level code testing strategy.

Review of Generalized Groundwater Models

The scientific and technical review process is qualitative in nature and comprises examination of model concepts, governing equations and algorithms, evaluation of documentation and general ease of use, inspection of program structure and program logic, error-prone analysis, and examination of the computer coding. If code testing has been performed by the code developers, the review process includes evaluation of the completeness of the tests and the significance of the test results (van der Heijde et al., 1985; Bryant and Wilburn, 1987). To facilitate the thorough review of the generalized model, detailed documentation of the model, its structure, operation, and developmental history are required, as is the availability of the source code for inspection. In addition, to ensure independent evaluation of the reproducibility of the code testing results, an executable version of the computer code should be available or be at least accessible for use by the reviewer, together with files containing the original test data and a description of the test problem set-up.

Model examination determines whether anything fundamental was omitted in the initial conceptualization of the prototype system. Such a procedure should determine whether the concepts underlying the generalized model adequately represent the nature of the prototype system, and should identify the processes and actions pertinent to the model's intended use. The examination should also determine whether the equations representing the various processes are valid within the range of the model's applicability, and whether these equations conform mathematically to the intended range of the model's use. Finally, model examination should determine the appropriateness of the selected initial and boundary conditions, and whether the selected solution approach is the most appropriate.

For complex models, detailed examination of the implemented algorithms is required to determine whether appropriate numerical schemes have been adopted to represent the model. This step should disclose any inherent numerical problems such as nonuniqueness of the numerical solution, inadequate definition of numerical parameters, incorrect or nonoptimal values used for these parameters, numerical dispersion, numerical instability such as oscillations or divergent solution, and problems regarding conservation of mass (ASTM, 1984). Consideration is also given to the ease with which the mathematical framework and the final results can be physically interpreted.

In *computer code inspection,* attention is given to the manner in which modern programming principles have been applied to code structure, compliance with programming standards, efficient use of programming languages, integrity of data handling, and internal documentation. This step might reveal undetected programming or logic errors not present in the documentation, or those errors hard to detect in code testing.

Another element of the review procedure is the *evaluation of model documentation.* Documentation of a generalized model is evaluated through visual inspection, comparison with existing documentation standards and guidelines, and through its use as a guide in preparing for and performing code testing runs. A discussion of what constitutes good documentation has been presented by van der Heijde and Elnawawy (1992).

FUNCTIONALITY ANALYSIS, PERFORMANCE EVALUATION, AND APPLICABILITY ASSESSMENT PROTOCOL

A systematic approach to code testing combines elements of error detection (i.e., *code verification* or *functionality analysis*), evaluation of the operational characteristics of the code (i.e., *performance evaluation*), and assessment of its suitability to solve certain types of management problems (i.e., *applicability assessment*), with well-designed test problems, carefully selected test data sets, and informative performance measures. Such a systematic approach is represented by the *functionality analysis, performance evaluation,* and *applicability assessment protocol,* developed by the IGWMC (van der Heijde et al., 1993). In this protocol, a systematic approach to the formulation of test objectives is combined with a comprehensive code testing strategy. The results of such testing are expressed in terms of correctness (e.g., in comparison with a benchmark), reliability (e.g., reproducibility of results, convergence and stability of solution algorithms, and absence of terminal failures), efficiency of coded algorithms (in terms of achieved numerical accuracy versus memory requirements and code execution time), and resources required for model set-up (e.g., input preparation time). The protocol requires that code testing be performed for the full range of parameters and stresses that the code be designed to simulate. Groundwater resource managers may be interested in further testing of a code to determine the consequences if such a code is used beyond its original design criteria, or beyond the

MODEL TESTING

range of applications for which it has already been tested. Through such extensive and systematic code testing and model evaluation, confidence in the applicability of the code will increase.

In software engineering, verification is the process of demonstrating consistency, completeness, and correctness of the software (Adrion et al., 1986). ASTM (1984) defines verification as the examination of the numerical technique in the computer code to ascertain that it truly represents the conceptual model, and that there are no inherent problems with obtaining a solution. In groundwater modeling, the objective of the code verification process is twofold: (1) to check the correctness and accuracy of the computational algorithms used to solve the governing equations, and (2) to assure that the computer code is fully operational. In code verification, the code is run using problems for which independently derived solutions are available and which are specifically designed to test individual code segments, subroutines, functions, and modules, as well as to test the overall program structure.

Since the early days of computer-based simulation of groundwater systems, verification of codes have been part of code development activities (e.g., Pinder and Bredehoeft, 1968; Prickett and Lonnquist, 1971; Pinder and Frind 1972). Recently, such code verification by the code developers has become quite elaborate (Ward et al., 1984; Reeves et al., 1986; Gupta et al., 1987; Sims et al., 1989, Faust et al., 1990).

In some cases, independent verification of codes and evaluation of their performance is needed (e.g., Watson and Brown, 1985; Beljin, 1988). Often, such third-party code testing is initiated by a regulatory agency to determine if the code is acceptable for use within a certain regulatory framework. The need for comprehensive verification of groundwater simulation codes is illustrated by the completion of a series of international cooperative programs: INTRACOIN, HYDROCOIN, and INTRAVAL (Larsson, 1992). These studies attempted to evaluate conceptual and mathematical models for groundwater flow and radionuclide transport in the context of performance assessment of repositories for radioactive waste (SKI, 1987, 1990). They were performed by participating modeling groups from different countries, and coordinated by the Swedish Nuclear Power Inspectorate. The first project, the International Nuclide Transport Code Intercomparison study, INTRACOIN, ran from 1981 to 1986 (INTRACOIN, 1984, 1986). It was succeeded by the Hydrologic Code Intercomparison study, HYDROCOIN, completed in 1990 (HYDROCOIN, 1988, 1990). A third project, the International Project to Study Validation of Geosphere Transport Models (INTRAVAL) was initiated in 1987 (Larsson, 1992). These studies tested the numerical accuracy of computer codes, the validity of the underlying conceptual models, and different techniques for sensitivity and uncertainty analysis.

In the mid-1980s the IGWMC developed a groundwater model testing strategy. Early versions of this strategy have been presented in van der Heijde et al. (1985), and applied by Huyakorn et al. (1984) and Beljin (1988) to two-dimensional flow and solute transport codes. The objective of the IGWMC test strategy was to provide a framework for evaluating a model's correctness.

A major limitation of these code testing approaches is the absence of information on the completeness of the testing performed. To address this concern, van der Heijde et al. (1993) discussed the application of an expanded version of the IGWMC testing strategy to three-dimensional flow and solute transport models. In this new version, three different code testing objectives are recognized: (1) functionality analysis; (2) performance evaluation; and (3) applicability assessment.

A code's functionality is defined as the set of functions and features that the code offers the user in terms of model framework geometry, simulated processes, boundary conditions, and analytical capabilities (see Table 1). Functionality analysis focuses on code error detection resulting in terminal failures, and incorrect, inconsistent, or inaccurate computational results. It consists of describing the code's functions and features in a systematic and comprehensive manner, and testing all individual functions and combination of functions. As a process functionality analysis covers many aspects of code verification discussed above, but as a term it has not been subject to so many different interpretations. To ensure that all code functions are addressed in the verification process, a *functionality matrix* is formulated combining the results of the functionality analysis with the test problem objectives in terms of tested functions and features (see Figure 1). The functionality matrix is used to check and illustrate the extent of the performed functionality analysis.

Performance evaluation is the operational characteristics of the code aimed at characterizing for a wide range of parameters and boundary conditions in terms of (1) convergence and stability; (2) sensitivity for grid orientation and resolution, and for time discretization; (3) reproducibility of results; (4) efficiency of coded algorithms; and (5) resources required for model setup. Results of the performance evaluation are expressed both quantitatively and qualitatively in tabular form (see Table 2). Reporting on performance evaluation should provide potential users information on both the actual performance obtained with the code, and a discussion of the setup of simulation control parameters, and the spatial and temporal discretization used.

Applicability assessment focuses on a code's suitability to solve certain types of management problems. Test problems are formulated in terms of representative hydrogeology, engineering designs, and management strategies. An *applicability matrix* is used to document the extent of the applicability assessment, comparable to the functionality matrix. Reporting on applicability assessment should include information on the optimal implementation of the test problems.

The functionality matrix, performance tables, and applicability matrix, together with the supporting test results, provide the information needed to select a model for a site-specific application, or to evaluate the appropriateness of a model used at a particular site.

CODE TESTING STRATEGY

The functionality analysis, performance evaluation, and applicability assessment protocol is implemented using a three-level code testing strategy. At Level 1, code testing is performed using *conceptual* tests and *benchmarks,* supporting both the functionality analysis and performance evaluation steps of the protocol. Conceptual tests are simple tests for which no known solution is available or necessary. They are aimed at testing a single or small number of code functions, and allow qualitative evaluation of code responses (see Figure 2). The essence of quantitative verification is the use of benchmarks, known solutions of a given problem.

Table 1. Functions and Features of a Typical Three-Dimensional Saturated Flow and Transport Model

General Model Capabilities
- uncoupled Darcian groundwater flow and nonconservative single-component solute transport in saturated porous medium
- distributed parameter discretization

Spatial Orientation
- 1-D horizontal
- 1-D vertical
- 2-D horizontal
- 2-D vertical
- quasi 3-D (layered)
- fully 3-D

Grid Design
- 1-D, 2-D, or 3-D block-centered finite difference grid with constant or variable cell size

Time Discretization
- steady-state flow
- transient flow
- transient transport
- variable time-step size
- multiple transport time-steps per flow time step
- multiple flow time-steps per stress period
- variable stress periods

Matrix Solvers
- SOR
- ADI
- PCG

Boundary Conditions for Flow
- fixed head
- prescribed time-varying head
- zero flow
- fixed boundary flux
- prescribed time-varying boundary flux
- areal recharge—variable in space, variable in time
- induced recharge from or discharge to stream; stream may not be directly connected to groundwater
- evapotranspiration dependent on distance surface to water table
- free surface

Solute Transport Processes
- advection
- hydrodynamic dispersion
- molecular diffusion
- linear equilibrium sorption
- first-order radioactive decay
- first-order chemical/microbial decay

Aquifer Conditions
- confined
- leaky-confined
- unconfined

Aquifer Systems
- single aquifer
- single aquifer/aquitard
- multiple aquifers/aquitards

Variable Aquifer Conditions in Space
- variable layer thickness
- confined and unconfined conditions in same aquifer
- aquitard pinchout
- aquifer pinchout

Changing Aquifer Conditions in Time
- desaturation of cells at water table
- resaturation of cells at water table
- confined/unconfined conversion

Parameter Representations
- hydraulic conductivity: heterogeneous (variable in space), anisotropic
- storage coefficient: heterogeneous
- longitudinal dispersivity: heterogeneous
- transverse dispersivity: heterogeneous
- sorption coefficient: homogeneous (single value for total model area)
- decay coefficient: homogeneous

Fluid Conditions
- density constant in time and space
- viscosity constant in time and space

Boundary Conditions for Solute Transport
- fixed concentration
- prescribed time-varying concentration
- zero solute flux
- specified constant or time-varying solute flux
- areal recharge of given (constant or time-varying) concentration
- induced infiltration of given (constant or time-varying) concentration
- concentration dependent solute flux

Sources/Sinks
- injection/production well with constant or time-varying flow rate
- injection well with constant or time-varying concentration
- injection well with constant or time-varying solute flux
- production well with aquifer concentration-dependent solute outflux
- springs with head-dependent flow rate and aquifer concentration-dependent solute flux

	functions				
test problem objective	function 1	function 2	function 3	function 4	function 5
test 1		X			X
test 2	X				
test 3		X			
test 4				X	X
test 5		X		X	
test 6				X	

Figure 1. Generic model functionality matrix; checked cells indicate that objective of test problem corresponds with a model function.

Table 2a. Example Performance Evaluation Table—Part 1

Test Case	Number of Nodes	Number of Time-Steps	Time-Step (days)	Convergence (Number of Iterations)	CPU Use (s)	RAM Use (Kbytes)
1	500	1	10	5	11	550
2	500	1	10	50 (maximum)	205	550
3	500	1	10	11	34	550
4	500	1	10	22	55	550
5a	500	1	10	7	21	550
5b	5000	1	10	9	309	3880
5c	500	10	1	21	80	550

Table 2b. Example Performance Evaluation Table—Part 2

Test Case	Sensitivity to Grid Size[a]	Sensitivity to Grid Orientation[b]	Sensitivity to Time Discretization[c]	Stability[d]	Reproducibility[e]
1	.1	.01	.1	satisfactory	100%
2	.02	.007	.2	unsatisfactory	100%
3	.03	.02	.1	satisfactory	90%
4	.001	.008	.3	satisfactory	80%
5a	.3	.04	.3	satisfactory	100%
5b	.25	.05	.25	satisfactory	100%
5c	.21	.045	.1	satisfactory	100%

[a] Sensitivity to grid size is determined by comparing the sum of absolute values of the differences in computed nodal values with the sum of computed nodal values divided by 2, employing two grid designs differing a factor 10 in number of active nodes.
[b] Sensitivity to grid orientation is determined by comparing the sum of absolute values of the differences in computed nodal values with the sum of computed values divided by 2, using two identical grid designs rotated 45° with respect to each other.
[c] Sensitivity to time discretization is determined by comparing the sum of absolute values of the differences in computed nodal values with the sum of computed values divided by 2, using for a constant period two time discretizations differing a factor 10.
[d] Stability is rated "unsatisfactory" if in one or more runs stability problems are encountered; otherwise stability is rated "satisfactory".
[e] Reproducibility is given as a percentage indicating the number of identical results in 10 tries.

In groundwater modeling, benchmarks are often represented by closed-form solution to the governing partial differential equation (i.e., analytical solutions). The numerical model to be tested provides solutions to the same equation at a limited number of discrete points in space and time. Assuming that the coding is correct, differences between the system responses described by the analytical solution and the numerical solution of the governing equation are due primarily to the approximate nature of the numerical method involved and to the limitations in computer accuracy, and are generally not randomly distributed. In many instances, the magnitude of these differences is related to the resolution in the discretization used in the computational scheme (Lapidus and Pinder, 1982). Theoretically, if the resolution increases such that the spatial and temporal step sizes approach zero, the differences between the numerical and the closed-form solution should disappear.

Designing and selecting appropriate test problems requires knowledge of the code's functionality. The selected conceptual and verification tests are designed and described in terms of test objectives (as related to code functions), problem description, and input data. It should be noted that if a computer code implementation of the analytical solutions has been used in this type of testing the resulting analytical modeling code should first be subject to appropriate testing. Verification of a coded analytical solution is restricted to comparison with independently calculated results using the same mathematical expression, (i.e., manual calculations, comparison with the results from computer programs coded independently by third-party programmers, or using general mathematical computer software systems such as Mathematica®[1] and Mathcad®[2]).

Implementing Level 1 testing to models designed to simulate groundwater flow in the saturated zone, tests may range from simple calculations based on gradient analysis, symmetry considerations, application of Darcy's law and mass balances to complex analytical solutions for piezometric head, groundwater flux, see page velocity, travel times, and capture zones. For solute transport, Level 1 tests may include gradient and symmetry analysis, mass balance and mass flux evaluations, and analytical solutions in terms of temporal and/or spatial concentration distributions. In some cases, complex benchmarks are derived by superposition of analytical solutions. Figure 3 provides an example of a Level 1 test problem.

The objective of Level 2 testing is threefold: (1) testing potentially problematic combinations of functions for which no Level 1 tests are available; (2) evaluating performance characteristics in addition to those established in Level 1 testing; and (3) demonstrating the code's applicability to typical real-world problems. Synthetic data sets are used to represent hypothetical, often simplified real-world groundwater systems. Due to the absence of an analytical solution to provide "ground truth", the code's test results are checked for unexpected or unexplained behavior and code intercomparison is used to obtain a relative measure of functionality and performance. This form of testing can be used to study the treatment of a number of naturally occurring conditions, including various hydrogeologic conditions (such as aquifer stratification and heterogeneities), physicochemical processes and ranges of their respective parameters, boundary and initial conditions, large variations in the gradient

---

[1] Registered trademark of Wolfram Research Inc, Champaign, Illinois.
[2] Registered trademark of Mathsoft, Inc., Cambridge, Massachusetts.

Program Name:	HOTWTR
Program Title:	Simulating Coupled Three-Dimensional Steady-State Groundwater Flow and Heat Transport in Saturated Media
Version:	1.1
Release Date:	September 1993
IGWMC Number:	FOS 67
Institution of Development:	U.S. Geological Survey, Denver, Colorado

TEST 03D: **Description**: multi-layer profile model; homogeneous aquifer of 13 by 1 cells horizontally, and 10 layers; aquifer parameters entered using zone option; internal and external heat conduction; no areal recharge; given heat flux condition at lower boundary (bottom); fixed temperature at opposite lateral boundaries; zero g.w. flux at lower, lateral, and upper boundaries.

Objective: to evaluate conductive heat flow through aquifer resulting from combined first and second type heat flow boundary conditions in the presence of heat in- or out-flux at the upper boundary (through conduction in overburden).

Results: Problem has zero g.w. flow; heat in-flux along lower and upper boundaries, and along upper part of high temperature boundary, and out-flux along lower part of high temperature boundary and along low temperature boundary.

Evaluation: results are conform expected behavior (qualitative conceptual test).

(A)

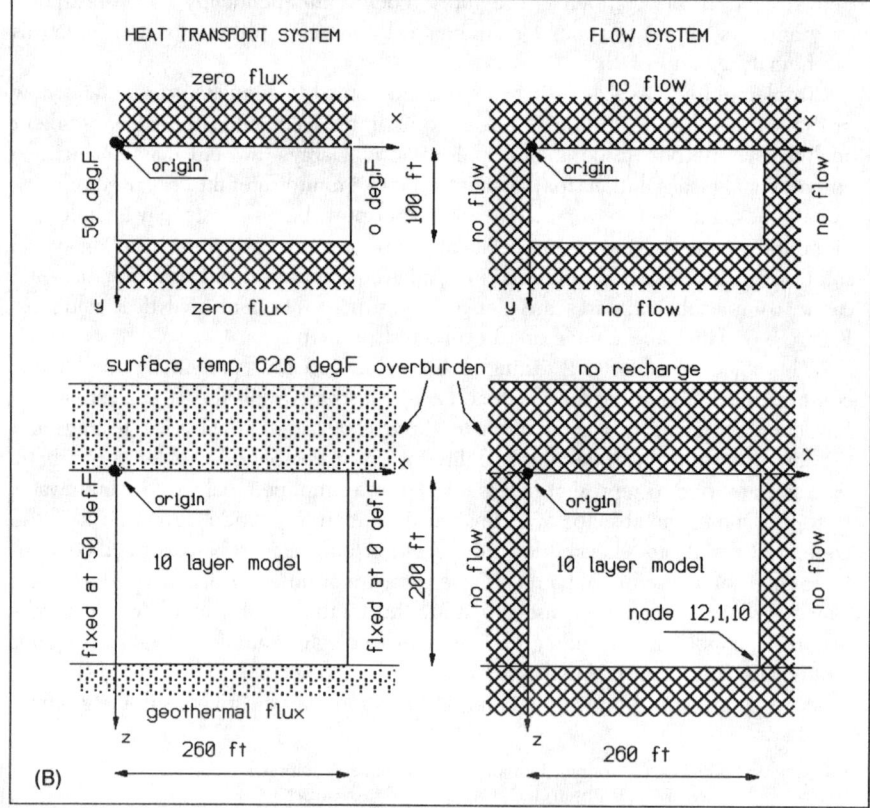

(B)

Figure 2. Example of a conceptual test problem. (A) Description of test; (B) test problem geometry; (C) resulting temperature distribution in °C.

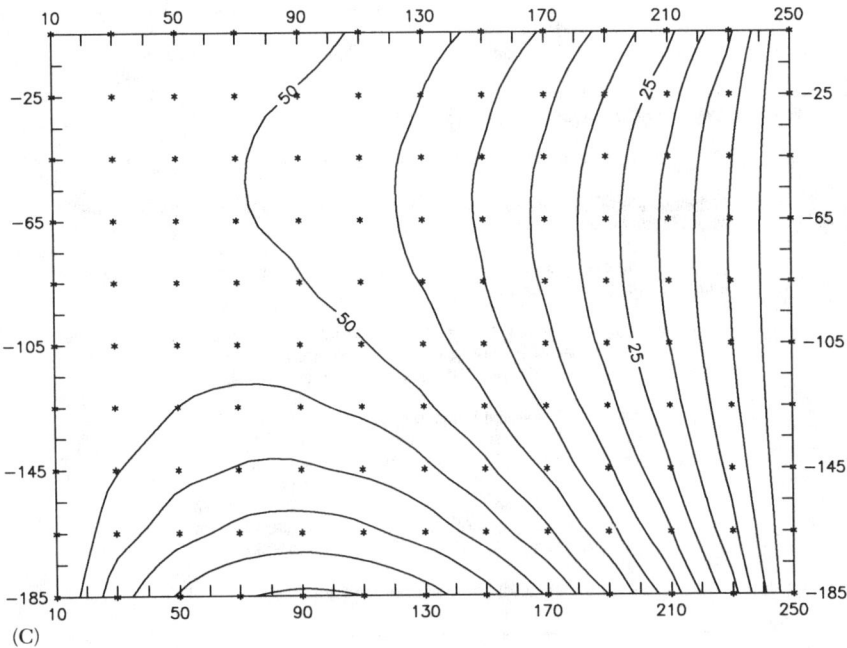
(C)

Figure 2. continued

of the dependent variable, (e.g., solute fronts), and sources and sinks. Some of the different conditions are summarized in Table 3.

The hypothetical problem used for a Level 2 test is defined by synthetic system parameters and system stresses and no independently observed system response is available. Therefore, testing takes place either by evaluating code behavior regarding such aspects as numerical consistency and stability, or by comparing the discretized predictions obtained with the numerical model tested with those obtained with another, preferably well-established, numerical model, using high-resolution spatial and temporal discretization schemes. As the absolute "truth" for these hypothetical problems is unknown, only a comparative verification of a model can be obtained. This approach provides a "relative" benchmark, which is in contrast with the "objective" benchmarks provided by Level 1 tests.

As is the case on Level 1, at Level 2 the grid orientation and resolution effects, as well as associated numerical dispersion and oscillations, are included in the evaluations. For this purpose, the test problems should be solved using a critical range of Peclet and Courant numbers. Accurate numerical solutions will be generated using codes that are known to effectively handle these critical conditions, high-resolution numerical grids, and small time steps. This approach is based on the idea that the smaller the discretization is in space and time, the better the approximate numerical solution will represent the real (unknown) solution of the governing partial differential equation (Huyakorn and Pinder, 1983). The resulting benchmarks are developed in a step-wise fashion, going from coarse resolution grids and large time steps to higher resolution grids and smaller time steps. After each run, computational differences will

```
Program name:            DYNFLOW
Program Title:           A Three-Dimensional Finite Element Ground Water Flow Model
Version:                 3.0
Release Date:            December 1993
Institution of Development: Camp Dresser and McKee

         TEST 26:   Radial Flow to Well, Conversion (Moench and Prickett, 1972)
```

<u>Description, assumptions, and limitations:</u> Transient, mixed phreatic and confined, two-dimensional radial uniform (horizontal) flow.

<u>Objectives:</u> Testing of no flow boundary conditions, 2-D horizontal flow, and storage management through conversion.

<u>Original reference(s):</u>   Moench, A. F. and T. A. Prickett. 1972. Radial Flow in an Infinite Aquifer Undergoing Conversion from Artesian to Water Table Conditions. Water Resources Research, 8(2):494-499.

<u>Test data:</u>
coefficient of storage, unconfined = 0.1
coefficient of storage, confined = .0001
aquifer thickness (ft) = 800 ft.
initial head (ft) = 801 ft.
aquifer transmissivity (ft^2/day) = 2 x 10^4 gpd/ft
flow of pumping well (ft^3/day) = 33636 ft^3/day
radius from well (ft) = 1000 ft

<u>Notes:</u> A full radial (360°) grid was used (see graph)

(A)

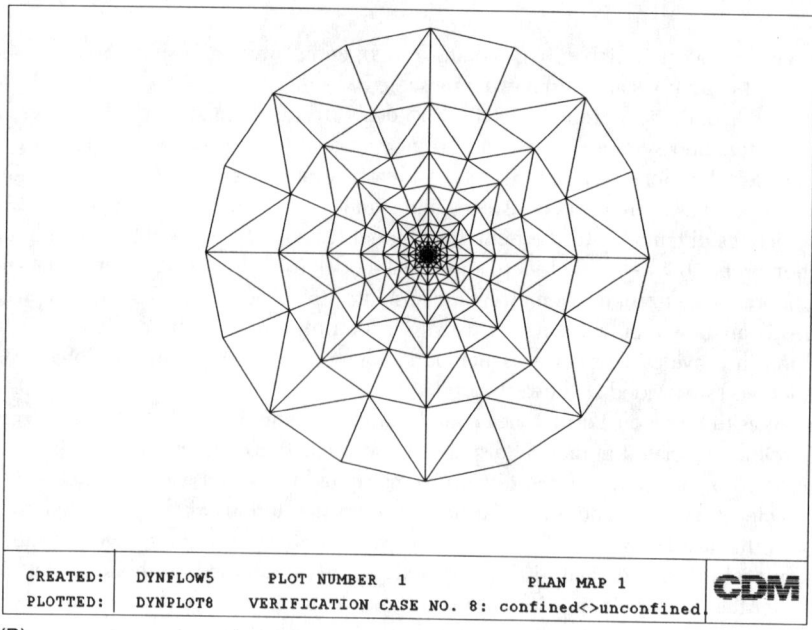

(B)

Figure 3. Example of a Level 1 test problem. (A) Description of test; (B) computational grid; (C) results in graphic and tabular form.

MODEL TESTING

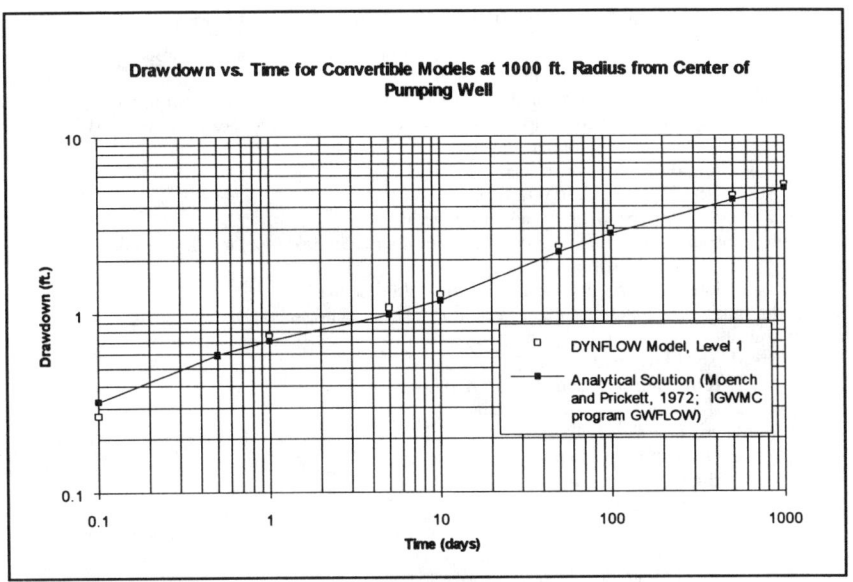

(C)

Figure 3. continued

be monitored. When further refinement, for example with a factor 2, does not provide significant changes in the computational results, the relative benchmark is established. If the simulation results in a Level 2 code intercomparison test do not deviate significantly, the "relative" or "comparative" test is considered successful. However, if significant differences occur, in-depth analysis of the results of simulation runs, performed with both codes, should occur. Figure 4 gives an example of a Level 2 test problem.

At Level 3 testing, the model (and its code) is compared with independently obtained field or laboratory data, determining the "degree of correlation" between calculated and independently observed cause-and-effect responses (van der Heijde and Elnawawy, 1992). This type of testing is sometimes referred to as "field or laboratory validation." The role of Level 3 testing in the protocol defined above is twofold: (1) determining how well a model's theoretical foundation and computer implementation describe actual system behavior; and (2) assessing of a code's applicability to real-world systems. The actual measured data of model input, system parameters, and

Table 3. Test Scenario for Three-Dimensional Solute Transport Codes

1. Solute transport in a steady-state uniform flow field in a large homogeneous isotropic aquifer (conceptual and analytical solutions are available):
    1.1. advection only (various boundary conditions, source locations, source strength)
    1.2. advection and dispersion (various boundary conditions, source locations, source strength, various ratios for longitudinal and transverse dispersion)
    1.3. advection, dispersion, and decay
    1.4. advection, dispersion, and retardation
    1.5. advection, dispersion, decay, and retardation
2. Solute transport to sink in a nonuniform steady-state flow field in a large homogeneous aquifer (analytical solutions available):
    2.1. advection and dispersion for various source/sink scenarios
3. Solute transport in a nonuniform flow field in a large homogeneous aquifer (analytical solutions not available):
    3.1. steady-state flow field:
        3.1.1. different solute source/sink conditions
        3.1.2. different boundary conditions
    3.2. nonsteady flow field with:
        3.2.1. constant source rates
        3.2.2. time-varying source rates
        3.2.3. time-varying boundary conditions
4. Nonuniform flow field in a heterogeneous anisotropic aquifer (no analytical solutions available):
    4.1. layered system:
        4.1.1. steady-state flow field:
            4.1.1.1. sources/sinks in various layers
            4.1.1.2. different boundary conditions
        4.1.2. nonsteady flow field with:
            4.1.2.1. sources/sinks in various layers
            4.1.2.2. different boundary conditions
    4.2. lens heterogeneities
    4.3. random heterogeneities

Source: van der Heijde, P. K. M. and Elnawawy, O. A., Quality assurance and quality control in the development and application of ground-water models, Report EPA/600/R-93/011, U.S. Environmental Protection Agency, Ada, OK, 1992.

system response are samples of the real system and inherently incorporate errors (NRC, 1990). An additional complexity is that often the data used for field validation are not collected directly from the field but are processed in an earlier study. Therefore, they are subject to inaccuracies, loss of information, interpretive bias, loss of precision, and transmission and processing errors, resulting in a general degradation of the data to be used in this type of testing.

The three-level code testing strategy is applied in a step-wise fashion using increasingly complex test problems. First, Level 1 testing (i.e., benchmarking or "objective" code verification) is conducted, and, if successfully completed, followed by Level 2 testing (i.e., code intercomparison or "subjective" verification). Eventually, it might gain further credibility by being subject to Level 3 testing (i.e., field or laboratory testing).

ASSESSMENT CRITERIA

An important aspect of code testing is the definition of informative and efficient measures for use as *evaluation* or *performance criteria*. Such measures should characterize quantitatively the results derived of the numerical model as compared to an

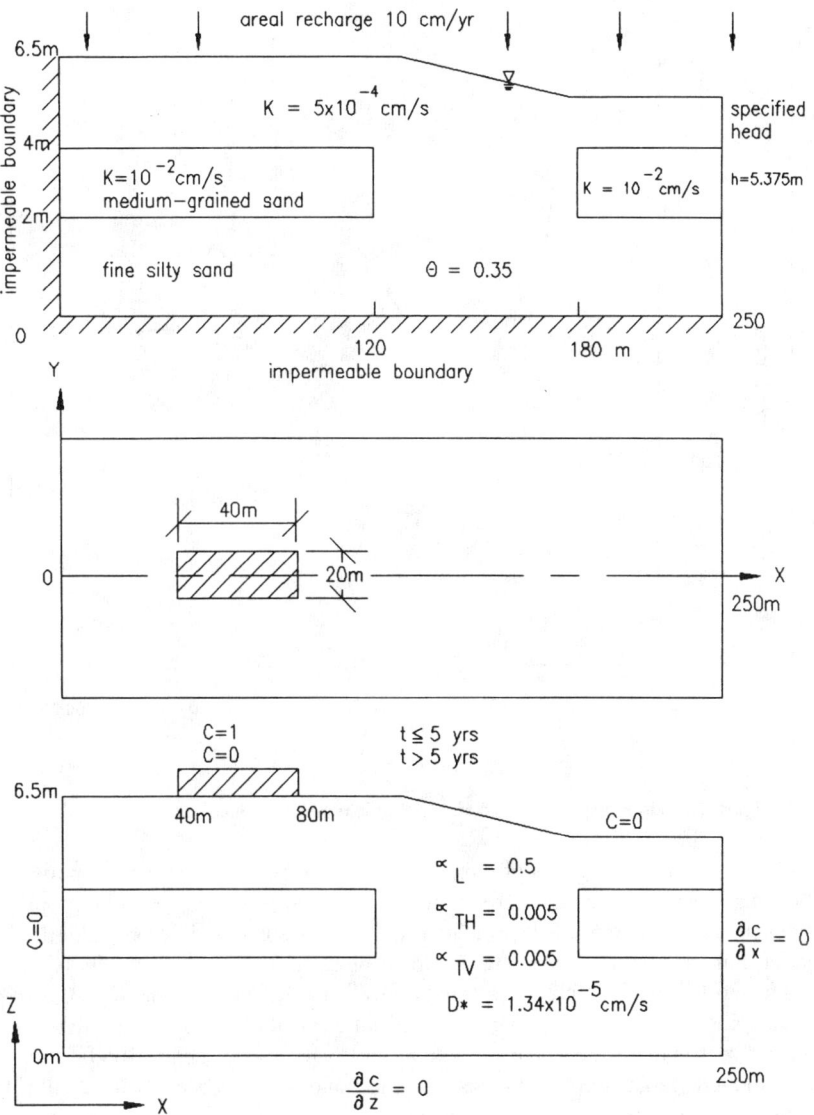

Figure 4. Geometry of a Level 2 example test problem (From van der Heijde, P. K. M. and Elnawawy, O. A., Quality assurance and quality control in the development and application of ground-water models, Report EPA/600/R-93/011, U.S. Environmental Protection Agency, Ada, OK, 1992).

established benchmark or the results obtained with a comparable numerical model (van der Heijde and Elnawawy, 1992).

To date, acceptance of code testing results has been primarily based on visual inspection of the graphical representation of the dependent variable computed with the numerical model and its benchmark (Figure 5). Graphical representation of test

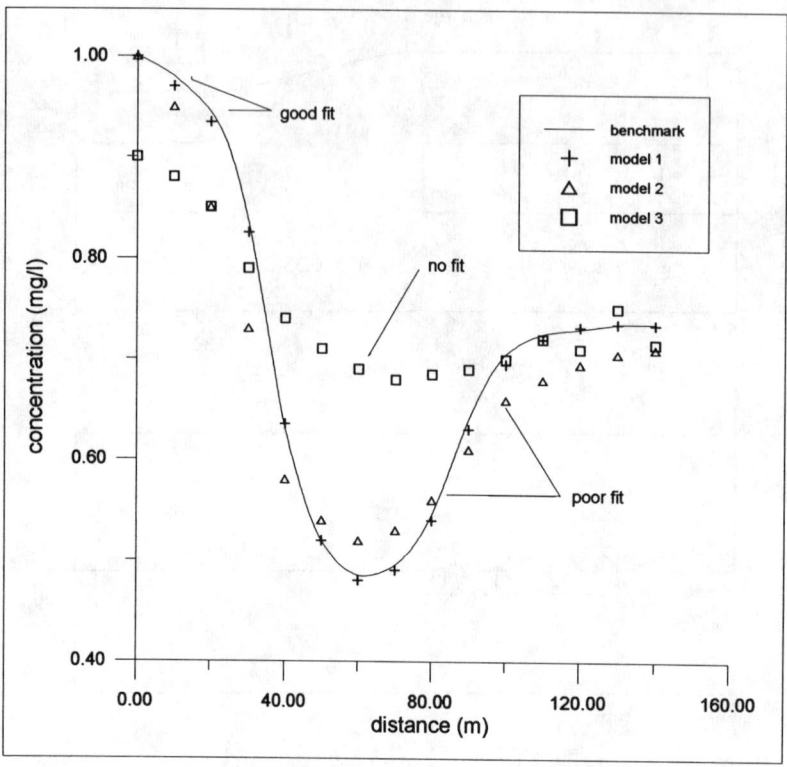

Figure 5. Visual inspection of goodness-of-fit between benchmark and tested models.

results should include graphs of the dependent variable(s) and other computed entities (e.g., mass balance, aquifer-stream fluxes) versus distance, and if appropriate, versus time. Two-dimensional spatial graphs depicting solution differences for the full model domain may also prove useful for evaluating code testing results (van der Heijde and Elnawawy, 1992). The conclusions from visual inspection of graphic representations of testing results may be described qualitatively by such attributes as "poor", "reasonable", "acceptable", "good", and "very good" (Beljin, 1988).

Although graphical comparison is an appropriate measure, acceptance should also be based on quantitative measures of the goodness-of-fit. There are three general procedures, coupled with standard linear regression statistics and estimation of error statistics, to provide such quantitative code performance assessment (Donigian and Rao, 1986):

- Paired-data performance—the comparison of simulated and observed data for exact locations in time and space;
- time and space integrated, paired-data performance—the comparison of spatially and temporally averaged simulated and observed data; and
- frequency domain performance—the comparison of simulated and observed frequency distributions.

Quantitative evaluation measures include mean error, mean absolute error, and root-mean-squared error comparable to those used in model calibration (Anderson and Woessner, 1992). The mean error (ME) of n data pairs is defined as the mean difference (i.e., deviation) between the dependent variable calculated by the numerical model h_c and the benchmark value of the dependent variable h_b.

$$\mathrm{ME} = \frac{\Sigma\,(h_c - h_b)}{n} \quad (1)$$

Because ME includes both positive and negative values which cancel each other, ME may not be the best indicator of an acceptable match. The mean absolute error (MAE) may provide a better indicator of agreement between model and benchmark, because it computes the absolute value of the deviations.

$$\mathrm{MAE} = \frac{\Sigma\,|(h_c - h_b)|}{n} \quad (2)$$

Also, the positive mean error (PME) and negative mean error (NME) may prove useful. The PME and NME are obtained by calculating the ME for the positive deviations and negative deviations, respectively.

Of special interest to model testing may be the direction of the deviations, indicating systematic overpredicting or underpredicting by the model. In that case, a composite measure may be appropriate, such as the mean error ratio (MER).

$$\mathrm{MER} = \frac{|\mathrm{ME}|}{\mathrm{ME}} \circ \frac{|\mathrm{NME}|}{\mathrm{PME}} \quad (\text{for } \mathrm{PME} < |\mathrm{NME}|) \quad (3a)$$

$$\mathrm{MER} = \frac{|\mathrm{ME}|}{\mathrm{ME}} \circ \frac{\mathrm{PME}}{|\mathrm{NME}|} \quad (\text{for } \mathrm{PME} > |\mathrm{NME}|) \quad (3b)$$

Another measure of model performance is the root-mean-squared error (RMSE). It is the average of the squared differences between the dependent variable calculated by the numerical model and its benchmark equivalent.

$$\mathrm{RMSE} = \frac{\Sigma\,(h_c - h_b)^2}{n} \quad (4)$$

For saturated flow codes, the focus of testing will be the computation of the distribution of hydraulic head (in space and time), head gradients, global water balance and individual boundary fluxes, flow velocity patterns (direction and magnitude), flow path lines, capture zones, and travel times. For solute transport codes, such evaluations will concern the concentration distribution in the aquifer (in space and time), global mass balance (per species), and breakthrough curves at observation points and sinks (wells, streams).

SUMMARY

Due to the rapid developments in groundwater modeling research and the demand for powerful analytical tools to evaluate alternative management strategies, an urgent need exists for comprehensive, systematic testing of all types of groundwater models aimed at establishing the operational characteristics and reliability of individual groundwater simulation codes. In recent years, the IGWMC has developed a methodology for model testing as part of a comprehensive QA program. This methodology consists of two elements: (1) a functionality analysis, performance evaluation, and applicability assessment protocol; and (2) a three-level code testing strategy. Various evaluation techniques and performance criteria are used to describe the results of testing. This methodology provides the rigor of a standardized procedure while offering the flexibility required by the wide variety in functionality of groundwater models. The methodology discussed in this paper is being implemented for two- and three-dimensional simulation of saturated flow, solute transport, and heat transport in porous media. Application to single-phase unsaturated zone flow models is still being prepared.

REFERENCES

Adrion, W. R., Branstad, M. A., and Cherniasky, J. C., Validation, verification and testing of computer software, in *Software Validation, Verification, Testing and Documentation*, Andriole, S. J., Ed., Petrocelli Books, Princeton, NJ, 1986, 81.

Anderson, M. P. and Woessner, W. W., *Applied Groundwater Modeling: Simulation of Flow and Advective Transport*, Academic Press, San Diego, CA, 1992.

ASTM, Standard Practices for evaluating environmental fate models of chemicals, in *Annual Book of ASTM Standards, E 978–84*, American Society for Testing and Materials, Philadelphia, 1984.

Beljin, M. S., Testing and validation of models for simulating solute transport in groundwater, Report GWMI 88–11, International Ground Water Modeling Center, Indianapolis, 1988.

Bredehoeft, J. D. and Konikow, L. F., Ground-water models: validate or invalidate, *Ground Water*, 31(2), 178, 1993.

Bryant, J. L. and Wilburn, N. P., Handbook of software quality assurance techniques applicable to the nuclear industry, Report NUREG/CR-4640, U.S. Nuclear Regulatory Commission, Washington, D.C., 1987.

Donigian, A. S., Jr. and Rao, P. S. C., Example model testing studies, in *Vadose Zone Modeling of Organic Pollutants*, Hern, S. C. and Melancon, S. M., Eds., Lewis Publishers, Chelsea, MI, 1986, 103.

Faust, C. R., Sims, P. N., Spalding, C. P., Andersen, P. F., and Stephenson, D. E., FTWORK: A three-dimensional groundwater flow and solute transport code, Report WRSC-RP-89–1085, Westinghouse Savannah River Company, Aiken, SC, 1990.

Gupta, S. K., Cole, C. R., Kincaid, C. T., and Monti, A. M., Coupled fluid, energy, and solute transport (CFEST) model; formulation and user's manual, Report BMI/ONWI-660, Battelle Office of Nuclear Waste Isolation, Columbus, OH, 1987.

Huyakorn, P. S. and Pinder, G. F., *Computational Methods in Subsurface Flow*, Academic Press, New York, 1983.

Huyakorn, P. S., Kretschek, A. G., Broome, R. W., Mercer, J. W., and Lester, B. H., Testing and validation of models for simulating solute transport in groundwater: development, evaluation, and comparison of benchmark techniques, Report GWMI 84-13, International Ground Water Modeling Center, Indianapolis, 1984.

HYDROCOIN Work Group, The International Hydrocoin Project, Level One: Code Verification, Swedish Nuclear Power Inspectorate, Stockholm, Sweden, 1988.

HYDROCOIN Work Group, The International Hydrocoin Project, Level Two: Model Validation, Swedish Nuclear Power Inspectorate, Stockholm, Sweden, 1990.

INTRACOIN Work Group, International Nuclear Transport Code Intercomparison (INTRACOIN) Study: Final Report Level One—Code Verification, Swedish Nuclear Power Inspectorate, Stockholm, Sweden, 1984.

INTRACOIN Work Group, International Nuclear Transport Code Intercomparison (INTRACOIN) Study: Final Report Levels Two and Three—Model Validation and Uncertainty Analysis. Swedish Nuclear Power Inspectorate, Stockholm, Sweden, 1986.

Konikow, L. F. and Bredehoeft, J. D., Ground-water models cannot be validated. *Adv. Water Resour.*, 15, 75, 1992.

Lapidus, L. and Pinder, G. F., *Numerical Solution of Partial Differential Equations in Science and Engineering*, John Wiley and Sons, New York, 1982.

Larsson, A., The international projects INTRACOIN, HYDROCOIN, and INTRAVAL, *Adv. Water Resour.*, 15, 85, 1992.

National Research Council (NRC), *Ground Water Models: Scientific and Regulatory Applications*, National Academy Press, Washington, D.C., 1990.

Pinder, G. F. and Bredehoeft, J. D., Application of the digital computer for aquifer evaluation, *Water Resour. Res.*, 4, 1069, 1968.

Pinder, G. F. and Frind, E. O., Application of Galerkin's procedure to aquifer analysis, *Water Resour. Res.*, 8, 108, 1972.

Prickett, T. A. and Lonnquist, C. G., Selected digital computer techniques for groundwater resource evaluation, Bulletin 55, Illinois State Water Survey, Champaign, IL, 1971.

Reeves, M., Ward, D. S., Johns, N. D., and Cranwell, R. M., Theory and implementation for SWIFT II, the sandia waste-isolation flow and transport model for fractured media; Release 8.84, Report NUREG/CR-3328, U.S. Nuclear Regulatory Commission, Washington, D.C., 1986.

Sims, P. N., Andersen, P. F., Stephenson, D. E., and Faust, C. R., Testing and benchmarking of a three-dimensional groundwater flow and solute transport model, in Proc. Conf. Solving Ground Water Problems with Models, National Water Well Association, Columbus, OH, 1989.

SKI, Proceedings of Symposium on Verification and Validation of Geosphere Performance Assessment Models, Swedish Nuclear Power Inspectorate, Stockholm, Sweden, 1987.

SKI, Proceedings of Symposium on Validation of Geosphere Flow and Transport Models, Swedish Nuclear Power Inspectorate, Stockholm, Sweden, 1990.

Tsang, C-H., Comments on model validation, *Transp. Porous Media*, 2, 623, 1987.

Tsang, C-H., The modeling process and model validation, *Ground Water*, 29, 825, 1991.

van der Heijde, P. K. M., Huyakorn, P. S., and Mercer, J. W., Testing and validation of ground water models, in Proc. Conf. on Practical Applications of Groundwater Models, National Water Well Association, Dublin, OH, 1985.

van der Heijde, P. K. M. and Park, R. A., U.S. EPA groundwater modeling policy study group; report of findings and discussion of selected groundwater modeling issues, Report GWMI 86-13, International Ground Water Modeling Center, Indianapolis, 1986.

van der Heijde, P. K. M. and Elnawawy, O. A., Quality assurance and quality control in the development and application of ground-water models, Report EPA/600/R-93/011, U.S. Environmental Protection Agency, Ada, OK, 1992.

van der Heijde, P. K. M., Paschke, S. S., and Kanzer, D. A., Ground-water flow and solute transport model functionality testing and performance evaluation, in Proc. Thirteenth AGU Hydrology Days, Morel-Seytoux, H. J., Ed., Fort Collins, CO, 1993.

Ward, D. S., Reeves, M., and Duda, L. E., Verification and field comparison of the sandia waste-isolation flow and transport model (SWIFT), Report NUREG/CR-3316, U.S. Nuclear Regulatory Commission, Washington, D.C., 1984.

Watson, D. B. and Brown, S. M., Testing and Evaluation of the SESOIL Model, Anderson-Nichols Company, Palo Alto, CA, 1985.

Wolfram, S., *Mathematica: A System for Doing Mathematics by Computer,* 2nd ed., Addison-Wesley, Redwood City, CA, 1991.

CHAPTER 4

The Value of Postaudits in Groundwater Model Applications

Leonard F. Konikow

INTRODUCTION

The U.S. Water Resources Council (1980) noted several significant attributes of aquifers that make these groundwater reservoirs important to water-resources planning. These attributes include (1) widespread occurrence, (2) general capacity to transmit water over great distances, (3) generally large storage capacity, (4) usually good chemical and bacterial quality, and (5) manageability.

About half the population of the United States uses groundwater for drinking or other domestic purposes, and nearly 40% of the nation's agricultural irrigation water is supplied by groundwater. Data on water use in the United States over 5-year intervals (Solley et al., 1993) show an increasing trend in groundwater withdrawals, which more than doubled from about 34 billion gallons per day in 1950 to about 79.4 billion gallons per day in 1990. The reported groundwater use in 1990 represents an increase of about 8% from 1985, but a slight decrease of about 4% from the historical peak in 1980. Most of the use, and most of the increase, are for irrigation supplies. The 1990 groundwater withdrawals represent about 23.5% of the total freshwater withdrawals (excluding instream use for hydroelectric power) in the United States.

The increase in groundwater use partly reflects the widespread recognition by local and regional water-resources planners and managers that groundwater must be considered as an integral part of the total water resource. It is also evident that issues of groundwater supply cannot be divorced from consideration of groundwater quality, and ultimately both issues must be reconciled with economic considerations. Because all water-supply sources are subject to both natural and human-induced variations in flow and storage, planners and managers often rely on predictions of future conditions as a partial basis for their decisions and actions. Matalas et al. (1982) state:

> Predictions of future and recurrent events enter into the risk-cost assessment on which water-management decisions are based. It is only after the fact that a prediction can be said to be right or wrong. However, predictions must be judged *a priori* to be good or bad and acted upon accordingly. The judgment is based on the credibility of predictions derived from their supporting scientific and philosophical arguments . . .

Process-simulating, deterministic, mathematical, and groundwater flow and transport models are being used with increasing frequency to predict changes in groundwater systems. The underlying philosophy is that given a high degree of understanding of the processes by which stresses on a system produce subsequent responses in that system, the system's response to any set of stresses can be defined or predetermined through that understanding of the governing or controlling processes, even if the magnitude of the new stresses falls outside of the range of historically observed stresses (Konikow and Patten, 1985). Predictions made this way assume an understanding of cause and effect relations. Typically, the conceptual model of the system is represented by partial differential equations that describe the interdependence of governing processes, driving forces, system properties, stresses, and boundary conditions, and the equations are solved numerically on digital computers.

The accuracy of such deterministic predictions thus depends on several factors, including: (1) how closely the concepts of the governing processes reflect the actual processes that are controlling the behavior of the system; (2) how accurately are the properties and boundaries of the system defined in the domain over which the processes and stresses are acting; (3) how well known is the state of the system at some point in time (either past or present); and (4) how reliable are estimates (or predictions) of the future stresses on the system. The prediction arises from the numerical solution of the governing equations, which may be expressions of physical and chemical laws. Therefore, if we can write the appropriate governing equations, and if we can accurately solve these equations, then we can determine mathematically the future state of the groundwater system *if* we know its properties and boundaries, future stresses, and present state of the system. Predictions made through deterministic modeling are inherently based on the belief that *the present is the key to the future* (Konikow and Patten, 1985). However, this presumption commonly proves invalid when applied to groundwater systems because typically too few data are available to define uniquely the current state of the system, and the system properties can be estimated only with a large degree of uncertainty. Furthermore, processes and boundary conditions that are negligible or insignificant under the past and present stress regime may become nontrivial or even dominant under a different set of imposed stresses. Thus, a conceptual model founded on observed behavior of a groundwater system may prove to be inadequate in the future, when existing stresses are increased or new stresses are added.

Predictive applications of deterministic groundwater models include assessments of groundwater availability, assessments of subsurface contaminant migration at toxic-waste sites, design of cleanup schemes for contaminated aquifers, prediction of saltwater intrusion in response to withdrawals from supply wells, and performance assessment for proposed radioactive-waste repositories. The latter might represent the most extreme example of reliance on groundwater model predictions, because regulators require the assurance of site safety for 10,000 years into the future. It is not uncommon for the accuracy and reliability of model predictions to be a major factor of contention in court cases concerning liability at contaminated sites or environmental impact assessments.

In light of the economic and legalistic importance of these types of model predictions, it may be asked whether there is any evidence that groundwater models can indeed accurately predict the future state of a groundwater system. One way to assess the predictive accuracy of groundwater models is by comparing the actual response

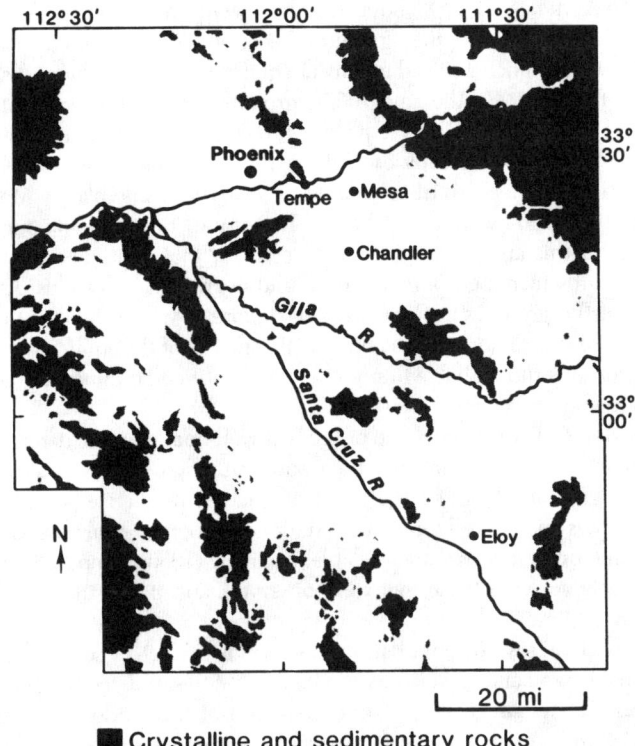

Figure 1. Map of study area in Arizona showing main geographic and physiographic features within the study area. (From Konikow, L. F., *Ground Water,* 24(2), 173, 1986. With permission.)

of a groundwater system with that predicted by the model, and performing such a comparison a sufficiently long time after the prediction was made so that the state of the system at the time of evaluation will not be dominated by its "memory" of conditions during the calibration period for the model (typically, this may require several years). This type of assessment of model reliability has been called a "postaudit" (Konikow, 1986; Alley and Emery, 1986; Konikow and Swain, 1990; Reichard and Meadows, 1992; and Anderson and Woessner, 1992). However, there are only a few examples in the literature in which this type of postaudit has been completed for areas for which calibrated models had been used to make predictions. The purpose of this paper is to review the results of postaudits performed on several different types of hydrogeologic problems, and to summarize the value and limitations of such an analysis.

EXAMPLE POSTAUDITS

Regional Water-Level Changes

The Salt River Valley and the lower Santa Cruz River basin are located near Phoenix, Arizona and are the two largest agricultural areas in Arizona (see Figure 1).

According to Anderson (1968), about 3240 km² were under cultivation at that time. Because of the arid climate, the agricultural economy depends on a reliable source of irrigation water. Since 1923, the alluvial aquifer system had been developed extensively in the area. By the mid-1960s, groundwater withdrawals totaled about 3.9×10^9 m³ (3.2 million acre-ft), equivalent to about 80% of the total annual water supply. Such withdrawals greatly exceed the rate of groundwater recharge and resulted in water-level declines of as much as 6 m/year in some places. Maximum declines from 1923–1964 were about 110 m. Because of the economic importance of groundwater in this area, there was much concern that continued declines would cause significantly increased pumping costs and decreased well yields. As part of an assessment of the groundwater resources of the area, Anderson (1968) constructed and calibrated a two-dimensional electric-analog model of the aquifer system, in part, ". . . to determine the probable future effects of continued groundwater withdrawals in central Arizona."

This model study represents one of the first well-documented, deterministic, distributed-parameter, model analyses of a groundwater system. It was also done sufficiently long ago that a long-term (10-year or greater) predictive period had passed since the analysis and related forecasts were done. Also, historical observations of the aquifer for the forecast period are available. Konikow (1986) compared the predicted water-level changes with those that were observed, and the results of his postaudit are summarized here.

The alluvial valleys are underlain by thousands of feet of sediments, including thick permeable sand and gravel units. Anderson (1968) simulated the uppermost 365 m of the aquifer system. He noted one limitation of the model was that values of transmissivity and storativity were not corrected with time to account for the interaction of vertical variations in properties and transient changes in water levels. The model was based on the assumption that prior to 1923 an equilibrium existed in the aquifer in which recharge balanced discharge. The model was calibrated by adjusting aquifer properties and boundary conditions to match observed historical changes in water levels during 1923–1964.

Although the model is used to predict future responses, the modeler must predict and impose the future stresses. In this case, past trends provided the basis for the simplifying assumption that the future amount and distribution of pumping would remain about the same as during the most recent 6-year period (1958–1964) (see Figure 2). Anderson cautioned, "The amount of water pumped probably will be less because the ever-increasing pumping lifts will make pumping increasingly expensive, and economically marginal lands may be withdrawn from cultivation." Thus, this assumed continuation of the recent pumpage patterns ". . . will cause the predicted water-level declines to be greater than are actually probable."

Water-level measurements are available for 77 wells within the study area for both 1964 and 1974. For these wells, the water-table decline was predicted to average about 25 m after 10 years and ranged from 4.5 to 65 m. However, measurements of the actual change in water level in the same wells showed an average decline of only 2.7 m, and the observed change ranged from a decline of 28 m to a rise of 45 m. The relation between observed and predicted water-level changes is shown in Figure 3. Data from all but three wells fall below the line connecting equal values of predicted and observed changes. This indicates poor predictive accuracy and the presence of a

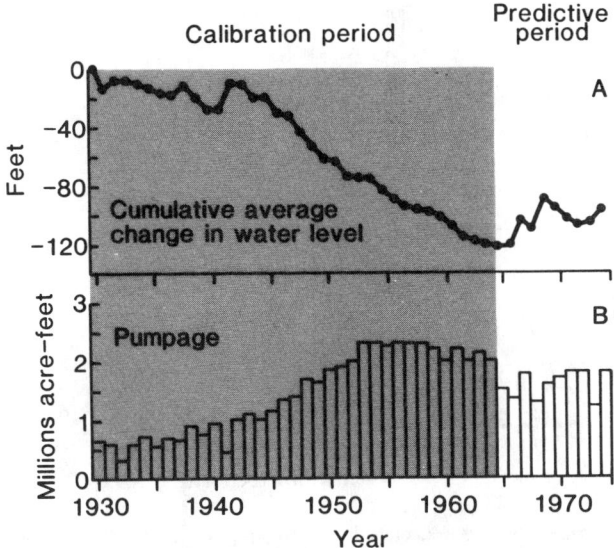

Figure 2. Average water-level change in the Tempe-Mesa-Chandler area (A) and annual groundwater withdrawals in the Salt River Valley, Arizona (B), 1930–1974. ((A) From Konikow, L. F., *Ground Water*, 24(2), 173, 1986. With permission. (B) Modified from Babcock, H. M., Arizona State Land Department Water Resources, Report 43, 1970.)

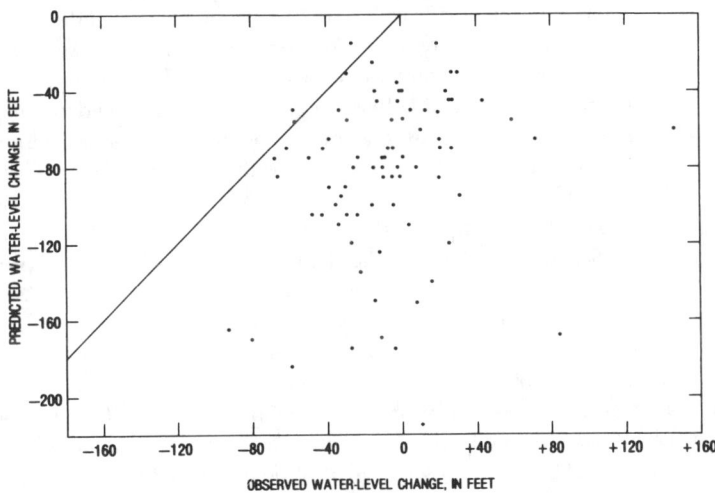

Figure 3. Relation between predicted and observed changes in water level in the Tempe-Mesa-Chandler area of the Salt River basin, Arizona, 1964–1974. Solid line shows where predicted equals observed values. (From Konikow, L. F., *Ground Water*, 24(2), 173, 1986. With permission.)

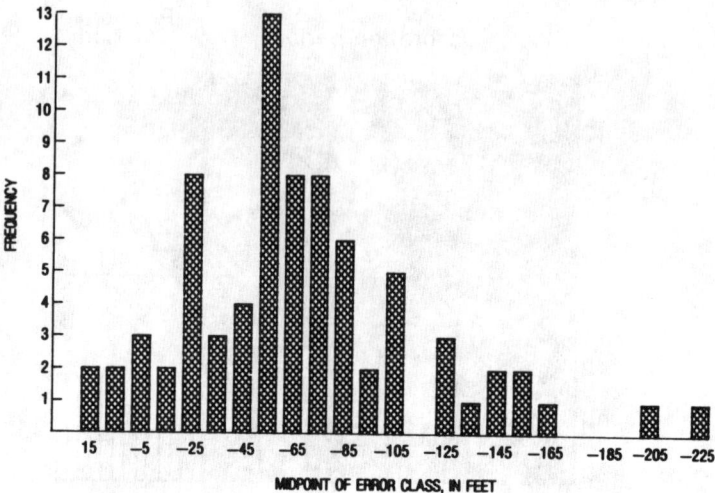

Figure 4. Histogram showing frequency distribution of errors for the flow model of the Tempe-Mesa-Chandler area of the Salt River basin, Arizona, 1964–1974. (From Konikow, L. F., *Ground Water*, 24(2), 173, 1986. With permission.)

bias in the model predictions. Also, the data show a relatively wide scatter, indicating that the model prediction is imprecise. A frequency distribution of the errors is shown in Figure 4. The errors are approximately normally distributed and have a mean of −22 m, a standard deviation of 14 m, and range from 3.4 to −69 m.

One obvious source of error is the error in the assumed future stress. Anderson (1968) proved accurate in forecasting that actual pumpage will likely be less than assumed (as shown on the right side of Figure 2B), as the actual withdrawals during 1965–1974 were 22% less than assumed. The volume of water represented by this error is equivalent to a saturated aquifer thickness of 14 to 27 m in the irrigated area of the basins. Therefore, it is possible that the gross error in assumed pumpage can account for a large part or all of the average bias in the predicted water-level changes.

However, removing the bias does not eliminate the spread in the error distribution. To help assess whether this lack of precision is related to errors in the assumed pumpage, Konikow (1986) compared the spatial distribution of errors in assumed pumpage for the Salt River Valley with the spatial distribution of errors in predicted water-level change. Figure 5 shows that there is little correlation between these two factors (r = −0.086), indicating that the relatively large spread in errors is probably not attributable to a variance in the accuracy of pumpage estimates. Hence, there are other significant sources of error.

The errors were mapped and contoured. As shown in Figure 6, it appears that the errors are not just randomly distributed in space, but rather exhibit some spatial correlation. In general, the errors seem greatest near the centers of the basins and least near their margins. However, with one partial exception, the error pattern does not correspond closely with the pattern for any single factor for which data are available. The one partial exception is land subsidence. Because significant subsidence is known to be occurring in parts of this study area (Schumann, 1974; Laney et al., 1978),

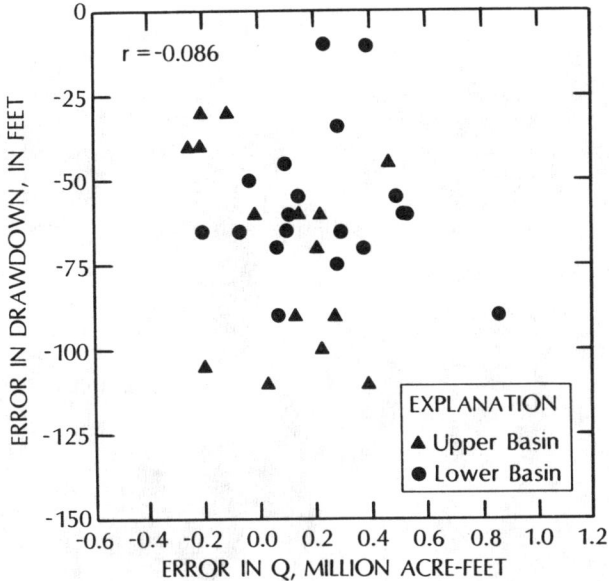

Figure 5. Relation between the error in the predicted water-level change and the error in the estimated pumpage per township in the upper (eastern) and lower (western) basins of the Salt River Valley, central Arizona, 1965–1974. (From Konikow, L. F., *Ground Water*, 24(2), 173, 1986. With permission.)

the possibility was considered that this process, which was not explicitly represented in the model, could account for some of the error in predicted water-level change. The land subsidence is caused by the compaction of sediments in the alluvial-fill basins. The compaction, in turn, is related to the compressibility of the sediments and to the decline in head in the aquifer. Laney et al. (1978) show that the greatest subsidence within the study area occurred in the lower Santa Cruz basin, where as much as 3.8 m of subsidence were observed from 1905–1977, whereas less than 1.5 m of subsidence are reported for the Salt River Valley. The greatest subsidence (2.1 to 3.8 m) occurred in two areas that total about 310 km^2. The largest of these two areas is about 285 km^2 and is located in the southwestern part of the basin near Eloy.

Hydrologically, the compaction that causes land subsidence also acts as a source of water to the aquifer system. If this fluid source were not accounted for in the model, then the water produced by compaction might cause the actual drawdowns to be less than would occur otherwise, which is indeed the nature of the predictive error observed here. However, if this hypothesis were correct, we would expect to see some correlation between the amount of subsidence and the magnitude of the error. Comparison of these two factors shows that the maximum subsidence zone near Eloy corresponds closely with a high error (> 30 m) zone in that same area, but that elsewhere there is no obvious association between patterns of errors and subsidence. As much as 1.2 m of subsidence were observed in the Eloy area during 1965–1974. As a first approximation, if we assume that 1.2 m of subsidence generates 1.2 m of water and if the specific yield of the aquifer in that area averages 0.15, then the subsidence may

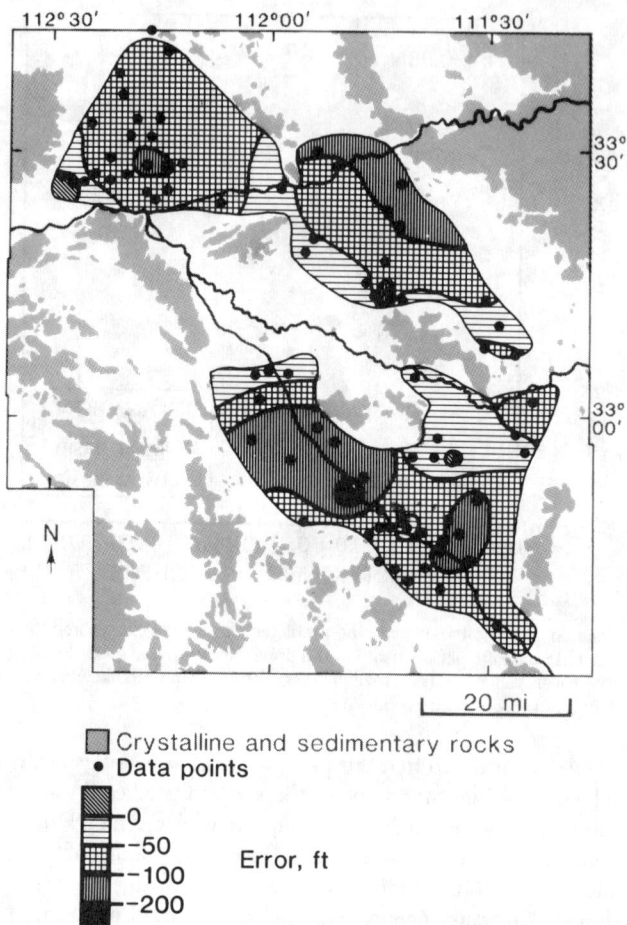

Figure 6. Map showing the spatial distribution of errors in predicted water-level change, 1965–1974, in the Salt River and lower Santa Cruz River basins, Arizona. (From Konikow, L. F., *Ground Water,* 24(2), 173, 1986. With permission.)

cause the water level to be 8.2 m higher than it would have been otherwise. In the remainder of the study area the subsidence during 1965–1974 averaged about 0.3 m, which may similarly be equivalent to about 2.1 m of head. These estimates are equivalent to about 20% of the error around Eloy and less elsewhere. The water produced by the compaction process would act as a delayed-yield or transient-leakage phenomena, which would cause long-term water-level declines to unit stresses to be less than they would otherwise if there were no leakage. It thus appears that, for the Eloy area, the lack of consideration of the land-subsidence process in the model contributed to the error in predicted water-level changes during 1965–1974. For the rest of the modeled area, there is no evidence to indicate that this could have been a significant factor.

In reexamining the history of the study area for the 1965–1974 period, it is believed that the following factors contributed to lower than anticipated net withdrawals:

(1) farmers in certain parts of the area ceased operations because of economic or other reasons, thereby eliminating their withdrawals for irrigation; (2) other farmers took measures to increase irrigation efficiency, thereby reducing their water requirements; (3) in some areas many wells were deepened to obtain water from deeper permeable zones, thereby reducing the drawdown relative to that resulting from equivalent withdrawals from wells that penetrate less of the aquifer; (4) additional surface water was available for irrigation, reducing the dependence on groundwater for supply; (5) in April 1965 an unusual flow event resulted in about 2.5×10^7 m^3 of recharge from infiltration in the channel of the Salt River, which is otherwise normally dry (Briggs and Werho, 1966); and (6) significantly greater than average precipitation occurred in the Salt River watershed in 1972–1973, and the subsequent unusually large runoff resulted in the direct recharge of about 6.2×10^8 m^3 of water along the river channels during 1973–1974 (Babcock, 1975). Local variations in these same factors also could have contributed to the variability in the error distribution.

A significant lack of uniformity with depth in the properties of the sediments would imply the existence of significant variations in hydraulic conductivity and specific storage with depth. This situation could induce significant vertical components of flow in places, which obviously could not be represented in the two-dimensional model of Anderson, and might thus be a contributor to the predictive error. In fact, Laney (R. L. Laney, U.S. Geological Survey, written commun., 1985) states that in much of this area at least three layers of differing transmissivity would be required to adequately describe the system. For example, the uppermost and highest transmissivity layer has gradually been dewatered since the 1940s and now is saturated only in parts of the basin (R. L. Laney, U.S. Geological Survey, written commun., 1985). This implies that the effective transmissivity may have changed significantly when and where dewatering has occurred. A rigorous test of this hypothesis would require the construction of alternative two- and three-dimensional models.

This example from central Arizona illustrates the weakness of basing a prediction of aquifer responses on a single set of assumed future stresses. Because the uncertainty of the 1965–1974 stresses was not assessed, we do not know whether the actual 1965–1974 responses fall within some associated confidence interval; hence, we cannot make a judgment based solely on these predictive errors as to whether the model is "good" or "bad". In cases like this, it would be preferable to assess the uncertainty in estimated (or assumed) future stresses and then present the forecasts as a range of responses with associated probabilities of occurrence or confidence intervals. Because Anderson had indicated (correctly) that the assumed stresses were probably greater than would occur, the predictions can be viewed as a "worst-case" estimate. From that perspective, the model predictions are reasonably accurate. From a water management perspective, the prediction of additional significant water-table declines was one factor contributing to water management decisions and actions leading to reduced groundwater withdrawals after 1964, which, in turn, was a major source of predictive error.

Point Source of Contamination

The Idaho National Engineering Laboratory (INEL) is located on 890 mi^2 of semiarid land in the eastern Snake River Plain of southeast Idaho. The facility, formerly called the National Reactor Testing Station, is now operated by the U.S.

Department of Energy for testing various types of nuclear reactors. Robertson (1974) reports that several facilities at the site generate and discharge low-level radioactive and dilute chemical liquid wastes to the subsurface through seepage ponds and disposal wells. The two most significant waste discharge facilities, the test reactor area (TRA) and the Idaho Chemical Processing Plant (ICPP), have discharged wastes continuously since 1952. This discussion of the INEL site and associated model is largely extracted and paraphrased from the reports of Robertson (1974) and Lewis and Goldstein (1982), to which the reader is referred for additional details.

As described in those two reports, the eastern Snake River plain is a large structural and topographic basin about 320 km long and 95 km wide. It is underlain by 600 to 3000 m of thin basaltic lava flows, rhyolite deposits, and interbedded alluvial and lacustrine sediments. These formations contain a vast amount of groundwater and comprise the major aquifer in Idaho, which is known as the Snake River plain aquifer. Groundwater flow is generally to the southwest at relatively high velocities (1.5 to 6 m/day), according to the reports on this area. The principal water-bearing zones occur in the basalts, the permeability fabric of which is highly heterogeneous, anisotropic, and complicated by secondary permeability features, such as fractures, cavities, and lava tubes.

Because of the concern about groundwater contamination resulting from waste discharge, in 1973 Robertson (1974) developed a digital solute-transport model to simulate the underlying aquifer system to help analyze groundwater flow and contaminant transport at the site. The numerical model was based on the method of characteristics (Bredehoeft and Pinder, 1973). Robertson first calibrated a flow model for a 6700 km^2 area, and then calibrated the transport model for a smaller part of that area in which contamination was of concern. The calibration of the transport model was based on a 20-year history of contamination, as documented by samples from about 45 wells near and downgradient from the known point sources of contamination. These data showed that chloride and tritium had spread over a 39 km^2 area and migrated as far as 8 km downgradient from discharge points. The distribution of waste chloride observed in November 1972 is shown in Figure 7. Robertson notes that the degree of observed lateral dispersion in the plumes is particularly large.

Robertson used the calibrated transport model to predict future concentrations of chloride, tritium, and strontium-90 for the years 1980 and 2000 under a variety of alternative possible future stresses. The scenario that came closest to what actually occurred for the chlorides included assumptions that disposal continues at 1973 rates and the Big Lost River recharges the aquifer in odd-numbered years. The projections indicated that by 1980 the leading edges of both the chloride (see Figure 8) and the tritium plumes would be at or near the INEL southern boundary.

Lewis and Goldstein (1982) report that eight wells were drilled during the summer of 1980 near the southern boundary to help fill data gaps and to monitor contaminants in groundwater flowing across the INEL boundary. They also used the data from the eight wells to help evaluate the accuracy of Robertson's predictive model—in effect, performing a postaudit. The distribution of waste chloride observed in October 1980 is shown in Figure 9. A comparison of this figure with Figure 7 indicates that the leading edge of the chloride plume had advanced 4 to 5 km during that 8-year period, and that the highest concentrations increased from about 85 mg/L to about 100 mg/L.

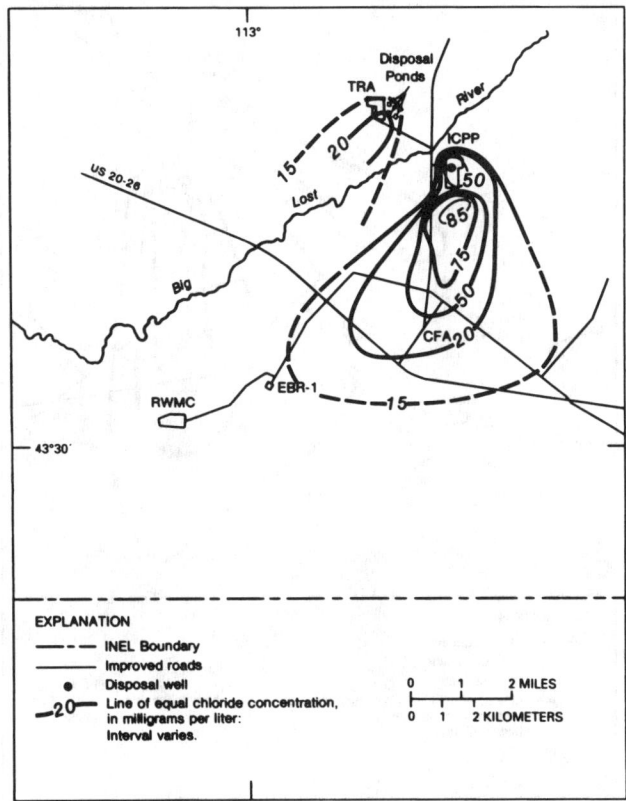

Figure 7. Map of ICPP-TRA vicinity showing observed distribution of waste chloride in the Snake River Plain aquifer water in 1972. (From Robertson, J. B., U.S. Geol. Survey Open-File Report IDO-22054, 1974.)

A comparison of Figures 8 and 9 indicates that although the observed and predicted plumes show general agreement in the direction, extent, and magnitude of contamination, there exist some apparently significant differences in detail. The observed plume is broader and exhibits more lateral spreading than was predicted, and has not spread as far south and as close to the INEL boundary as was predicted. Also, the predicted secondary plume north of the Big Lost River, emanating from the TRA, was essentially not detected in the field.

Lewis and Goldstein (1982) presented a number of factors that they felt contributed to the discrepancy between predicted and observed results, including (1) dilution from recharge decreased during 1977–1980 because of below-normal river flow, (2) chloride disposal rates at the ICPP facility increased during the several years preceding 1980, (3) the model grid may have been too coarse, (4) the model calibration incorporated inaccurate hydraulic and transport parameters, (5) vertical components of flow and transport may be significant in the aquifer but cannot be evaluated with the two-dimensional areal model, (6) there may be too few wells to map the actual plumes accurately, and some existing wells may not be constructed properly to yield

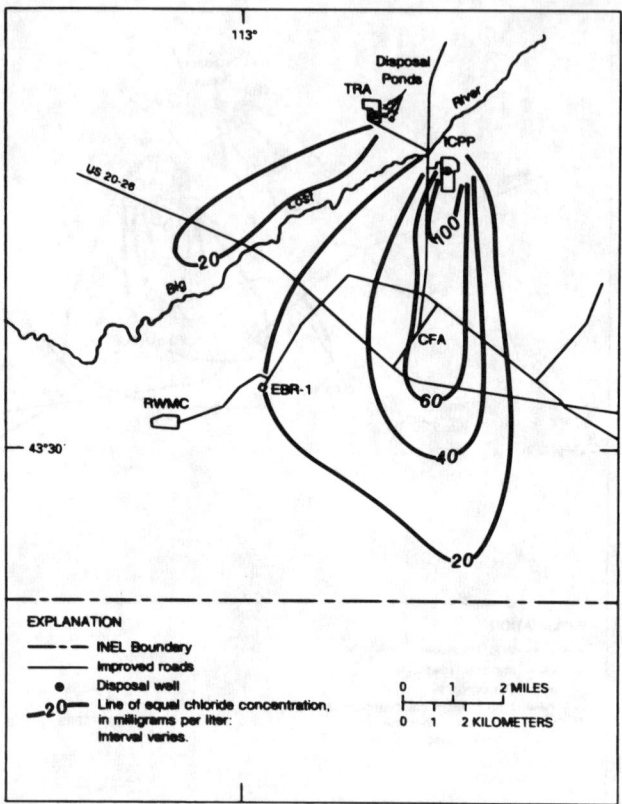

Figure 8. Model-projected distribution of waste chloride in the Snake River Plain aquifer for 1980, ICPP-TRA vicinity, assuming disposal continues at 1973 rates and the Big Lost River recharges the aquifer in odd numbered years. (From Robertson, J. B., U.S. Geol. Survey Open-File Report IDO-22054, 1974.)

representative measurements, and (7) the numerical method introduced some errors. Although these factors can be expanded upon, and additional factors added, it is extremely difficult to assess the contribution of any single factor to the total error. Recalibration of the earlier model using the now extended historical record could be used to test some of these hypotheses. Other factors could only be tested if new models are developed that incorporate additional or more complex concepts, such as density differences and three-dimensional flow. Such a recalibration and model revision should lead to a model that has greater predictive power and reliability.

Goode and Konikow (1990) recalibrated a transport model of the INEL site to evaluate the possible contribution of transient flow to the apparent dispersivity. Their calibration was based on a root-mean-squared error (RMSE) criteria for the point observations. They found that, under transient flow, lowest calibration errors were achieved using larger dispersivities than those used by Robertson, and using a dispersivity tensor that has the typical characteristic $\alpha_L > \alpha_T$. However, there was little

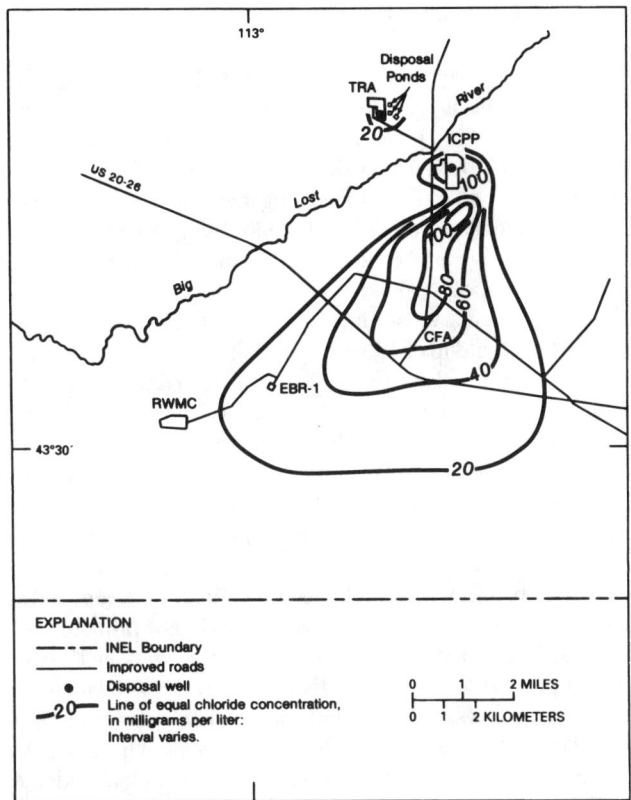

Figure 9. Distribution of waste chloride in the Snake River Plain aquifer, ICPP-TRA vicinity, October 1980. (From Lewis, B. D. and Goldstein, F. J., U.S. Geol. Survey Water-Resources Inv. 82–25, 1982.)

difference between the RMSE for steady-flow and transient-flow simulations, and explicitly accounting for the effects of transient flow conditions did not significantly improve model fit.

Whether the errors in the model of Robertson (1974) were significant in relation to the overall problem can be best (or perhaps, only) answered by those who sponsored the model study in light of (1) what they expected, (2) what actions were taken or not taken because of these predictions, and (3) what predictive alternatives were available. It is clear, however, that the model predictions represented one hypothesis of future contaminant spreading that could be tested in the field, and indeed the 1980 test drilling was designed, to a large extent, to test that very prediction. The process of collecting data seems most efficient when guided by an objective of hypothesis testing. Thus, a major value of the model so far has been to help optimize the data collection and monitoring process; that is, the predictive model offers a means to help decide how frequently and where water samples should be collected to track the plume.

Diffuse Source of Contamination

Groundwater and surface water are interrelated in stream-aquifer systems in which groundwater in the floodplain alluvium is in hydraulic connection with the stream. In the arid to semiarid regions of the western United States the fertile floodplain soils are commonly irrigated with both diverted surface water and pumped shallow groundwater. Much of the applied irrigation water is lost by evapotranspiration, but some of it recharges the alluvial aquifer and provides return flow to the stream. Dissolved solids become concentrated in the recharged water (and hence in the aquifer) because evapotranspiration consumes some of the water but has little effect on the mass of chemical constituents dissolved in the water. The down-valley reuse of water causes a buildup of salts to levels that may be intolerable to many crops.

Because crop yields are related to the quality of applied irrigation water, successful water management in an irrigated area can be aided by the capability to predict and evaluate the impact of any proposed changes in irrigation practices or water management on both the quantity and quality of surface water and groundwater. Consequently, an accurate hydrologic and water-quality simulation model is a desirable management tool for predicting responses and optimizing the use of the total water resource.

Increases in salinity (or dissolved solids content) of groundwater and surface water in the Arkansas River valley of southeastern Colorado are primarily related to irrigation practices. Konikow and Bredehoeft (1974a) simulated an 18-km reach of the valley for a 1-year period (March 1971 to February 1972) that included one complete irrigation season. All inflows, outflows, and changes in aquifer storage of both water and dissolved solids within the study reach were determined from detailed field measurements, which served as a basis for calibrating the simulation model. A more complete description of the study is presented by Konikow and Bredehoeft (1974a). The model used was an early version of the two-dimensional solute-transport model of Konikow and Bredehoeft (1978). It was assumed that the dissolved-solids concentration (or salinity) would act as a conservative tracer.

The valley is cut into relatively impermeable bedrock consisting of Cretaceous shale and limestone. The alluvium consists of moderately permeable but inhomogeneous deposits of gravel, sand, silt, and clay. In the study area the Arkansas River flows in a sandy channel and is in good hydraulic connection with groundwater in the floodplain alluvium. During the study period the mean flow of the Arkansas River at La Junta, near the upstream end of the study reach, was approximately 3.77 m^3/s, which is very close to the median of the annual flows (3.82 m^3/s) during identical periods from 1955 to 1982.

Variations of hydrological and water-quality parameters with time were simulated on a monthly basis. The model was calibrated by adjusting several parameters within a narrow range of values until a best fit was obtained between the observed data and the simulation results. During the study period the measured water levels in 23 observation wells varied by an average of about 1 m. The model reproduced the water-table elevation within 0.3 m of the observed value more than 90% of the time. The dissolved-solids concentration observed in the aquifer during the 1-year study period ranged from about 800 mg/L to greater than 3800 mg/L. Comparisons were made between the observed and calculated spatial salinity patterns, areas of change, and variations of salinity over time. These data indicate that the model can successfully reproduce observed changes in salinity in the aquifer. The computed dissolved-solids concentration was within 10% of the observed value approximately 80% of the time.

The observed change in the salinity in the river from the upstream to downstream ends of the study reach was an increase that averaged 475 mg/L during the study period. This represents an increase of about 40% over the 1175 mg/L average value observed in the river entering the study reach. The calculated increases are in close agreement with the observations. The model results indicated that approximately one-half of the total increase is attributable to the salt load contributed by tributary inflow, and the other half is attributable to the discharge to the river of high-salinity groundwater.

Despite some limitations, it was felt that the model was sufficiently well calibrated to help evaluate the impact of alternative decisions or policies regarding problems of water-planning, water-management, and water-quality control. The application of the model to management problems in the Arkansas River valley was demonstrated by evaluating the impact of several possible changes (Konikow and Bredehoeft, 1974b). The calibrated model was used to predict longer-term (5 year) changes that might occur within the stream-aquifer system, both with and without any changes in irrigation practices. It was assumed that all stresses that occurred during the first year would recur in each succeeding future year. Extending the simulation period on the basis of the stresses and irrigation practices observed during the 1-year study period provides data to serve as a basis of comparison for evaluating the effects of changing management decisions.

The results of such a 5-year simulation are presented in Figure 10. Two hydrological variables, the volume of groundwater in storage and the stream gains and losses, showed marked seasonal variations during each year but little change from one year to the next. On the other hand the mass of salt stored in the aquifer showed seasonal trends superimposed on a long-term increase. During the 10 years following the end of the 1-year study period, land-use, water-use, and irrigation practices remained approximately the same in the study area. The study area was resampled during February 1982 to help evaluate actual long-term trends in groundwater salinity and to help assess the accuracy of the model predictions (Konikow and Person, 1985). Figure 11 shows a comparison between the mean dissolved-solids concentration observed in 1971, 1972, and 1982 and that predicted by the model (the model predictions were extended by extrapolation from 1976–1982). The predictions indicate that a gradual long-term increase in groundwater salinity of about 2% per year would occur under a continuation of stresses observed during 1971–1972. The observed groundwater salinity increased about 6% between the winters of 1971 and 1972, which is a statistically significant increase (at the 0.025 level). However, the observed average salinity increased only about 1% from 1972–1982, which is not a statistically significant increase (at the 0.025 level). The model also predicted (see Figure 10) that there would be no long-term changes in the water levels in the aquifer. This prediction was borne out by the water-level measurements in 1982, which indicated that the water levels in the observation wells were always within a few tenths of a foot of the earlier measurements.

Figure 11 also shows the results of a preliminary recalibration of the model, in which only some of the parameters affecting the estimates of the amount, timing, and

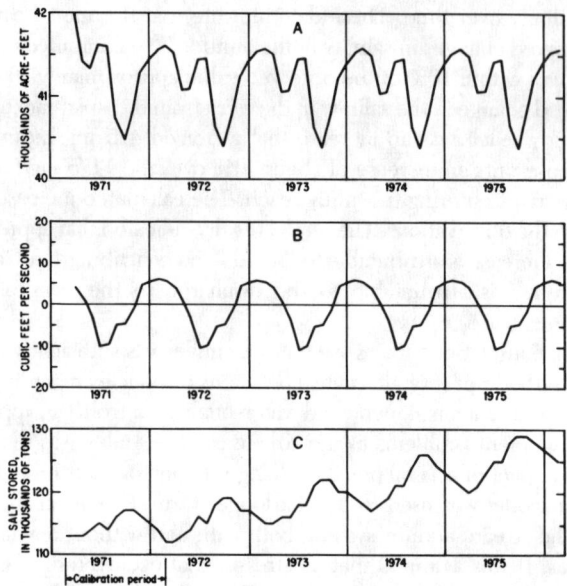

Figure 10. Results of extending 1-year calibration period to a 5-year simulation with the solute-transport model of the Arkansas River valley, Colorado. (A) volume of groundwater in storage; (B) change in streamflow; and (C) mass of salt in aquifer. (From Konikow, L. F. and Bredehoeft, J. D., A water quality model to evaluate water management practices in an irrigated stream-aquifer system, in *Salinity in Water Resources,* Flack, J. E. and Howe, C. W., Eds., 36–59, Merriman, Boulder, CO, 1974.)

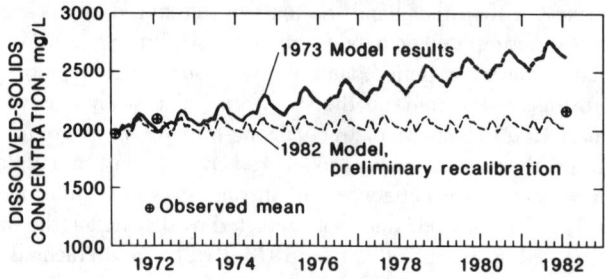

Figure 11. Mean dissolved-solids concentration in alluvial aquifer, Arkansas River valley, Colorado, 1971–1982 (From Konikow, L. F. and Person, M. A., *Water Resour. Res.,* 21(11), 1611, 1985.)

salinity of the recharge are adjusted within the relatively small range of uncertainty associated with them. This illustrates that the model is sufficiently sensitive to errors in recharge estimates such that these factors alone can account for the inaccuracy in the long-term prediction. Person and Konikow (1986) used the 1982 data to recalibrate the original model further. They found that the model was improved by incorporating a recharge lag related to salt transport through the unsaturated zone. The long-term salinity changes were then more accurately reproduced by imposing the actual river salinities observed during 1972–1982 (as opposed to extrapolating the 1971–1972 observations). Figure 11 also illustrates a "lever" effect, in that the parameter adjustment has very little effect on early-time responses, particularly during the original 1-year calibration period, but the effect grows with time so that the differences after 11 years (for 1982) are large. It is thus clear that the reliability of the forecast decreases with time.

An examination of historical streamflow records indicates that the 1-year study period occurred in the middle of a 3-year short-term annual trend (1971–1973) of decreasing streamflow. This would correspond with a trend of increasing salinity in the river, as well as in the irrigation canals, which supplied about 35% of the total applied irrigation water during the study period. Because this short-term (3-year) trend was not indicative of long-term trends in the Arkansas River, it was concluded that the 1-year study period provided too short a historical basis for calibrating a solute-transport model to make long-term predictions. An analysis of the few available historical data on groundwater salinity in the area (mostly 1959–1961) supports the hypothesis that groundwater salinity in this irrigated area has reached a long-term dynamic equilibrium in response to modern irrigation practices. Because the model was calibrated to a period during which a significant trend of increasing salinity was observed, it is logical that the model predicts that an increase will continue in time.

Because of the sensitivity of the solute-transport model to river salinity, Person and Konikow (1986) estimated how long of a period of record would be needed to provide a reliable basis for calibrating the transport model in this area. Based on stream salinity records, they calculated that from 4 to 20 years of data would be needed to estimate the mean salinity within 10 to 5%, respectively.

In this case, the forecasts pertaining to groundwater flow proved to be more accurate than those pertaining to groundwater salinity. For the range and types of stresses existing in this area, and for the purpose of long-range forecasting, the 1-year calibration period was adequate for groundwater flow but too short for solute transport. That is not to say that the solute-transport model was of no value. On the contrary, its development and application led to a significantly improved understanding of the interrelations among the factors affecting salinity changes in that stream-aquifer system, and clearly was and is a valuable tool for hypothesis testing. The recalibrated model extends its historical basis and thus tends to stabilize and increase the reliability of new forecasts to be made with the model. But even if the recalibrated model based on the 11-year period of record is used to predict the impacts of new or major changes in stresses that are not typical or outside the range of stresses observed during the historical period of record, the predictions must be interpreted cautiously.

CONCLUSIONS

It is fairly common for comprehensive and intensive hydrogeologic investigations to include the development, application, and calibration of a simulation model, as well as to use that model to make predictions. When model parameters have been adjusted during calibration to obtain a "best fit" to historical data, there is a bias towards extrapolating existing trends when predicting future conditions, in part because predictions of future stresses are often based on existing trends. Thus, although one advantage of deterministic models is that they represent processes and thus have cause-and-effect relations built into them, careful attention must be paid to the accuracy with which future "causes" (stresses) can be estimated, because that can be the major source of error in the predictions of future "effects" (system responses). Furthermore, concepts inherent in a given model (for example, two-dimensional flow and vertically averaged parameter values) may be adequate over the observed range of stresses, but may prove to be oversimplified or invalid approximations under a new type of stress or for a larger magnitude of stress.

Data collection and monitoring efforts in a study area tend to be curtailed after the project has ended. This will inevitably result in a future deficiency in data on actual stresses and responses during the prediction period. However, the collection of new data after a prediction has been made provides the basis for a strict test of model accuracy—the postaudit. If a model is to be used for prediction of responses in a system subject to continuing water management constraints, it should be periodically postaudited and recalibrated to incorporate new information, such as changes in imposed stresses or revisions in the assumed conceptual model.

Despite the inconclusiveness in pinpointing the exact sources of error, the postaudit in the Arizona example pointed out the large predictive error that occurred and the change in withdrawal trends that occurred immediately after the end of the calibration period. Although in this case it is possible that errors in assumed stresses can account for most of the predictive error, the postaudit analysis helped in formulating hypotheses about how errors in conceptualization and in estimates of hydraulic parameters may have contributed to the low predictive accuracy. The original forecasts had been extended to 1984, and subsequent to this type of postaudit, the extended prediction could have been revised to more accurately account for the change in pumping patterns and the occurrence of occasional but significant recharge events.

In general, results from all postaudits show that predictive accuracy is typically not high, and that perhaps the predictive capability of models should not be considered their primary value. Instead, groundwater models should be viewed as powerful analytical tools that encourage quantification and integration of theory and site-specific data. Model use tends to lead to an increased understanding of a system faster than could be achieved without the use of a model. The postaudits also show that predictions should be accompanied by an assessment of their reliability that is based on the uncertainty in all model parameters, including stresses and boundary conditions. A major value of the postaudit is that the evaluation of the nature and magnitude of predictive errors may itself lead to a large increase in the understanding of the system and in the value of a subsequently revised model. Revised predictions can then be made with greater reliability.

There is no sure way to reliably predict the future, but, because management decisions must be made, predictions of future conditions are needed and will be made in one manner or another. To make the most reliable prediction for a given groundwater problem, all relevant information should be considered and evaluated in order to arrive at the best estimate of the future behavior of the system. Deterministic simulation models can help accomplish this quantitatively by providing a format to integrate and synthesize all available information in a manner consistent with theories describing the governing processes. Our present understanding of the many processes affecting groundwater is sufficiently adequate to allow us, in theory, to predict responses in a groundwater system. In practice, we are severely limited by the inadequacy of available data to describe aquifer properties and historical stresses and responses, and by an inability to predict future stresses. Overall, extreme caution is required in making, presenting, and accepting predictions of future groundwater behavior.

An aquifer-simulation model is no more than an approximation of a complex field situation. Improvements in the approximation are always possible; thus, models should be considered as dynamic representations of nature, subject to further refinement and improvement. As new information becomes available, previous forecasts could and should be modified. Feedback from preliminary models not only helps an investigator to set improved priorities for the collection of additional data, but also helps test hypotheses concerning governing processes in order to develop an improved conceptual model of the system and problem of concern. In summary, the primary value of deterministic groundwater models in many analyses is in providing a disciplined format to improve one's understanding of the aquifer system. This knowledge, in turn, should allow better management of groundwater resources of an area, regardless of the predictive accuracy of the model.

REFERENCES

Alley, W. M. and Emery, P. A., Groundwater model of the Blue River basin, Nebraska—twenty years later, *J. Hydrol.*, 85, 225, 1986.

Anderson, M. P. and Woessner, W. W., *Applied Groundwater Modeling: Simulation of Flow and Advective Transport*, Academic Press, San Diego, CA, 1992, 321 pp.

Anderson, T. W., Electric Analog Analysis of Ground-Water Depletion in Central Arizona, U.S. Geol. Survey Water-Supply Paper 1860, 1968, 21 pp.

Babcock, H. M., Annual Report on Ground Water in Arizona, Spring 1969 to Spring 1970, 44. Arizona State Land Department Water Resources Report 43, 1970.

Babcock, H. M., Annual Report on Ground Water in Arizona with Special Emphasis on Gila Bend Basin and McMullen Valley and the Southeast Part of the Harquahala Plains, Spring 1973 to Spring 1974, 45. Arizona Water Commission Bulletin 9, 1975.

Bredehoeft, J. D. and Pinder, G. F., Mass transport in flowing groundwater, *Water Resour. Res.*, 9(1), 194, 1973.

Briggs, P. C. and Werho, L. L., Infiltration and Recharge from the Flow of April 1965 in the Salt River near Phoenix, Arizona, 12. Arizona State Land Dept. Water Resources Report 29, 1966.

Goode, D. J. and Konikow, L. F., Re-evaluation of large-scale dispersivities for a waste chloride plume: Effects of transient flow, in *ModelCARE 90: Calibration and Reliability in Groundwater Modeling*, Kovar, K., Ed., IAHS Publ. no. 195, 1990, 417.

Konikow, L. F. Predictive accuracy of a ground-water model—Lessons from a postaudit, *Ground Water*, 24(2), 173, 1986.

Konikow, L. F. and Bredehoeft, J. D., Modeling flow and chemical quality changes in an irrigated stream-aquifer system, *Water Resour. Res.*, 10(3), 546, 1974a.

Konikow, L. F. and Bredehoeft, J. D., A water quality model to evaluate water management practices in an irrigated stream-aquifer system, in *Salinity in Water Resources,* Flack, J. E. and Howe, C. W., Eds., Merriman, Boulder CO, 1974b, 36.

Konikow, L. F. and Bredehoeft, J. D., Computer model of two-dimensional solute transport and dispersion in ground water, 90. U.S. Geol. Survey Techniques of Water-Resources Inv., Book 7, Chap. C2, 1978.

Konikow, L. F. and Patten, E. P., Jr., Groundwater forecasting, in *Hydrological Forecasting,* Anderson, M. G. and Burt, T. P., John Wiley & Sons, 1985, 221.

Konikow, L. F. and Person, M. A., Assessment of long-term salinity changes in an irrigated stream-aquifer system, *Water Resour. Res.,* 21(11), 1611, 1985.

Konikow, L. F. and Swain, L. A., Assessment of predictive accuracy of a model of artificial recharge effects in the Upper Coachella Valley, California, in *Selected Papers on Hydrogeology,* Simpson, E. S. and Sharp, J. M., Jr., International Assoc. of Hydrogeologists, Proc. 1989 IGC Meeting, Washington, D.C., 1, 443, 1990.

Laney, R. L., Raymond, R. H., and Winikka, C. C., Maps Showing Water-Level Declines, Land Subsidence, and Earth Fissures in South-Central Arizona, 2 sheets. U.S. Geol. Survey Water-Resources Inv. 78–83, 1978.

Lewis, B. D. and Goldstein, F. J., Evaluation of a Predictive Ground-Water Solute-Transport Model at the Idaho National Engineering Laboratory, Idaho, 71. U.S. Geol. Survey Water-Resources Inv. 82–25, 1982.

Matalas, N. C., Landwehr, J. M., and Wolman, M. G., Prediction in water management, in *Scientific Basis of Water Resources Management,* Geophysics Study Committee, Ed., 118–127, National Academy Press, Washington D.C., 1982.

Person, M. A. and Konikow, L. F., The recalibration and predictive reliability of a solute-transport model of an irrigated stream-aquifer system, *J. Hydrol.* 87, 145, 1986.

Reichard, E. G. and Meadows, J. K., Evaluation of a Ground-Water Flow and Transport Model of the Upper Coachella Valley, California, 101. U.S. Geol. Survey Water-Resources Inv. Rept. 91–4142, 1992.

Robertson, J. B., Digital Modeling of Radioactive and Chemical Waste Transport in the Snake River Plain Aquifer at the National Reactor Testing Station, Idaho, U.S. Geol. Survey Open-File Report IDO-22054, 1974.

Schumann, H. H., Land Subsidence and Earth Fissures in Alluvial Deposits in the Phoenix Area, Arizona, 1 sheet. U.S. Geol. Survey Misc. Inv. Ser. Map I-845-H, 1974.

Solley, W. B., Pierce, R. R., and Perlman, H. A., Estimated Use of Water in the United States in 1990, 76. U.S. Geol. Survey Circular 1081, 1993.

U.S. Water Resources Council, Essentials of Ground-Water Hydrology Pertinent to Water-Resources Planning, Bulletin 16 (revised), 1980.

CHAPTER 5

Groundwater Modeling in the 21st Century

Mary P. Anderson

INTRODUCTION

We stand close to the dawn of a new century—a century that is sure to bring a new era of groundwater modeling. As we prepare for the future, it is fitting to review progress achieved during the present century and to anticipate the challenges ahead.

The science of groundwater modeling has been closely bound to advances in computer technology. Numerical models executed on digital computers quickly replaced early electric analog models as digital computers became more affordable and more powerful. PLASM, the first widely used flow code, was introduced in 1971 by Prickett and Lonnquist (1971); a multitude of groundwater flow codes quickly followed. While there are still many different flow codes in use today (van der Heijde and Elnawawy, 1993), it appears that the USGS code MODFLOW (McDonald and Harbaugh, 1988) is becoming an industry standard. The availability of increasingly sophisticated computers, which allow the capability of simulating complex systems, has encouraged the development of pre- and postprocessing software for visualization of input and output.

Advances in the science of groundwater modeling were driven by the need to solve water supply problems in the 1960s and later to confront problems involving transport of contaminants in groundwater. The challenges of quantifying solute transport processes brought groundwater modelers into closer collaboration with chemists, microbiologists, geologists, statisticians, and mathematicians (Abriola, 1987). The complexities of quantifying solute transport will continue to challenge modelers into the 21st century.

MILESTONES OF THE 20TH CENTURY

Milestones in groundwater modeling are listed in Figure 1. When we trace developments in flow and transport modeling, we find that some of the same researchers were involved in both types of modeling. C. S. Slichter was a professor of mathematics at University of Wisconsin–Madison who did important early work in groundwater flow modeling and also conducted what may have been the first subsurface tracer experiments. Later C. V. Theis, who is noted primarily for his contributions in well hydraulics, demonstrated that he understood that recognition of aquifer heterogeneity is critical to the description of dispersion.

MILESTONES IN GROUNDWATER MODELING

FLOW MODELING

1856   Darcy's law

1886/89   Forchheimer/Slichter apply LaPlace eqn.

1935   Theis equation

1964-65   1st computer modeling applications

TRANSPORT MODELING

1905   Slichter's tracer experiments

1964   Bachmat & Bear's dispersion theory

1967   Theis emphasizes heterogeneity

1973   1st computer modeling applications

1975
Freeze calls attention to uncertainty

1977 Penrose Conference
Geostatistical Concepts and Stochastic Methods in Hydrogeology

Figure 1.   Significant events in groundwater modeling.

The origins of groundwater flow modeling go farther back than the start of this century. The foundation of groundwater modeling is, of course, Darcy's law, which was published in 1856. Freeze (1994) recently presented an enlightening account of Darcy's life. Among other things, Freeze reports that the famous sand column experiments were performed near the end of Darcy's life in the courtyard of the hospital in Dijon, his home town. Recognition that the Laplace equation is relevant to groundwater flow came later when Forchheimer (1886) in Europe and Slichter (1899) in the United States independently applied the Laplace equation to groundwater problems. Later, Theis (1935) borrowed a solution from the heat flow literature to develop a model for transient flow to a well and presented an inverse procedure for calculating transmissivity and storage coefficient from drawdown data.

Numerical modeling was introduced in the mid 1960s and quickly became a standard tool for analyzing groundwater flow problems. One of the first applications of numerical methods to a field problem was presented by Remson et al. (1965), who published a short computer program to solve the Laplace equation using finite difference techniques. They then developed a numerical model to solve for changes in water levels as a result of the construction of a dam. Later Remson et al. (1971) published the first textbook devoted to numerical methods for subsurface flow.

Interest in transport processes in this century started with Slichter (1905) who performed both field and laboratory experiments to investigate dispersion (Wang, 1987). In his field tracer studies, he obtained the now familiar S-shaped breakthrough curves, which he correctly attributed to dispersion. He also conducted laboratory tracer experiments in a sand tank to study the dispersion phenomenon. The fundamentals of dispersion theory were developed some 60 years after Slichter's work and comprehensively articulated by Bachmat and Bear (1964). The theory invoked the assumption that dispersion is a Fickian process and therefore can be represented by an expression similar to Fick's law of diffusion, using a coefficient of dispersion in place of the diffusion coefficient. This expression is embedded in what is now commonly known as the advection-dispersion equation, but researchers are still struggling to find a practical way to measure the dispersion coefficient in the field.

Skibitzke and Robinson (1963) conducted a laboratory tracer experiment to show the effects of dispersion in a heterogeneous medium. After viewing the results of this experiment, Theis (1967) made a remarkable observation that anticipated the problem future researchers would have in trying to make the theory of dispersion match the realities of a heterogeneous world. He said that: "It seems obvious that mixing processes are involved in real aquifers that are not reproduced in dispersion experiments in the laboratory. It also seems obvious that the heterogeneous character of clastic sediments and other porous rocks must be involved in these mixing processes." This comment appeared in an obscure paper published in a proceedings volume in a conference that was filled with papers on well hydraulics. Theis was probably the first person fully to appreciate the complexity of the dispersion process. He went on to say that in order to explain the dispersion process " . . . we need a new conceptual model, containing the known heterogeneities of the natural aquifer."

One of the first attempts at numerical simulation of transport on a regional scale was by Bredehoeft and Pinder (1973) who investigated saltwater contamination of the groundwater system at Brunswick, Georgia. This simulation appeared to produce an acceptable match between measured and simulated conditions. However, later transport simulations by researchers who attempted to reproduce the movement of other contaminant plumes (e.g., Robertson, 1974; Konikow, 1977) clearly established the difficulties involved in getting simulated concentrations to match observed values.

The parallel lines of flow and transport investigations came together in 1975 (Figure 1), when Allan Freeze called attention to uncertainty in flow models, just when others were beginning to question the way in which contaminant transport simulations were being done. The seminal paper by Freeze (1975) launched a new age of investigations into stochastic analysis of groundwater systems. A Penrose Conference in 1977, which Freeze helped organize, set the stage for new directions in groundwater modeling that are likely to continue into the 21st century.

DIRECTIONS FOR THE 21ST CENTURY

The 1977 Penrose Conference on geostatistical concepts and stochastic methods in hydrogeology brought together a number of researchers who were already thinking about uncertainty and the use of stochastic methods. The roots of the major research directions in groundwater modeling for the last quarter of the 20th century and for the coming century grew out of this conference.

Parameter Estimation and Model Reliability

Although inverse modeling had been explored earlier by Stallman (1956) and Nelson (1960), the Penrose Conference gave this methodology new impetus. Groundwater flow modeling developed with a sense of euphoria over the realization that with computers we could now solve complex problems. In the 1970s, skepticism about modeling arose because contaminant transport models did not live up to expectations and calibration of flow models was now recognized to be a highly uncertain process owing to uncertainty over parameter values. In the 1990s, we are increasingly questioning the reliability of our modeling results.

The usual calibration procedure is by trial-and-error adjustment of parameter values until simulated heads are in some sense close to measured heads. However, calibration procedures frequently are not well documented; in most calibrations the justification for the final selection of parameter values is not well defined. Hence, there is a lot of uncertainty in most calibrated models. Parameter estimation models for solving the inverse problem guide the modeler through the calibration process and help the modeler make informed decisions during calibration, leading to better calibrations. The embryos of parameter estimation codes were discussed at the 1977 Penrose Conference. One of these has now emerged as MODFLOWP (Hill, 1990) and is based on codes (e.g., INVFD by Cooley and Naff, 1990) pioneered by Dick Cooley, who was a participant at the 1977 conference. MODFLOWP is a parameter estimation code linked to the popular flow code MODFLOW (McDonald and Harbaugh, 1988). It is likely that the use of codes like MODFLOWP will become common during the next century.

Current interest in parameter estimation models as a way of improving model reliability was evident during a recent conference in The Netherlands (Kovar, 1990), which featured papers on the application of inverse models for both flow and transport problems. Modelers have always been aware of the utility of flux data in calibrating flow models. Recently, investigators have pointed out that calibration of a flow model is enhanced by the use of concentration data (Kauffmann et al., 1990; Medina et al., 1990; Krabbenhoft et al., 1990; Gailey and Gorelick, 1991; Billings and Woessner, 1993), or thermal data (Woodbury and Smith, 1988). Hence, modelers must be alert to the possibilities of collecting these kinds of data and become more actively involved in field investigations.

Regulators are also becoming increasingly concerned about model reliability and asking for model validation (Voss, 1990). One definition of validation states that, "A conceptual model and the computer code derived from it are validated when it is confirmed that the conceptual model and the computer code provide a good representation of the actual processes occurring in the real system. Validation is thus carried out by comparison of calculations with field observations and experimental measurements. A model cannot be considered validated until sufficient testing has been performed to ensure an acceptable level of predictive accuracy. (Note that the acceptable level of accuracy is judgmental and will vary depending on the specific problem or question to be addressed by the model.)" (IAEA, 1982)

Defined in this way, validation requires a postaudit in which field measurements are used to check the results of a prediction, some 20 or more years after the model has been run. A number of such postaudits have been performed recently (Anderson and Woessner, 1992; Konikow and Bredehoeft, 1992). They demonstrate that model predictions were not accurate owing to an incorrect or incomplete conceptual model of the system or to the input of incorrect future stresses. These failings are not necessarily the modeler's fault, but can be attributed to a lack of field data and the inability to know with certainty the magnitude and timing of future stresses. Hence, the inevitable conclusion is that a valid model is an unattainable goal of model validation

(NRC, 1990). Oreskes et al. (1994) use philosophical arguments to show that validation of models of natural systems is impossible.

Anderson and Woessner (1992) recommended that regulatory demands for validation should shift to demands for good modeling protocol that include a complete description of model design, thorough assessment of model calibration, and an uncertainty analysis. Freeze et al. (1988) point out that the "iterative process by which sites are measured and modeled constitutes both a calibration and a validation. Viewed as calibration, the process provides quantitative reduction in parameter uncertainty; viewed as validation, it provides qualitative reduction in model uncertainty." A similar sentiment was expressed by McCombie and McKinley (1993) and by Oreskes et al. (1994), who pointed out that calibration does not demonstrate the veracity of a model but only supports the probability that the model is an accurate representation of the field system.

Geological Heterogeneity

Another major theme of the 1977 Penrose Conference was the use of stochastic methods in hydrogeology as a way of dealing with uncertainties, including uncertainty in the geological description of aquifers. A rational approach for quantifying dispersivity values (Gelhar et al., 1992) requires information on geological heterogeneity. Proponents of the stochastic approach represent heterogeneity using random hydraulic conductivity fields with specified statistical properties in the hope of capturing the relevant features in the subsurface that govern contaminant movement. There are four main approaches including reliance on effective parameters, geostatistics, Monte Carlo simulation, and conditional simulation (Yeh, 1992). In addition, Frind et al. (1988) and Ababou et al. (1989) used a single realization approach.

The use of effective parameters is an extension of the equivalent homogeneous porous medium approach, whereby effective parameter values are defined statistically. It seems that relatively homogeneous aquifers like the one that has been studied for many years at the Borden site (Sudicky, 1986) may be characterized statistically using effective parameters (Hess et al., 1992; Dagan, 1984). In this approach variograms are used to describe the hydraulic conductivity distribution in terms of mean, variance, and correlation length. It is less likely that heterogeneous aquifers such as the one at the MADE site in Columbus, Mississippi (Boggs et al., 1992) can be adequately characterized in this way, although the effective parameter approach has been applied to this site as well (Adams and Gelhar, 1992).

The geostatistical approach relies on interpolation techniques such as kriging. Monte Carlo simulations utilize multiple realizations of random conductivity fields to help define average transport behavior. Conditional simulations additionally use point field measurements to fix values at known points. Conditional simulation models that rely on facies analysis to provide input data for the simulation are used by engineers to simulate heterogeneity in petroleum reservoirs (Haldorsen and Damsleth, 1990; Ravenne and Beucher, 1988). The single realization approach (Ababou et al., 1989) relies on small grid spacing and definition of a set of length scales to generate one statistically meaningful realization. This approach requires large numbers of nodes (of the order of 10^6) and solution using a supercomputer.

Anderson (1987, 1989) pointed out the need to incorporate geological information into the stochastic description of aquifers, and introduced the possibility of using conceptual geological facies models to define trends in hydraulic conductivity. Facies refers to materials in a geological deposit that have similar physical characteristics. Facies models have been used for years by geologists to help locate petroleum deposits. Facies mapped on a local scale may provide a way of defining units of heterogeneity (Anderson, 1989, 1991a,b). Phillips and Wilson (1989) and Phillips et al. (1989) also discuss the use of soft geological information in stochastic descriptions of aquifers. Fogg (1986, 1990) pointed out the importance of interconnection of geological units and Anderson (1991a) demonstrated how interconnected units of high hydraulic conductivity cause channeling of contaminants. Silliman and Wright (1988) used Monte Carlo analysis to investigate the importance of paths of high hydraulic conductivity within low conductivity media. They found that for the cases they examined, there was at least one path crossing the three-dimensional grid along which the hydraulic conductivity was greater than the effective conductivity of the medium. The implication is that contaminants would preferentially follow the path of high hydraulic conductivity. Desbarats (1990) used a geostatistical description of a dual permeability sand/shale system combined with particle tracking to study preferential flow. He found that some of his systems exhibited pronounced channeling of particles and bimodal breakthrough curves.

An extension of a geologically based approach is the use of geological process simulation models, which allow the direct simulation of geological structures (e.g., Tetzlaff and Harbaugh, 1989). An early paper by Price (1974), who used a random walk model to simulate alluvial fan deposition, anticipated the interest hydrogeologists would later show in process simulation models. Recently, Koltermann and Gorelick (1993) used a much more sophisticated process simulation model to reproduce the geological evolution of an alluvial fan in northern California over the past 600,000 years.

Webb (1994) developed a geological simulation model based on superposition of geomorphological surfaces to form sediment packages in a braided stream system. Results show that high conductivity preferential flow paths are present in the simulated deposit and cause funneling of contaminants (Webb and Anderson, 1994). When a random distribution of hydraulic conductivities is used to represent variation within each facies unit, the high conductivity units are not so obvious but still form preferential flow paths. Hence, even in seemingly random distributions of hydraulic conductivity, geological structures control flow and cause funneling. Preferential flow is a key feature of contaminant transport that cannot be described by effective parameters or by a Fickian model of dispersion. Preferential flow paths are difficult and perhaps impossible to identify in the field. Detailed field measurements will be needed to estimate the probability of occurrence of interconnected paths in different types of deposits.

It is clear that hydrogeological analysis requires a quantitative description of the geological setting. This entails taking geological information, which is often qualitative, and developing ways of using it to produce realistic distributions of hydraulic conductivity. Groundwater hydrologists are also finding that information and methods in the traditional geological literature are not directly transferable to hydrogeological investigations owing to discrepancies in scale between traditional geological investigations and contaminant studies. Given that geological structures are an important part of heterogeneity and that uncertainty will always dictate a stochastic approach, it seems that some combination of methods is required. Conditional simulations using geological simulation models provide a way to combine stochastic simulations with geological input.

A number of researchers (e.g., Johnson and Dreiss, 1989; Poeter and Gaylord, 1991; Davis et al., 1993) have applied traditional geostatistical methods involving variograms and kriging to geological field data in order to quantify aquifer heterogeneity. Others (e.g., Brannan and Haselow, 1993; McKenna and Poeter, 1993; Wen and Kung, 1993) explored the possibility of using imprecise geological information to condition flow and transport simulations in which the aquifer is described stochastically. Rubin et al. (1992) used geophysical data in the form of seismic measurements to condition an hydrologic inverse model to solve for hydraulic conductivity.

Plume Characterization

A good description of geological heterogeneity is necessary for accurate simulation of the movement of a contaminant plume, but so is careful and detailed sampling of the plume itself as well as an understanding of chemical reactions, and the history of the source. Tracer experiments at the Borden site (Freyberg, 1986), the MADE site (Boggs et al., 1992), and at the Cape Cod site (LaBlanc et al., 1991) demonstrate the difficulties of tracking plumes in field experiments. These experiments utilized elaborate arrays of monitoring wells, but such dense coverage is not typical of most field studies of contaminant plumes. Sampling is especially difficult when dealing with multiphase plumes. For example, Essaid et al. (1991) obtained 146 core samples along a transect through a plume originating from a crude-oil spill in Bemidji, Minnesota, and found poor correspondence between oil-layer thicknesses as measured in wells and the oil distribution determined from core samples (Figure 2).

To date, most applications of modeling to regulatory problems involving aquifer remediation utilize either particle tracking codes or simple solute transport codes. A particle tracking code (e.g., PATH3D by Zheng, 1989; MODPATH by Pollock, 1989) is used as a postprocessor to a flow code, thereby allowing the delineation of flow paths; retardation can be simulated but dispersion and decay are not considered. Simple solute transport codes such as USGS MOC (Konikow and Bredehoeft, 1978), MT3D (Zheng, 1990), and FTWORK (Faust et al., 1990) allow for dispersion and simple chemical reactions including retardation and first order decay. However, chemical parameters like the retardation factor and a first order decay constant are inadequate to describe many chemical reactions. Chemical complexities within contaminant plumes are caused by interactions between two or more chemicals, biodegradation, and nonequilibrium effects, none of which are described by simple transport codes. Abriola (1987) reviewed ways of modeling complex chemistry by discussing three broad classes of models: geochemical models, sorption models, and biological transformation models.

Modeling of realistic chemically complex plumes is itself complex. Chen et al. (1992) described a one-dimensional numerical model to simulate the transport of benzene and toluene. Their model includes the following governing equations: five nonlinear partial differential equations describing component transport in the bulk pore fluids, five nonlinear algebraic equations governing interphase mass exchange,

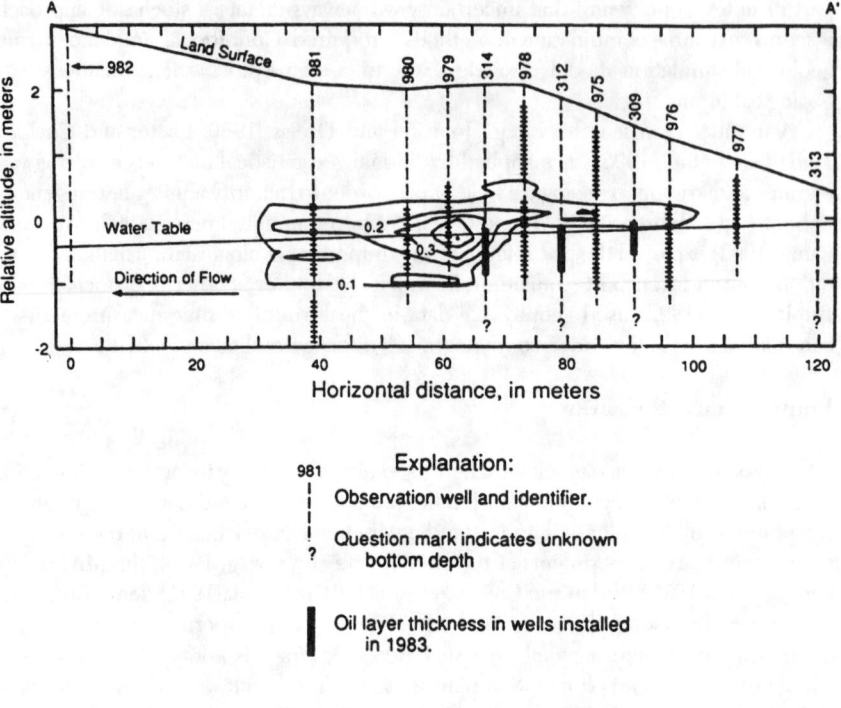

Figure 2. Transect through a multiphase plume of air, oil, and water at Bemidji, Minnesota, showing contours of oil saturation and oil thickness measured in wells. (From Essaid et al., Water-Resources Investigation Report 91–4034, 1991, 614.)

and two ordinary differential equations governing microbial growth. Providing input parameters to such sophisticated contaminant transport models is a difficult task. Chen et al. (1992) used 39 parameters in their simulations. These ranged from relatively mundane parameters such as porosity, soil bulk density, and dispersivity to more elusive parameters such as initial biomass, oxygen use coefficient, and maximum specific substrate utilization rate.

The history of the source is typically poorly known in terms of the amount of contaminant introduced into the system and the timing of the release. Hence, the release rate is sometimes used as a calibration parameter (NRC, 1990). Another modeling approach commonly used in remediation simulations when the history of the source is unknown, is to input field measured concentrations as the initial conditions and then to proceed directly to predictive simulations without performing a calibration. It would, of course, be preferable to start a predictive simulation using a model calibrated to the historical development of the plume. Optimization schemes (Gorelick et al., 1983), statistical pattern recognition (Datta et al., 1989), inverse models (Wagner, 1992), and analytical solutions (Skaggs and Kabala, 1994) have been proposed as ways to recover the release history of a groundwater contaminant.

Simulating the transport of dissolved contaminants in continuous porous media is difficult enough but representing contaminants in fractured porous media (Wang, 1991) and simulating multiphase flow (Parker, 1989) or density effects is more complicated, requiring the input of parameters that are even more difficult to estimate.

The challenges ahead for modelers of the 21st century are evident in an example from Bemidji, Minnesota (Essaid et al., 1991). The site was contaminated in 1979 when a pipeline ruptured, releasing crude oil into a glacial outwash aquifer. In October 1989, core samples were collected along a transect (Figure 2) through the plume. Oil layer thickness was also measured in monitoring wells. It is noteworthy that the thickness of oil measured in the wells did not accurately reflect the distribution of the plume as recorded in the core samples (Figure 2). The spread of the plume was simulated using a multiphase flow model. The results of simulating 10 years of oil movement under three different assumptions about the hydraulic conductivity distribution are shown in Figures 3a–c and can be compared with the field observations shown in Figure 2. Results are sensitive to the distribution of heterogeneities and the characteristic functions relating pressures, saturations, and relative permeabilities. Similar conclusions regarding multiphase flow modeling were reported by Kueper and Frind (1991). None of the simulated plumes accurately reproduce the spreading of oil within the capillary fringe, nor do the models accurately reproduce the details of the oil distribution, although they do capture the general features of the observed plume.

SUMMARY AND CONCLUSIONS

The major directions for groundwater modelers in the 21st century include using parameter estimation codes to help with calibration, using good modeling protocols to improve model reliability, and developing field techniques to help with geological characterization of heterogeneity and plume characterization.

Given a complex geological setting, we are faced with the daunting prospect of defining the nature and distribution of heterogeneities beneath the site. This requires general geological information at a regional scale larger than the site as well as detailed site-specific borehole data and informed correlation of the stratigraphy by a trained geologist. The task of correlating geological units between boreholes is aided by geological intuition and conceptual facies models (Anderson, 1989). But these models cannot predict the location of high conductivity preferential flow paths and units of low permeability, both of which are critical features of the hydrogeological setting for purposes of aquifer remediation. It is likely that some type of statistical analysis such as conditional simulation using geological information as input or conditional simulation with a geological simulation model will become an essential part of the protocol for hydrogeological site characterization in the 21st century.

In order to model the site using a solute transport code, we need to estimate values of hydrogeological parameters including hydraulic conductivity, storativity, effective porosity, and dispersivity, at appropriate scales for each geological unit, in addition to parameters to describe chemical reactions in the subsurface. When addressing regulatory problems, modelers of the 20th century typically ignored the

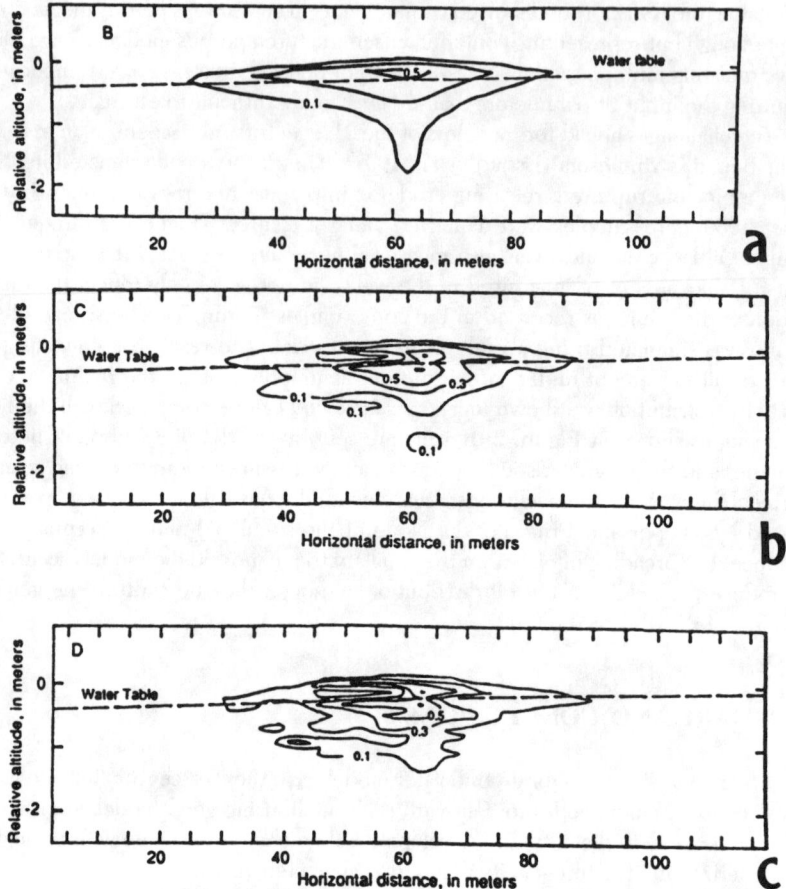

Figure 3. Simulated oil saturations after ten years of oil movement. (a) Uniform mean hydraulic properties; (b) heterogeneous permeability distribution with correlation lengths of 0.5 m (vertical) and 25 m (horizontal); (c) heterogeneous permeability distribution with correlation lengths of 0.25 m (vertical) and 12.5 m (horizontal). (From Essaid et al., Water Resources Investigation Report 91-4034, 1991.)

complications of complex chemistry and tried to avoid the use of even a simple transport model by relying on particle tracking codes to trace out contaminant pathways. It may be that modelers of the 21st century will choose that route as well. Researchers will continue to attempt to improve the simulation of basic transport processes such as dispersion and simple chemical reactions and to incorporate more sophisticated chemical transport processes into models of both continuous and fractured porous media. But the difficulties associated with the measurement of field parameters required by these types of codes will plague modelers until at least the end of the 21st century and probably beyond.

REFERENCES

Ababou, R., McLaughlin, D., Gelhar, L. W., and Tompson, A. F. B., Numerical simulation of three-dimensional saturated flow in randomly heterogeneous porous media, *Transp. Porous Media*, 4, 549, 1989.

Abriola, L. M., Modeling contaminant transport in the subsurface: an interdisciplinary challenge, *Rev. Geophys.*, 25, 125, 1987.

Adams, E. E. and Gelhar, L. W., Field study of dispersion in a heterogeneous aquifer, 2, spatial moments analysis, *Water Resour. Res.*, 28, 3293, 1992.

Anderson, M. P., Treatment of heterogeneities in ground water flow modeling, in *Solving Ground Water Problems with Models*, National Ground Water Association, Columbus, OH, 1987, 444.

Anderson, M. P., Hydrogeologic facies models to delineate large-scale spatial trends in glacial and glaciofluvial sediments, *Geol. Soc. Am. Bull.*, 101, 501, 1989.

Anderson, M. P., Aquifer heterogeneity—a geological perspective, in *Parameter Identification and Estimation for Aquifer and Reservoir Characterization*, B. Hitchon, Ed., National Ground Water Association, Columbus, OH, 1991a, 3.

Anderson, M. P., Comment on "Universal scaling of hydraulic conductivities and dispersivities in geologic media" by S. P. Neuman, *Water Resour. Res.*, 27, 1381, 1991b.

Anderson, M. P. and Woessner, W. W., The role of the postaudit in model validation, *Adv. Water Res.*, 15, 167, 1992.

Bachmat, Y. and Bear, J., The general equations of hydrodynamic dispersion, *J. Geophys. Res.*, 69, 2561, 1964.

Billings, J. G. and Woessner, W. W., The use of natural gradient tracer test data to refine the three-dimensional hydraulic conductivity distribution of a heterogeneous unconfined aquifer, in *Proc. 1993 Ground Water Modeling Conf.*, International Ground Water Modeling Center, 1, 1993.

Boggs, J. M., Young, S. C., Beard, L. M., Gelhar, L. W., Rehfeldt, K. R., and Adams, E. E., Field study of dispersion in a heterogeneous aquifer, 1, overview and site description, *Water Resour. Res.*, 28, 3281, 1992.

Brannan, J. R. and Haselow, J. S., Compound random field models of multiple scale hydraulic conductivity, *Water Resour. Res.*, 29, 365, 1993.

Bredehoeft, J. D. and Pinder, G. F., Mass transport in flowing groundwater, *Water Resour. Res.*, 9, 194, 1973.

Chen, Y-M, Abriola, L. M., Alvarez, P. J. J., Anid, P. J., and Vogel, T. M., Modeling transport and biodegradation of benzene and toluene in sandy aquifer material: comparisons with experimental measurements, *Water Resour. Res.*, 28, 1833, 1992.

Cooley, R. L., and Naff, R. L., Regression modeling of ground-water flow, U.S. Geological Survey Techniques of Water-Resources Investigations O3-B4, 232, 1990.

Dagan, G., Solute transport in heterogeneous porous formations, *J. Fluid Mech.*, 145, 151, 1984.

Datta, B., Beegle, J. E., Kavvas, M. L., and Orlog, G. T., *Development of an Expert System Embedding Pattern Recognition Technique for Groundwater Pollution Source Identification*, National Technical Information Service, Springfield, VA, 1989.

Davis, J. M., Lohmann, R. C., Phillips, F. M., Wilson, J. L., and Love, D. W., Architecture of the Sierra Ladrones Formation, central New Mexico: depositional controls on the permeability correlation structure, *Geol. Soc. Am. Bull.*, 105, 998, 1993.

Desbarats, A. J., Macrodispersion in sand-shale sequences, *Water Resour. Res.*, 26, 153, 1990.

Essaid, H. I., Herkelrath, W. N., and Hess, K. M., Air, oil and water distributions at a crude-oil spill site, Bemidji, Minnesota, in U.S. Geological Survey Toxic Substances Hydrology Program Meeting, Mallard, G. E., and Arsonson, D. A., Eds., U.S. Geological Survey, Water-Resources Investigation Report 91–4034, 1991, 614.

Faust, C. R., Sims, P. N., Spalding, C. P., and Andersen P. F., FTWORK: Groundwater Flow and Solute Transport in Three Dimensions, Version 2.4, Computer Code Documentation, GeoTrans, Sterling, VA, 1990, 172.

Fogg, G. E., Groundwater flow and sand body interconnectedness in a thick, multiple-aquifer system, *Water Resour. Res.*, 22, 679, 1986.

Fogg, G. E., Architecture of low-permeability geologic media and its influence on pathways for fluid flow, in *Hydrogeology of Low Permeability Environments*, Neuman, S. P., and Neretnieks, I., Eds., International Association of Hydrogeologists, 1990, 19.

Forchheimer, P., Uber die ergiebigkeit von brunnen, anlagen and sickerschlitzen, *Zeitsch. Archit. Ing. Ver. Hannover*, 32, 539, 1886.

Freeze, R. A., A stochastic conceptual analysis of one-dimensional groundwater flow in nonuniform homogeneous media, *Water Resour. Res.*, 11, 725, 1975.

Freeze, R. A., Henry Darcy and the fountains of Dijon, *Ground Water*, 32, 23, 1994.

Freeze, R. A., De Marsily, G., Smith, L., and Massmann, J., Some uncertainties about uncertainty, in *Geostatistical, Sensitivity and Uncertainty Methods for Groundwater Flow and Radionuclide Transport Modeling*, Battelle Press, 1988, 231.

Freyberg, D. L., A natural gradient experiment on solute transport in a sand aquifer, 2, spatial moments and the advection and dispersion of nonreactive tracers, *Water Resour. Res.*, 22, 2031, 1986.

Frind, E. O., Sudicky, E. A., and Schellenberg, S. L., Micro-scale modelling in the study of plume evolution in heterogeneous media, in *Groundwater Flow and Quality Modelling*, Custodio, E., et al., Eds., D. Reidel, Dordrecht, 1988, 439.

Gailey, R. M. and Gorelick, S. M., Coupled process parameter estimation and prediction uncertainty using hydraulic head and concentration data, *Adv. Water Res.*, 14, 301, 1991.

Gelhar, L. W., Welty, C., and Rehfeldt, K. R., A critical review of data on field-scale dispersion in aquifers, *Water Resour. Res.*, 28, 1955, 1992.

Gorelick S. M., Evans, B., and Remson, I., Identifying sources of groundwater pollution: an optimization approach, *Water Resour. Res.*, 19, 779, 1983.

Haldorsen, H. H. and Damsleth, E., Stochastic Modeling, Society of Petroleum Engineers, SPE 20321, Distinguished Author Series, 1990, 404.

Hess, K. M., Wolf, S. H., and Celia, M. A., Large-scale natural gradient tracer test in sand and gravel, Cape Cod, Massachusetts, 3, hydraulic conductivity variability and calculated macrodispersivities, *Water Resour. Res.*, 28, 2011, 1992.

Hill, M. C., MODFLOWP: a computer program for estimating parameters of a transient, three-dimensional ground-water flow model using nonlinear regression, U.S. Geological Survey Open-File Report 91–484, 1990, 317.

International Atomic Energy Agency (IAEA), *Radioactive Waste Management Glossary*, IAEA-TECDOC-264, International Atomic Energy Agency, Vienna, 1982.

Johnson, N. M. and Dreiss, S. J., Hydrostratigraphic interpretation using indicator geostatistics, *Water Resour. Res.*, 25, 2501, 1989.

Kauffmann, C., Kinzelbach, W., and Fried, J. J., Simultaneous calibration of flow and transport models and optimization of remediation measures, in *Calibration and Reliability in Groundwater Modeling*, Kovar, K., Ed., IAHS Publ No. 195, 1990, 159.

Koltermann, C. E. and Gorelick, S. M. Paleoclimatic signature in terrestrial flood deposits, *Science*, 256, 1775, 1992.

Konikow, L. F., Modeling chloride movement in the alluvial aquifer at the Rocky Mountain Arsenal, Colorado, U.S. Geological Survey Water-Supply Paper 2044, 1977.

Konikow, L. F. and Bredehoeft, J. D., Computer model of two-dimensional solute transport and dispersion in ground water, U.S. Geological Survey Techniques of Water-Resources Investigations, Book 7, Chap. C2, 1978, 90.

Konikow, L. F. and Bredehoeft, J. D., Ground-water models cannot be validated, *Adv. Water Resour.*, 15, 75, 1992.

Kovar, K., Ed., *Calibration and Reliability in Groundwater Modeling*, IAHS Publication No. 195, 1990, 539.

Krabbenhoft, D. P., Bowser, C. J., Anderson, M. P., and Valley, J. W., Estimating groundwater exchange with lakes, 1, the stable isotope mass balance method, *Water Resour. Res.*, 26, 2445, 1990.

Kueper, B. H. and Frind, E. O., Two-phase flow in heterogeneous porous media, 1, model development; 2, model application, *Water Resour. Res.*, 27, 1049, 1991.

LeBlanc, D. R., Garabedian, S. P., Hess, K. M., Gelhar, L. W., Quandri, R. D., Stollenwerk, K. G., and Wood, W. W., Large-scale natural gradient tracer test in sand and gravel, Cape Cod, Massachusetts, 1, experimental design and observed tracer movement, *Water Resour. Res.*, 27, 895, 1991.

McCombie, C. and McKinley, I., Validation—another perspective, *Ground Water*, 31, 530, 1993.

McDonald, M. G. and Harbaugh, A. W., A modular three-dimensional finite-difference ground-water flow model, Techniques of Water-Resources Investigations O6-A1, U.S. Geological Survey, 1988, 576.

McKenna, S. A. and Poeter, E. P., Conditioning of hydrofacies simulations with imprecise data, in *Proc. 1993 Ground Water Model. Conf.*, International Ground Water Modeling Center, Golden, CO, 1993, 4.

Medina, A., Carrera, J., and Galarza, G., Inverse modeling of coupled flow and solute transport problems, in *Calibration and Reliability in Groundwater Modeling*, K. Kovar, Ed., IAHS Publ. No. 195, 1990, 185.

National Research Council (NRC), *Ground Water Models: Scientific and Regulatory Applications*, National Academy Press, 1990, 303.

Nelson, R. W., In-place measurement of permeability in heterogeneous media, 1, theory of a proposed method, *J. Geophys. Res.*, 65, 1753, 1960.

Oreskes, N., Shrader-Frechette, K., and Belitz, K., Verification, validation, and confirmation of numerical models in the earth sciences, *Science*, 263, 641, 1994.

Parker, J. C., Multiphase flow and transport in porous media, *Rev. Geophys.*, 27, 311, 1989.

Phillips, F. M. and Wilson, J. L., An approach to estimating hydraulic conductivity spatial correlation scales using geological characteristics, *Water Resour. Res.*, 25, 141, 1989.

Phillips, F. M., Wilson, J. L., and Davis, J. M., Statistical analysis of hydraulic conductivity distributions: a quantitative geological approach, in *Proc. New Field Techniques for Quantifying the Physical and Chemical Properties of Heterogeneous Aquifers*, National Ground Water Association, Columbus, OH, 1989.

Poeter, E. and Gaylord, D. R., Influence of aquifer heterogeneity on contaminant transport at the Hanford Site, *Ground Water*, 28, 900, 1991.

Pollock, D. W., Documentation of computer programs to complete and display pathlines using results from the U.S. Geological Survey modular three-dimensional finite-difference ground-water model, U.S. Geological Survey Open-File Rept. 89–381, 1989, 81.

Price, W. E., Simulation of alluvial fan deposition by a random walk model, *Water Resour. Res.,* 10, 263, 1974.

Prickett, T. A. and Lonnquist, C. G., Selected Digital Computer Techniques for Groundwater Resource Evaluation, *Ill. State Water Surv. Bull.* 55, 62, 1971.

Ravenne, C. and Beucher, H., Recent development in description of sedimentary bodies in a fluvio deltaic reservoir and their 3D conditional simulations, *Soc. Pet. Eng. SPE 18310,* 1988, 463.

Remson, I., Appel, C. A., and Webster, R. A., Ground-water models solved by digital computer, Journal of the Hydraulics Division, *Proc. of the Am. Soc. Civil Eng.,* HY 3, 133, 1965.

Remson, I., Hornberger, G. M., and Molz, F. J., *Numerical Methods in Subsurface Hydrology,* Wiley-Interscience, New York, 1971, 389.

Robertson, J. B., Digital modeling of radioactive and chemical waste transport in the Snake River Plain aquifer at the National Reactor Testing Station, Idaho, U.S. Geological Survey Open File Report ID0–22054, 1974, 41.

Rubin, Y., Mavko, G., and Harris, J., Mapping permeability in heterogeneous aquifers using hydrologic and seismic data, *Water Resour. Res.,* 28, 1809, 1992.

Silliman, S. E. and Wright, A. L., Stochastic analysis of paths of high hydraulic conductivity in porous media, *Water Resour. Res.,* 24, 1901, 1988.

Skaggs, T. H. and Kabala, Z. J., Recovering the release history of a groundwater contaminant, *Water Resour. Res.,* 30, 71, 1994.

Skibitzke, H. E. and Robinson, G. M., Dispersion in groundwater flowing through heterogeneous materials, U.S. Geological Survey Professional Paper 386-B, 1963.

Slichter, C. S., Field measurements of the rate of movement of underground waters, U.S. Geological Survey Water-Supply and Irrigation Paper No. 140, 1905, 122.

Slichter, C. S., Theoretical investigation of the motion of groundwaters, 19th Annual Report, Part II, United States Geological Survey, 295, 1899, reprinted in *Ground Water,* 23, 396, 1985.

Stallman, R. W., Numerical analysis of regional water levels to define aquifer hydrology, EOS *Trans. Am. Geophys. Union,* 37, 451, 1956.

Sudicky, E. A., A natural gradient experiment on solute transport in a sand aquifer: spatial variability of hydraulic conductivity and its role in the dispersion process, *Water Resour. Res.,* 22, 2069, 1986.

Tetzlaff, D. M. and Harbaugh, J. W., *Simulating Clastic Sedimentation,* Van Nostrand Reinhold, New York, 1989, 202.

Theis, C. V., The relation between the lowering of the piezometric surface and the rate and duration of discharge of a well using groundwater storage, *Trans. Am. Geophys. Union,* 2, 519, 1935.

Theis, C. V., Aquifers and models, in *Proc. Symp. Ground-Water Hydrol.,* M. A. Marino, Ed., American Water Resources Association, 1967, 138.

van der Heijde, P. K. M. and Elnawawy, O. A., Compilation of Ground-Water Models, EPA/600/R-93/118, 1993, 87.

Voss, C. I., A proposed methodology for validating performance assessment models for the DOE Office of Civilian Radioactive Waste Management Program, in *High Level*

Radioactive Waste Management, Vol. 1, American Nuclear Society Inc., La Grange Park, IL, and American Society of Civil Engineers, New York, 1990, 359.

Wagner, B. J., Simultaneous parameter estimation and contaminant source characterization for coupled groundwater flow and contaminant transport modelling, *J. Hydrol.,* 135, 275, 1992.

Wang, H. F., Charles Sumner Slichter—an engineer in mathematician's clothing, in *The History of Hydrology,* American Geophysical Union, 1987, 103.

Wang, J. S. Y., Flow and transport in fractured rocks, in *Contributions in Hydrology,* U.S. National Report 1987–1990, American Geophysical Union, 1991, 254.

Webb, E. K., Simulating the three-dimensional distribution of sediment units in braided stream deposits, *J. Sediment Res.,* B64, No. 2, 219, 1994.

Webb, E. K. and Anderson, M. P., Simulation of preferential flow in three-dimensional, heterogeneous, random fields with realistic internal architecture, *Water Resour. Res.,* in review, 1994.

Wen, X-H. and Kung, C-S., Stochastic simulation of solute transport in heterogeneous formations: a comparison of parametric and nonparametric geostatistical approaches, *Ground Water,* 31, 953, 1993.

Woodbury, A. D. and Smith, L., Simultaneous inversion of hydrogeologic and thermal data, 2, incorporation of thermal data, *Water Resour. Res.,* 24, 356, 1988.

Yeh, T-C. Jim, Stochastic modelling of groundwater flow and solute transport in aquifers, *Hydrol. Process.,* 6, 369, 1992.

Zheng, C., PATH3D, S. S. Papadopulos & Assoc., Rockville, MD, 1989.

Zheng, C., MT3D, A Modular Three-Dimensional Transport Model, S. S. Papadopulos & Assoc., Rockville, MD, 1990.

CHAPTER 6

Needs for the Next Generation of Models

Editor's Note

An open forum was held on the last day of the "Assessment of Models for Groundwater Resources Analysis and Management" conference, which was held in Hawaii, March 21–23, 1994. The objective of the discussion was to identify weaknesses in models and modeling and to suggest needs for the next generation of models. Discussion was led by a convener and panelists, with active participation from the audience. The convener was Steven M. Gorelick of Stanford University. The panelists were as follows.

Research:
    Lynn W. Gelhar, Lawrence Berkeley Laboratory, Berkeley, California; on leave from Massachusetts Institute of Technology
    Joel W. Massmann, University of Washington, Seattle, Washington
    R. J. Wagenet, Cornell University, Ithaca, New York

Regulation/Decision Making:
    Philip Berger, U.S. Environmental Protection Agency, Washington, D.C.
    Leonard F. Konikow, U.S. Geological Survey, Reston, Virginia

Model Use:
    Thomas A. Prickett, Thomas A. Prickett and Associates, Urbana, Illinois
    Daniel B. Stephens, Daniel B. Stephens and Associates, Albuquerque, New Mexico

Quality Assurance/Quality Control:
    Paul K. M. van der Heijde, International Ground Water Modeling Center, Golden, Colorado

The following summary of the discussion that took place in the forum represents a near consensus among the participants. Complete agreement was evident on many issues related to process simulation, parameter evaluation, and model application. However, it should be realized that complete consensus of all participants cannot be possible.

RESEARCH

Groundwater models can be classified, in general, as flow and transport/fate models. Flow models have progressed to reach what can be termed a second-generation status with improved modeling capabilities. In such a class, it is now possible to model three-dimensional flow problems on faster computers. The success is mainly attributed to the validity of the macroscopic viewpoint of flow in porous media. However, success is not as sound regarding modeling fracture or preferential flow where the continuum principle is not valid, such as flow in karstic or volcanic formations. A detailed account of the various hydraulic properties of the medium is not generally possible. Techniques for measuring such properties, as well as conceptual models for heterogeneity analysis, are very much needed. Although research models have addressed uncertainty assessment, an acceptable rigorous approach is not readily available. Another difficulty exists in modeling nonlinear flow problems, such as those related to unsaturated and density-dependent flow. Numerical solutions may require the use of a very fine grid to achieve acceptable accuracy. Efficient solution techniques are essential here to solve large-scale field problems. The third generation of models additionally requires the ability to use visualization and to have interactive and complete control of the modeling process.

Although transport/fate models have advanced in their capabilities, they have not overcome their initial first generation status in terms of their limited ability to solve field problems. Difficulties include the failure to describe the flow field on the appropriate scale and the absence of accurate conceptualization of the chemical and biological activities involved. The next generation of transport/fate models needs to overcome the major specific problems summarized below.

Field Techniques

Computer technology has outpaced field technology. The lack of appropriate data causes significant barriers in the modeling process. Many parameters of concern cannot be independently estimated, and a type of inverse technique is needed in this regard. The estimated parameters depend thus on the conceptual model used in the inversion process and the scale of measurements. There is a need to develop new field techniques to characterize subsurface properties. The techniques should be able to characterize the heterogeneities involved in the physical, chemical, and biological parameters.

Conceptual Models

Many modeling processes and their interaction are not well understood. Conceptual models having realistically practical data requirements are lacking in such areas as chemical, biological, and multiphase-flow modeling, especially under nonequilibrium conditions. Modeling coupled processes, particularly under complicated field conditions, is an area that requires much attention. There is a need to consider heterogeneity of biological and chemical processes. Such processes have been assumed as spatially invariant, which may not be an accurate conceptualization, due to the interrelation among chemical, biological, and hydrological parameters. The problem of parameter variability needs to be studied on both the small and large scale. Scale-up theories, which will allow better interpretation of field measurements at the correct scale, need to be advanced.

Of immediate need also are efforts to estimate various transport/fate parameters, such as dispersivities, independently from chemical concentration data. Although Fickian transport theory has been proven, through several controlled field experiments, to be adequate in describing the dispersion phenomenon, there is still a need to develop other vehicles for modeling contaminant transport.

Models to quantify uncertainty and risk should be advanced, most appropriately within a stochastic framework that links the modeling process to available data. Finally, interdisciplinary research efforts should be encouraged to include cooperation between the various branches of earth sciences. For example, it is possible to integrate advances in oil engineering, soil physics, and hydrogeology in addressing a certain problem from an environmental perspective.

Computation

Software and hardware computer technology has advanced greatly over the last decade, outgrowing our conceptual and data collection abilities. Three-dimensional modeling is now possible; yet, because such models are not flexible, more efficient solution techniques are needed. For example, nonlinear flow and transport problems require the use of an extremely fine grid, on the order of a few centimeters, which would make a field-scale problem intractable. The new class of "mega-models", which is promoted as "general", is probably not useful at this stage because of the degree of complexities involved and the need for experienced users.

APPLICATIONS

Modeling

The availability of models that are too user-friendly can lead to their misuse. New advances in computer software and hardware provide the opportunity for developing a relatively easy modeling process with a user-friendly interface. However, because many models in such systems are too simple and are based on severe assumptions, the user has to be aware of the limitations and restrictions involved. User interaction, intuition, and common sense are important parts of the modeling process and should not be substituted by the machine and its software. Decision or policy models are generally generic with many built-in parameters and are intended for screening purposes. Policies or decisions based on these models should be analyzed carefully and their interpretation based on the assumptions included within a comparative framework. In this and other cases, modeling goals and objectives should be set clearly as early as possible because they constitute an important factor in choosing a model.

It is essential that modelers reveal their subjectivity and personal judgment in the study report. The basic element in any report should be the ability of the modeler to defend his or her effort and justify any modeling decisions. One of these decisions concerns model choice with the appropriate level of complexity. Although complex models are generally more accurate, their use requires extensive data sets which may not be suitable for the problem at hand.

Successful model use requires the availability of experienced model users and good model documentation that adopt acceptable standards. Well-trained modelers should have the knowledge and expertise necessary to reduce any chance of model misuse. Documentation should clearly describe model limitations and restrictions. There is also a need to close the gap between model researchers and users and to improve the usability of research models. Modeling conferences and workshops that involve the two groups should be held regularly. Many research models are not suitable for use mainly because of their extensive data requirements, the absence of documentation, and their experimental, unfinished status.

Model Validation

A clear and consistent modeling vocabulary needs to be used. The use of many fitting parameters that cannot be independently defined has deemed model validation a useless concept. A need exists to fully examine this issue and to standardize techniques to gain confidence in the predictive capability of models. It should be realized that general modeling standards can be useful although they may not be appropriate for all conditions. Standards cannot and should not substitute for a modeler's justification for a specific model use and for specific results interpretation. Although post-audits deal with site-specific problems, they can be useful. Such studies are concerned with assessing model predictions based on actual outcomes and therefore, can enhance the understanding of processes involved and add to the modeler's practical experience.

SECTION 2

On Models as Management Tools

CHAPTER 7

If It Works, Don't Fix It: Benefits from Regional Groundwater Management

John D. Bredehoeft, Eric G. Reichard, and Steven M. Gorelick

INTRODUCTION

The question of managing water resources in a democratic society, such as the United States, is one of implementing institutions that regulate the use of the resource. There are a small set of published papers that consider the problem of "optimal" (meaning economically efficient) groundwater management on a regional scale. Most, although not all (O'Mara and Duloy, 1984), of these papers deal with the arid western United States. The results of these studies suggest some general conclusions about the institutions that society uses to manage groundwater. We recognize that these generalizations are speculative; however, to the extent that they are valid, we believe that they have interesting implications for managing groundwater.

One issue is the ability of a free market to allocate the resource efficiently. To examine this issue, we look at the utility of optimization models used to analyze groundwater quantity and quality management. We inspected a subset of the published problem formulations and categorized these studies in terms of their objective functions. Based on the small number of investigations that have generated empirical data, we ask the question: Of what potential use are the optimization methods in evaluating institutions for groundwater management? In attempting to answer this question we consider the form of the resulting objective.

An underlying assumption of most analyses is that existing water rights and government subsidies are taken as given. The impact of altering these assumptions may be significant. Gisser (1983) suggested that the task of interest is to investigate how modifying water rights will improve allocation rather than computing optimal allocation schemes within the existing right structure. We agree that examining the impact of water rights is an important issue; yet, few studies have treated both water rights and government subsidies as decision variables. Of interest are quantitative investigations that consider the impact of changing the entire institutional framework under which water is allocated. The feasibility of changing these institutions is a subject for debate.

Lefkoff and Gorelick (1990) investigated the potential economic benefits of a local water market in the Arkansas Valley of Colorado. A local rental market would relax the established water rights structure. They indicated substantial benefits both in economic terms as well as improvements in water quality. The Colorado system of stream-flow augmentation by pumpers involves a water market (Young et al., 1985). California is finally moving toward a rental market for water, which is a major change in institutions.

Model Classification

Banks (1993) classified models into two types: (1) *consolidative* which he defined as consolidating known facts into a single package and then using it as a surrogate for the real system; and (2) *exploratory* which he also defined as the use of a series of computation experiments to explore the implications of varying assumptions and hypotheses. Consolidative models neglect uncertainty; they lead to single valued *predictions*. This kind of modeling was done a decade or more years ago, partly because computers in those years were much smaller. Some problems elucidated by postaudits are the result of single valued predictions made by the consolidative models of an earlier era.

Today, most of us recognize that there are real uncertainties in both the conceptual models and the model parameters. This leads us to exploratory modeling; most of us espouse this idea. Banks (1993) went on to define three types of exploratory modeling. Bank's background at the Rand Corporation is in global scale, war games. We have modified his definitions to better fit groundwater modeling:

1. Data driven exploratory modeling—history matching.
   A. Conceptual model identification—is the basic model correct? For example: two- or three-dimensional; transient or steady-state; correct boundary condition; etc.
   B. Parameter identification (inverse problem)—adjust the parameters until a satisfactory history match is achieved.
2. Policy question driven modeling—This type of modeling searches among plausible scenarios in an effort to examine the system response to human actions or to illuminate policy choices.
3. Conceptually driven exploratory modeling—This exploration compares the output of different conceptual models.

Most groundwater model analyses are with the data driven exploratory models—the history matching activity. However, our focus in this paper is on a different set of models—the policy driven models. These are the set of groundwater models that are used to examine questions of regional groundwater management. The published literature on this set of models is small. We will attempt to review the relevant published models.

The problems addressed in the literature are divided into two types (Gorelick, 1983): (1) regional scale problems in which management is an institutional issue; and (2) smaller scale, design problems in which management is an operational problem. An underlying issue, especially at the regional scale, is the difference between managing water quality versus water quantity in a *free market*.

Optimization techniques were applied to the small scale operational problems such as: well field management for aquifer dewatering (Aguado and Remson, 1974; Danskin and Freckleton, 1992); maximizing *safe yield* (Maddock, 1973; Larson et al., 1977; Heidari, 1982); hydraulic gradient control for contaminant removal (Molz and Bell, 1977; Remson and Gorelick, 1980; Atwood and Gorelick, 1985); aquifer restoration (Gorelick et al., 1984); and pollutant source management (Willis, 1979; Gorelick et al., 1979). Recently, Gorelick (1990) reviewed optimization techniques applied to problems of subsurface contamination. Yeh (1992) also presented a review of optimization applied to groundwater problems. These models have an important role in design and control of both groundwater hydraulics and the movement of contaminants. The small-scale design and operation problems are not the focus of this discussion; our concern is with the application of groundwater management models to regional scale problems.

POTENTIAL PROBLEMS ASSOCIATED WITH LARGE-SCALE GROUNDWATER DEVELOPMENT

Large groundwater developments are a recent phenomena. They have occurred since World War II and are the product of improved technology, both the invention of efficient pumps and the availability of cheap energy. To a large extent, the development of groundwater has gone on almost unregulated. Land owners who perceived that it was in their economic interest to drill wells have done so. Often the only serious effort to control development came late, usually when very substantial pumping capacity was already in place in the system.

Four potential problems are associated with this development:

1. It is possible for a single user (or group) to pump a disproportionally large quantity that lowers water levels for one's neighbors such that they must pump from unduly large depths. There is little, or no, incentive for individuals to reduce their pumpage in an effort to conserve water (Alley and Schefter, 1987). There are analogous problems with all common property, or nonexclusive, resources such as fisheries (Gordon, 1954) or common land used for grazing or hunting.
2. In a system that uses both surface and groundwater conjunctively, groundwater use can divert water from associated surface water bodies thereby depleting the available surface water, especially for downstream users.
3. Some developments are a mining operation with the resource depleted over time.
4. Water quality will be changed by aquifer use. The action of one individual can adversely change water quality in the system. This can have long-term, adverse impacts on his/her neighbor's groundwater.

Because of the way groundwater developed historically, especially for irrigation, the problems associated with disproportionately large withdrawals by a single pumper (or group) are usually not serious. In the largely uncontrolled mode in which the resource has traditionally been managed, groundwater has been available to almost anyone willing to drill a well. The economic incentives to pump groundwater have been almost equal for most users. One usually finds that development is rather uniformly distributed throughout the system.

Conjunctive use in systems where groundwater development can adversely impact the availability of surface water has proved to be a difficult management problem. The economic advantages of using groundwater in conjunction with surface water are usually large. Young and Bredehoeft (1972) showed a doubling of net benefits in the South Platte River Valley in Colorado from using groundwater in conjunction with surface water. It is not surprising that many conjunctive use irrigated agricultural systems developed in the western United States. Often the development of groundwater in these systems has gone on practically unregulated.

The impacts of groundwater withdrawals on streamflow during the growing season are not as large as might be expected. Bredehoeft and Young (1983) showed that the depletion of streamflow in the South Platte system in Colorado is probably less than 10% of the total rate of groundwater pumping, even during periods of extremely low stream flows. An innovative institution designed to lessen the adverse impacts of pumping in a conjunctive use groundwater/surface water system is the augmentation scheme currently used in Colorado. The Colorado system requires a groundwater users' collective to make provisions to augment streamflow during low flow periods. Augmentation is accomplished either by setting water aside in reservoirs, or by drilling wells that are used solely for augmenting depleted streamflow. A modest pumping tax provides the cost of augmentation.

Commonly regulations restrict groundwater mining so that the system will have a life of at least several decades. In many places, however, the mining of groundwater has gone on almost unchecked. In large systems, depletion of the resource may take many years. As water levels fall, the costs of lifting groundwater from great depths provide an economic constraint on continued use. In some areas of Arizona, for example, approximately one-third of the land has gone out of production because water levels have fallen and pumping has become expensive. In other areas farmers have shifted to higher value crops, such as citrus, to compensate for the higher costs of groundwater.

In the past, groundwater quality has gone largely unregulated; in many areas groundwater quality has suffered from inadequate management. As discussed below, there are characteristics of groundwater quality deterioration that require groundwater management.

GROUNDWATER MANAGEMENT OBJECTIVES

We consider groundwater management to involve an objective (or possibly multiple objectives) and a series of constraints. It may be difficult to formulate problems in this way. The distinction between constraints and objectives is often unclear. In addition, some management objectives are hard to quantify.

There are two principal objectives: (1) the maximization of net revenue, and (2) the minimization of the cost of achieving some goal. The first category includes surrogate economic objectives. In surrogate objectives, economic benefits are not explicitly stated; the implied economic goal can frequently be identified. Examples of surrogate objectives include maximizing the waste disposal capacity (a surrogate for minimizing treatment costs), and the minimization of total pumping (a surrogate for minimizing pumping costs). Surrogate objectives are used because they may be easier to formulate quantitatively, or the economic implications of the problem may not be recognized.

There are substantive philosophical differences between maximizing net benefits and minimizing costs. Maximizing net benefits exploits the difference between revenues generated by water use and costs, including the cost of water. In maximizing net benefits, demand is not fixed; the quantity to pump is one of the decisions. Water use depends on the level of economic activity; for example, water use for irrigation will depend on the choice of crops planted.

In minimizing costs the demand for water is assumed fixed; one minimizes the cost of meeting a fixed demand. One either explicitly or implicitly assumes that the quantity demanded is insensitive to cost. Again surrogate measures, such as drawdowns or pumping lift, may be used in the objective function. From a policy perspective, the more interesting problems of regional agricultural-hydrological management were formulated in terms of maximizing net benefits. Minimizing costs is the principal concern in the small-scale, design problems. Another category of economically unquantifiable (or at least difficult to quantify) objectives often involves human health, or environmental objectives. For example, groundwater contamination can be a potential threat to human life, or to the lives of plants and wildlife. One way to formulate such objectives is in terms of risk management. The literature only treats extensively the first category—quantifiable economic objectives. This category is the primary focus of our discussion. We wish to emphasize the need for further study of the noneconomic objectives because they are important and cannot be dismissed.

Two additional factors further categorize groundwater management: (1) the spatial scale; and (2) the time domain of the problem. The spatial scale often dictates the objectives of groundwater management. The objective for a given problem can be formulated differently depending on whether one's concern is with regional water management, or the concerns of a local management district, or the interests of an individual user. Problems viewed from all three perspectives invariably result in multiple objectives.

Other factors influence the time domain of management. Problems that involve mining groundwater have a finite life. In conjunctive use, on the other hand, one often expects the development to operate indefinitely. The time response of the groundwater system to hydraulic versus quality stresses is different. Our focus is on the regional scale problems, not the small-scale operational problems.

Quantity Management: Maximizing Net Benefits

For large agricultural areas, where the net benefits relate directly to the cost of obtaining water and revenues come from irrigated crops, the objective function near the point of optimality may be essentially nonunique. Often there is a broad plateau where a wide range of water use and differing cropping patterns will yield near optimum solutions.

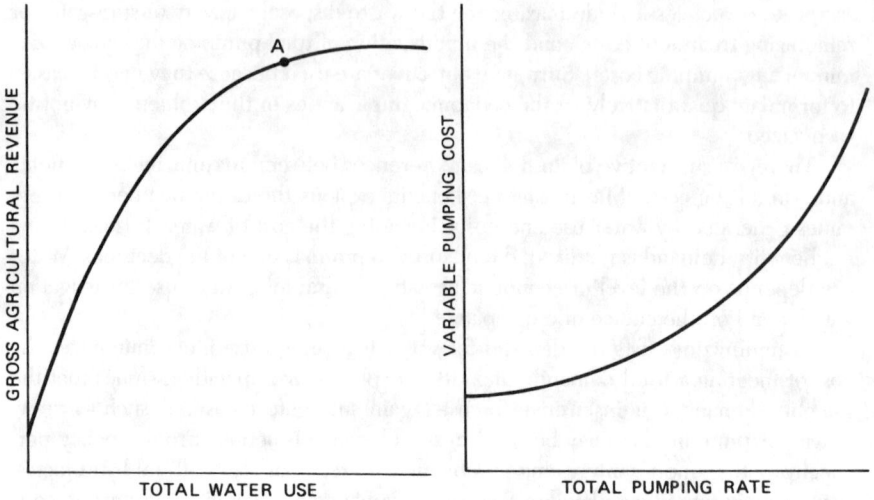

Figure 1. Typical agricultural production function for water.

Figure 2. Typical pumping cost function.

The shape of the objective function will depend on both the shape and relative magnitude of two components: (1) the agricultural revenue function (revenues excluding water costs) and (2) the pumping cost function. The agricultural revenue functions will generally have a concave shape such as shown in Figure 1. The concave form results from the fact that marginal returns diminish as more water is used. Pumping costs will vary as a function of the product of pumping rate and pumping lift. As one uses more water, both the rate of pumping and the pumping lift increase. For this reason, the variable pumping cost function will have a convex shape similar to that shown in Figure 2. Since the net revenue is the difference between these two functions its overall shape will depend upon their magnitudes. Capital costs are not included in this analysis; including them complicates the functions.

Given current energy costs, the gross agricultural revenue associated with a given amount of water use is likely to be much larger than the pumping costs; for example, in the Salinas Valley of California gross agricultural revenue is three to four times the cost to pump groundwater (Lemoine, 1984; Reichard, 1987). In this case, the shape of the net revenue function is dominated by agricultural revenue. As long as a certain minimum amount of water is used (to the right of point A in Figure 1), the net revenue function will be relatively flat.

We present a review of the literature relating to the maximization of agricultural revenue below. In reviewing this literature, we consider both lumped parameter (i.e., treating the groundwater system as a single, homogeneous, infinitely permeable reservoir) and distributed parameter systems (i.e., actually solving the appropriate partial differential equations over the relevant spatial and temporal domain) simulation-management models. We wish to emphasize that the available literature is limited and uses for most of its prototype systems are in the western United States. The conclusions gleaned from this set of examples are based on our intuition and speculation; we do not pretend that our sample of the literature is all inclusive.

A number of studies employed a single-cell groundwater basin formulation to test how net agricultural revenues can be increased by optimizing groundwater use. Gisser and Sanchez (1980) and Allen and Gisser (1984) stated that given more properly defined water rights, the increase in agricultural revenues resulting from optimization of groundwater use will be relatively small in areally extensive aquifers with high storage coefficients. Where there are uncertainties in estimating demand functions, they suggest that attempts to optimize groundwater use over time may actually result in a reduction of revenues.

Nieswiadony (1985) used a similar analytic formulation to compare time paths of water use under optimal control in the Texas High Plains. He estimated that the potential percentage benefits from groundwater management in this scenario would be small. Howitt (1979) considered the benefits of optimizing groundwater use in four subbasins of the Central Valley in California. The computed present value of benefits ranged from $4 to $65 per acre. Feinerman and Knapp (1983) did a sensitivity analysis for various parameters in their single-cell model. For the various scenarios considered, the increase in revenues from more efficient groundwater ranged from 7 to 28%. Their example system was a conjunctive surface water/groundwater system, and involved much direct surface water use.

The simplifications and assumptions inherent in treating the groundwater system as a single cell need to be considered when analyzing the results of these studies. In addition, both Howitt (1979) and Feinerman and Knapp (1983) assumed a fixed recharge rate that determined the steady-state basin yield. The problems of such an assumption are discussed by Bredehoeft et al. (1982). The first work to include distributed parameter groundwater simulation in the optimization of agricultural revenues was carried out by Bredehoeft and Young (1970). Their approach involved iterating between the groundwater simulator and a linear program that represented agricultural production.

The studies of Bredehoeft and Young (1970, 1983) and Young and Bredehoeft (1972) indicated that an agricultural system can operate under a variety of pumping taxes, pumping quotas, and installed well capacities with only small changes in total revenue. Figure 3 shows net expected benefit (mean annual income) per acre versus well capacity from their 1983 study. The relatively flat portion of the curves beyond a capacity of 200 cfs is apparent.

In the South Platte Valley the installed capacity to pump groundwater is sufficient to irrigate completely the system (400 cfs on Figure 3). The farmers have discounted the availability of surface water. The groundwater capacity ensures the water supply during drought. A plateau in the benefit function leads to this result. There is little or no diminution of average annual income associated with installing excess well capacity. There is the additional benefit of insuring against a drought. Tsur (1990) presented an application in Israel in which groundwater provided insurance against the variability of the available surface water; in this case this was the major benefit of the groundwater development.

Using a similar simulation approach, Daubert and Young (1982) computed agricultural revenues under two water rights scenarios. In Daubert and Young's first case, groundwater pumping was prohibited to protect downstream surface water user rights. In the second case uncontrolled groundwater pumping was allowed. The second policy yields twice as much total revenue under average streamflow conditions

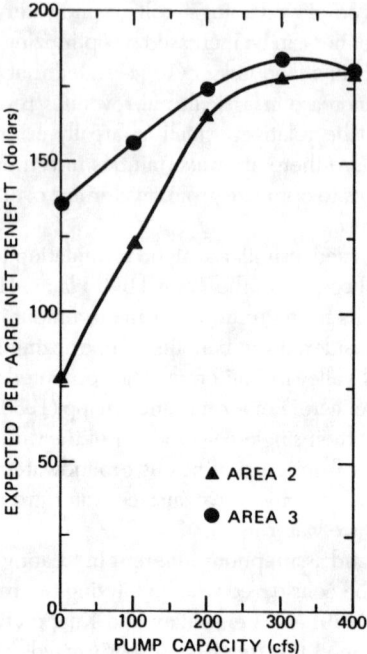

Figure 3. Expected per acre income versus well capacity. (From Bredehoeft, J. D., and Young, R. A., *Water Resour. Res.*, 19, 1111, 1983.)

and three times as much revenue under drought conditions; however, downstream users incurred severe revenue losses. These results, which indicate a doubling of revenue when groundwater is used in conjunction with surface water are consistent with Young and Bredehoeft's (1972) (Bredehoeft and Young, 1983) results.

Young et al. (1985) investigated the economic efficiency of the Colorado augmentation scheme. They used a distributed parameter model in which effects were simulated using a discrete kernel mathematical approach pioneered by Morel-Seytoux (Morel-Seytoux and Daly, 1975; Illangasekare et al., 1984). Their analysis indicated that an augmentation capacity of 25% of the groundwater pumping capacity was optimal. This capacity is larger than the 5% augmentation required by the state of Colorado. Their analysis again indicated that conjunctive use in the South Platte Valley increased the average annual output by a factor of two—100%. Once pumping capacity was introduced into the system the objective function exhibits a broad plateau on which there is no sharply defined optimum point, a result also found by Young and Bredehoeft (1972) (Bredehoeft and Young, 1983).

Recent studies have incorporated aspects of distributed parameter groundwater simulation directly into an agricultural optimization model. This situation is in contrast to the simulation approach where the analysis involves iterating back and forth between the agricultural and hydrologic models. Noel et al. (1980) presented a quadratic optimization model for part of California's Central Valley. They found that dynamic optimization would increase the social benefits of water use by a little more than 20%. O'Mara and Duloy (1984) constructed a hierarchical water management model for the Punjab Basin in Pakistan. They tested the impact of several alternatives regarding pricing and allocation of water, as well as capital improvements. All of their scenarios increased per capita income by approximately 20%.

Lemoine (1984) and Reichard (1987) looked at the management of groundwater in the Salinas Valley in California. Their approach involved coupling agricultural production functions and groundwater response functions in a single optimization model. For the various scenarios they considered, basin-wide management of groundwater was found to increase total revenues from 3 to 16%.

Peralta has done a number of studies (Peralta, 1985; Peralta and Killian, 1985; Yazdanian and Peralta, 1986) of optimal development using the conjunctive groundwater/surface water system in the Grand Prairie of Arkansas as his prototype. He set as a constraint in a linear or quadratic program water levels in the aquifer system that maintain a viable saturated thickness of aquifer throughout the area. Given this constraint he minimizes the cost of delivering water. The results show improvements in revenues of 10%, or less.

An informative study is that of Lefkoff and Gorelick (1990a,b). They examined the economic effects of water quality on agricultural production along a reach of the Arkansas River Valley in Colorado. In this portion of the valley the groundwater quality is severely degraded; the degradation is such that it affects crop yields. They used a distributed parameter simulation model that incorporated the farm decisions. Some farmers had groundwater from the alluvial aquifer that underlies parts of the valley; others were in portions of the valley that did not have the aquifer and did not have access to groundwater.

Lefkoff and Gorelick (1990b) showed that a local rental market had several beneficial results. The market allowed water users with groundwater to rent their surface water during dry years. It increased the average annual benefits from $80 per acre to $94 per acre (17%); it decreased the standard deviation in per acre income from $28 to $13 providing income insurance; and it improved the water quality of the aquifer. It was especially effective in improving income during dry years when farmers without a well were very short of water. This suggests that modest changes in the water rights structure, especially a local rental market can improve system output. However, the local water market invoked by Lefkoff and Gorelick (1990b) may have adverse quality impacts on the Arkansas River downstream from the area investigated that was unaccounted for in their analysis.

A rental market is being looked at more and more to provide a solution to the allocation problem, especially as demand in the urban sector of the economy increases and competes for agricultural irrigation water. Even in California, irrigated agriculture currently consumes 85% of the water. In other western states the percent consumed by agriculture is larger than in California. There is also a growing recognition that the ecosystem within a water shed needs protection including some right to water. This growing concern impedes the large scale transfer of water.

Quantity Management: Minimizing Costs

The second category of regional scale groundwater quantity management problems involves minimizing the costs of water supply. The problem of meeting a prespecified demand for water in the least costly manner has been investigated with several different approaches. Denninger (1970), Maddock (1972), Schwarz (1976), Willis and Newman (1977), and Wanakule et al. (1986) had as their objective minimizing pumping costs, or an equivalent surrogate measure, to satisfy a given demand. Yu and Haimes (1974), Maddock (1974), Gupta and Goodman (1986), and Li et al. (1987) considered the minimization of water supply costs in conjunctive use systems. These studies have generally not involved sensitivity analyses.

Danskin and Gorelick (1985) studied the simultaneous management of groundwater and surface water in a complex system. Their results indicated that the current operations were far from optimal. Their simulation/management model of the Livermore Valley in California showed that the water authority was spending twice as much as they should have to achieve a variety of goals involving hydraulic, water quality, and water allocation. They suggested that the primary overexpenditure related to the purchase and transport of expensive surface water that was released along one stream reach in an effort to recharge the aquifer. The management model indicated that this practice was both ineffective and wasteful. The problem addressed by Danskin and Gorelick (1985) shows that groundwater/surface water management models can improve basin operation by identifying inefficient practices. This study indicates that inefficiencies occur when market forces are neglected for one reason or another.

Lall and Lin (1991) considered the problem of minimizing municipal and industrial pumping costs in a basin in Utah that had several supply agencies. They explicitly considered constraints that restricted flow across the boundaries of the supply agencies. Their results suggested that the redistribution of pumping within the boundaries of the agency supply areas could reduce the basin-wide pumping costs by more than 50%. In addition there were potential advantages to transferring water between agencies. However, the capital costs of constructing additional distribution facilities were not included in their analysis.

Regional Quantity Problems: Summary

Our survey of the different studies concerned with maximizing agricultural revenue indicates that the maximum potential increase in revenues without dramatic changes in the existing institutions is generally less than 20%. Table 1 summarizes our review. One can argue that the free market which has generally operated in allocating groundwater has lead to efficient systems. Admittedly the literature deals mostly with irrigated agriculture because it is the dominant water consumption in the arid western United States. It also seems that approximately the same revenues can be obtained by several different patterns of water use and cropping. This suggests that the objective functions are relatively flat near the optimum—the optimum is essentially non-unique. Often, a broad plateau exists where all policies yield approximately the same benefits. It is essential that one reach the plateau; for example, in Figure 3 it is important to install at least 200 cfs of pumping capacity. At any lesser capacity there is a significant decrease in annual income. Beyond 200 cfs it does not matter much; other considerations such as minimizing the variation in annual income come into play. This broad plateau near the optimum makes the problem in water resources management easier.

When considering the management of groundwater used for agriculture, it may be important to consider factors other than percentage changes in revenue. First, the

Table 1. Summary of Net Benefits from Groundwater Quantity Management Models

Model	Benefits	Comments
1. Central Valley, CA Howitt (1979)	$4–65/acre	Steady-state flow
2. Conjunctive use Feinerman and Knapp (1983)	7–28%	Steady-state flow
3. Conjunctive use South Platte Valley, CO Young and Bredehoeft (1972) Bredehoeft and Young (1983)	Less than 10%	Large plateau benefit function (Figure 3)
4. Conjunctive use South Platte Valley, CO Young, et al. (1985)	Less than 10%	Large plateau benefit function 25% augment
5. Central Valley, CA Noel et al. (1980)	20%	
6. Punjab, Pakistan O'Mara and Duloy (1984)	20%	Hierarchical model
7. Salinas Valley, CA Lemoine (1984) Reichard (1987)	3–16%	
8. Grand Prairie, Arkansas Peralta (1985) Peralta and Killian (1985) Yazdanian and Peralta (1986)	10% or less	Constrained water levels
9. Arkansas Valley, CO Lefkoff and Gorelick (1990a,b)	17%	Local market

total dollar value of revenue changes may be important. If total revenues are large, small percentage changes may involve large amounts of money. For an area such as the Salinas Valley in California with annual net agricultural revenues of nearly $50 million, a 10% improvement in annual revenues represents $5 million. Second, the distribution of revenue may be important. For example, O'Mara and Duloy (1984) found that areas with groundwater salinity problems had potentially larger revenue increases than did the areas where the groundwater was fresh. Their work and Reichard's (1987) work indicated that relatively large revenue increases can be obtained in specific areas of a groundwater basin without penalizing others.

The flatness of the objective function also suggests a potential institutional flexibility in the management of groundwater for agricultural use. Specifically, it may be possible to maximize or minimize multiple objectives without greatly reducing the optimum revenue. Such additional objectives could decrease the variance in annual income (Bredehoeft and Young, 1983; Lefkoff and Gorelick, 1990), the volume of water used (Reichard, 1987), or the level of employment (O'Mara and Duloy, 1984).

A final inference can be drawn regarding the seeming flatness of the objective function. Maknoon and Burges (1978) suggested that, in cases where the response surfaces are relatively flat, simulation models might be preferable to optimization since much more detail can be incorporated into simulation models without optimization. With the advances in both computer capacity and methodology of the last several years, optimization models can now be used to model complicated hydrologic systems. Danskin and Gorelick's (1985) work suggests a different result; the actual conditions were the result of inefficient institutional controls. Water allocation decisions were made by a water authority that did not have a market incentive to identify inefficiency. A groundwater management model identified this inefficiency. Lall and Lin (1991) reached a similar conclusion.

REGIONAL QUALITY PROBLEMS

For the regional-scale groundwater quantity problems considered above a reasonable argument can be made that market forces drive water users to maximize profits and water suppliers to minimize the costs of delivery. It is difficult to apply this argument to problems of groundwater quality management. A brief episode of groundwater contamination will generally have long-lasting effects. The long-lasting nature of this problem has been at the heart of the nuclear waste disposal issue (Gorelick et al., 1979). The period for contamination problems is usually from tens to hundreds of years, often longer, much longer than for most problems of groundwater quantity. The impact of contamination may not be felt until sometime in the future. For this reason these impacts are often ignored in decision making. Often groundwater contamination is unknown. Put another way, the benefits of groundwater quality management usually require a long planning horizon; adverse quality impacts may have few effects on short-term profits. Health and safety considerations are often outside the decisions of individual users.

Regional pollution damage is often irreversible over any reasonable planning horizon. Once an aquifer is contaminated it is generally a lengthy and expensive process to restore it so that water can be used again without treatment (Sharefkin et al., 1984). This situation is unusual for groundwater quantity. Except for mining, once the stress is discontinued, water levels recover quickly (see Burt 1966, 1967 for discussions of groundwater *mining*). An example illustrates the difference in the physics governing groundwater quality and quantity management. Suppose a water user pumps for years causing a detrimental impact to a distant water user—an increase in pumping costs due to declining water levels. This effect quickly begins to reverse when the *guilty party* stops pumping. Alternatively, let us assume that the individual disposes of toxic waste into the aquifer. Even if this person stops disposal, the distant water user may continue to suffer from contaminated groundwater for a long period.

Other differences between the groundwater quality and quantity problems are illustrated by this example. For the water quantity problem, the first user is causing a drawdown in water levels; significant withdrawals from the aquifer are readily identified by monitoring water levels. The fact that the first individual has polluted the aquifer may not be discovered because of the difficulty of monitoring groundwater quality. The contaminant plume may not be identified until it has migrated away from the source or mixed with other contaminants.

For the water quantity problem, water levels for the distant users decline steadily over time; however, the resource remains usable, albeit at a higher cost, until the aquifer is dewatered. The arrival of contamination at the distant water user's well could render the water unusable (at least from a legal standpoint) or could be a factor to diminished agricultural production or increase health risks. Some contamination does not render water unusable; it simply increases the costs of water. Individual contaminants have differing levels of toxicity and require different treatment. Chlorides, for example, are different from man-made organic compounds, such as organic pesticides.

Groundwater versus Surface Water Quality

Many problems discussed above have been recognized by economists considering the problems of market externalities. Randall (1983) describes an externality as a situation in which the affected party is influenced by one or more activities under the control of another (or others). Mitnick (1980) describes externalities that are not adequately handled by a market. First there are those externalities for which there is a general lack of knowledge or lack of knowledge regarding their severity. The cost to individuals of obtaining information is often prohibitively expensive. Collective action will often be required to remove the externality. Finally, permanent resolution of certain externality problems may require the institution of new or additional regulations. The modification of water rights could also resolve externality problems. Randall (1983) argues that most externalities can be handled properly by existing markets. Only in situations where the transaction costs of removing the externality are prohibitively high will markets break down. Externalities related to a nonexclusive resource such as groundwater can present such a situation.

There is a body of literature dealing with surface-water pollution that is particularly relevant to the discussion of groundwater quality. The fact that an upstream polluter may cause damages to downstream water users is widely recognized. From a basin-wide perspective, economically efficient operation occurs when the marginal cost of controlling the upstream discharge equals the marginal damage to downstream users. Herfindahl and Kneese (1974) suggest that the likelihood of reaching this optimum through private exchange is small. Transaction costs make it difficult for the generally large number of damaged parties to organize effectively.

There is also the *free-rider* problem—parties can benefit from improvements in water quality without sharing any of the costs, since water is generally nonexclusive. Under these circumstances, free market mechanisms will usually not drive the system toward optimum conditions; some form of collective action or management is necessary. Often, as indicated by Kneese and Schultze (1975), waste assimilation capacities of air and water do not command a price . . . , the price system works with marvelous efficiency, but in the wrong direction. There are exceptions to this generalization. For example, in unsewered residential areas soil suitability for septic systems will have an impact on land value. The same reasons for the market's inability to regulate surface water contamination apply to the problem of groundwater pollution. The situation is further complicated in groundwater because:

1. The exact direction of flow is not always apparent, and can change as the location and magnitude hydrologic stresses change;
2. physical parameters are highly uncertain, distributed in three dimensions and may be anisotropic, and difficult to measure; and
3. dispersive as well as advective transport occurs accompanied by chemical and biological reactions.

Regional Quality Models

There are only a few investigations that discuss the regional management of water quality. The theoretical work of Cummings (1971) and Cummings and McFarland (1971) incorporate functions that represent the impact of seawater intrusion and salinity build-up into a management model. They discuss the need for central control to address groundwater quality concerns. However, they do not include any applications in their analysis.

Shamir et al. (1984) present a multiobjective linear programming model for optimal management of a coastal aquifer. Two of the objectives incorporate quality concerns: minimizing the difference between the actual and the desired location of the seawater–freshwater interface and minimizing the total sum of chloride concentrations in all locations. Willis and Finney (1988) developed an optimization model for controlling seawater intrusion in Taiwan. They linked nonlinear optimization with a sharp interface model. They used a composite objective function that included the location of the toe of the interface, target quantities of both pumping and costs, and artificial recharge. Their results suggested that the resulting objective function had a large plateau, much like the quantity management model results discussed above.

Gardner and Young (1988) investigated the economics of reducing the salinity load in Colorado using as their prototype irrigation in the Grand Valley of Colorado. Return flow from irrigation in the Grand Valley is high in dissolved salts. The salts are leached from the Cretaceous Mancos Shale that underlies the irrigated fields. The cost of removing salt ranges from $3 to $12 per ton depending on the policy implemented. The policy question is further complicated by the questions of equity about who should pay these costs. Congress currently subsidizes removing the salt load from the Colorado River; all taxpaying Americans pay these costs.

As mentioned above, Lefkoff and Gorelick (1990a,b) used a distributed parameter flow and quality management model to investigate the impacts of differing policies on the surface and groundwater quality in the Arkansas Valley of Colorado. They found that a local water market improved the supply during drought years and improved the local groundwater quality of the system. The local market improved the total economic output of the local system.

The west side of the San Joaquin Valley in California has been the focus of recent efforts to improve water management. Several years ago selenium was found in irrigation drain discharges from this part of the valley. Irrigation leaches selenium from the soil; it is then discharged with groundwater into field drains (Benson et al., 1991; White et al., 1991). Part of the area of concern is in the Westlands Project, the largest irrigation project of the Bureau of Reclamation. The occurrence of selenium at Kesterson Reservoir, a wildlife preserve, caused a great public outcry. Birds, especially ducks, living at the preserve had serious birth defects caused by the selenium. The identification of selenium in the irrigation drain flow caused a restriction in irrigation drainage. This in turn lead to a concern about the water logging of land in the lower San Joaquin Valley. Water logging usually involves a high water table accompanied by a build-up in salt; it is usually both a quantity and a quality problem.

There have been several management studies of the agricultural drainage problem in the San Joaquin Valley (San Joaquin Valley Drainage Program, 1987,

1989). One interesting effort is the modeling of the U.S. Geological Survey (Belitz and Phillips, 1992). They demonstrated that by decreasing surface-water applications and increasing the groundwater pumping, the water logged area can be greatly reduced. Instead of approximately 50% of the 550 square miles studied becoming water logged by the year 2040, it can be reduced to less than approximately 25% of the area, approximately 140 square miles—still a serious problem, but reduced significantly by the management modeling.

Bernardo et al. (1993a) assembled a regional groundwater quality management model, that used the USGS MODFLOW groundwater flow model coupled to a programming model that simulated irrigation, to look at the economic impacts of imposing water quality changes. They used the Ogalalla Aquifer in Texas, Oklahoma, and Kansas as their prototype. They looked in particular at the economic consequences of imposing nitrogen fertilizer limits to improve the quality of the runoff from the system (Bernardo et al., 1993b). The nitrogen restriction decreased farm revenues; decreases range from 20 to 28%. The also investigated restricting pesticide use. This too decreased farm revenues; however, the decrease was less than 10%.

Bouzaher et al. (1993) argue for the use of *metamodels* for studying nonpoint pollution from agriculture. They define a metamodel as a statistical method to eliminate unneeded detail for regional analysis by approximating the outcomes of a complex process model through statistically validated parametric forms. They studied the impact of various policies of herbicide applications on both groundwater and surface water.

These studies indicate that improving the quality of regional groundwater will have an associated cost. Table 2 summarizes our review. Some land will have to be retired from further agricultural use in the San Joaquin Valley because of water logging. The study by Belitz and Phillips (1992) suggests that this will be more than 100 square miles. The study of nitrates in the Ogalalla Aquifer indicates that decreasing nitrate in the aquifer will reduce revenues somewhere between 20 and 30% (Bernardo et al., 1993b). The study of salt loading in the Grand Valley by Gardner and Young (1988) indicates that salt can be reduced at a cost of approximately $5 to $10 per ton. All these studies indicate that there is no free lunch in improving water quality; it will cost.

CONCLUDING REMARKS

Aquifer management models that combine simulation with optimization help both the hydrologist and the water resource analyst understand how social and economic forces interact with the water resource allocation. Just as a hydrologic simulation model is a tool to understand the physical/chemical behavior of an aquifer system, a management model is a tool to provide insight into the economic and social consequences of institutional changes. Just as the usefulness of a physical simulation model depends on the degree to which it simulates the essentials of the system modeled, the usefulness of management models depends on assumptions made regarding economics and institutions and on the accuracy of the model parameters. From our analysis of the literature, several speculative remarks can be made about the relevance of management models.

Table 2. Summary of Costs Associated with Improving Water Quality as Derived from Management Models

Model	Cost	Comments

1. Seawater intrusion and agricultural salinity
  Cummings (1974)
  Cummings and McFarland (1971)
	NA	

2. Coastal aquifer
  Shamir et al. (1984)
  Willis and Finney (1988)
	NA	sharp interface objective plateau

3. Salt load to the Colorado River Grand Valley, CO Gardner and Young (1988)	$3–12/ton	? of equity: who pays
4. Arkansas River Valley, CO Lefkoff and Gorelick (1990b)	benefit	Local market
5. San Joaquin Valley, CA Belitz and Phillips (1992)	100+/− mi^2 640,000 acre	waterlogged land ($2000/acre)

6. Ogalalla Aquifer—KA, NM, TX
  Nitrogen Control
  Pesticide Control
  Bernardo et al. (1993b)
	20–28%
	10%
		large distributed model—uses MODFLOW

7. Metamodels—Statistical Models
  Herbicide control
  Bouzaher et al. (1993)
	NA	Simplified models

Regional aquifer management models are used to investigate the effectiveness of institutions and policies for allocating water in agricultural areas. Irrigated agriculture is of interest because it consumes more than 85% or more of the water used in all western states. The models are also used to study how to provide regional water supplies at a minimum cost. The results of the various analyses suggest that a large percentage increase in economic efficiency through *better* management is unlikely. The market forces that currently operate generally allocate groundwater quantity to maximize net revenue. In nearly every instance reviewed the improvement indicated by optimal management was less than 20%. It should be remembered, however, that a small percentage improvement in a large agricultural system may mean large amounts of money.

Regional agricultural management models can prove useful in suggesting alternate policies that are near optimal and provide other social or environmental benefits at a small cost in benefits. A plateau in the objective surface near the optimum suggests that other objectives ranging from the distribution of employment to ensuring against drought can be achieved with little impact on net benefits. Models may be useful in identifying institutions that better account for economic externalities. They also may help identify what the important system variables are as well as those portions of the modeling that require more detailed investigation (Maknoon and Burges, 1978). In terms of water supply, the literature indicates that groundwater management models can be useful in indicating inefficiencies that exist in a complex water system, such as conjunctive use. Experience suggests that these inefficiencies occur where the market forces are ineffective for one reason or another.

We expect that the nonexclusive nature of groundwater will make it difficult for problems of water quality to be efficiently handled via traditional markets. Water quality degradation has historically been a problem that was not considered by individual groundwater users. The short-term market is not usually effective in preserving long-term health and safety.

While both groundwater quantity and quality can be viewed as nonexclusive properties, there are fundamental differences between them. Problems of quantity involve the impact of users upon each other. This impact is generally identifiable, and rapidly reversible (except for mining). This is usually not so when one inspects the impacts of even one water polluter on other water users. The nature of groundwater transport exacerbates the problems of economic efficiency associated with water quality degradation. A brief episode of contamination may impact the system at distant times and unexpected locations. For these reasons, management to protect groundwater quality is important. The study of Lefkoff and Gorelick (1990b) illustrated the benefits of a local water market. This is a clear instance where improving the institution can have a significant beneficial impact. Moreover, the local market relaxes the traditional water right structure.

Given the difficulty of defining and measuring the appropriate social objective function, groundwater management models cannot define a precise solution that represents a single policy of choice. For problems in which a plateau exists in the objective function, the expectation of a single optimal policy is particularly ludicrous. The output computed by a groundwater management model may be helpful in evaluating the efficiency of current operations. The optimum should be considered a starting point in the exploration of differing policies and institutions. Identification of the optimal value of the objective is useful even if the optimal set of decision variables is largely ignored. Determining the optimal value provides a standard with which to compare the efficiency of current and alternative operations. Frequently, the intricacies of a management problem are unclear to the hydrologist, economist, and planner. When a problem is difficult to formulate, simulation is one means of experimentation. A second approach was suggested by Rogers et al. (1983) for resource planning in which resilient near optimal solutions are found for a set of different objectives. We support the logic of this concept, but both the data requirements and its implementation are unclear.

Given the nature of the objective in which there is no sharp optimal peak [rather there is broad plateau where several solutions provide nearly equal benefits (Figure 3)], additional objectives or constraints may be considered in reformulating the problem. Multiobjective analysis (Cohon and Marks, 1975; Hippel, 1992) may push decisions to one region on the plateau. If one is interested in limiting the dimensions of the decision space, then an objective might include minimizing the number of decision variables. We expect the new analyses to account explicitly for parameter uncertainty in the selection of optimal policies.

The Bottom Line

Our analysis indicates that many existing groundwater developments are currently operated at near optimal economic efficiency, especially when we consider quantity alone. None of the analyses suggest an improvement in net benefits of more than 20%—although 20% may be a large amount of money (Nieswiadony, 1985). Market forces are effective in managing groundwater. As a result, one should proceed cautiously in changing current institutions. The management of many groundwater systems is not broken; they do not need fixing. Where the free market is ignored there may be significant savings in identifying efficient management strategies. In contrast, regional management of water quality is likely to have large costs associated with it.

ACKNOWLEDGMENT

We gratefully acknowledge the support of NSF grant BCS-8957186.

REFERENCES

Alley, W. M. and Schefter, J. E., External effects of irrigators' pumping decisions, High Plains Aquifer, *Water Resour. Res.*, 23, 1123, 1987.

Aguado, E. and Remson, I., Ground water hydraulics in aquifer management, *J. Hydraul., ASCE,* 100(HY1), 103, 1974.

Allen, R. C. and Gisser, M., Competition versus optimal control in ground water pumping when demand is nonlinear, *Water Resour. Res.*, 20, 752, 1984.

Atwood, D. F. and Gorelick, S. M., Hydraulic gradient control for ground water contaminant removal, *J. Hydrol.*, 76, 85, 1985.

Banks, S., Exploratory modeling for policy analysis, *Oper. Res.*, 41, 435, 1993.

Belitz, K. and Phillips, S. P., Simulation of water-table response to management alternatives, central part of the Western San Joaquin Valley, California, U.S. Geological Survey Water Resources Investigation Report 91–4193, 1992.

Benson, S. M., White, A. F., Halfman, S., Flexser, S., and Alavi, M., Ground water contamination at the Kesterson Reservoir, California: 1. Hydrologic setting and conservative solute transport, *Water Resour. Res.*, 27, 1071, 1991.

Bernardo, D. J., Mapp, H. P., Sabbagh, G. J., Galeta, S., Watkins, K. B., Elliot, R. L., and Stone, J. F., Economic and environmental impacts of water quality protection policies: 1. Framework for regional analysis, *Water Resour. Res.*, 29, 3069, 1993a.

Bernardo, D. J., Mapp, H. P., Sabbagh, G. J., Galeta, S., Watkins, K. B., Elliott, R. L., and Stone, J. F., Economic and environmental impacts of water quality protection policies: 2. Application to the Central High Plains, *Water Resour. Res.*, 29, 3081, 1993b.

Bredehoeft, J. D., Papadopulos, S. S., and Cooper, H. H., Ground water: the water budget myth, National Research Council Studies, in *Geophysics: Scientific Basis of Water-Resource Management*, 51, 1982.

Bredehoeft, J. D. and Young, R. A., The temporal allocation of ground water: a simulation approach, *Water Resour. Res.*, 6, 3, 1970.

Bredehoeft, J. D. and Young, R. A., Conjunctive use of ground water and surface water for irrigated agriculture: risk aversion, *Water Resour. Res.*, 19, 1111, 1983.

Burt, O. R., Economic control of ground water reserves, *J. Farm Econ.*, 48, 632, 1966.

Burt, O. R., Temporal allocation of ground water, *Water Resour. Res.*, 3, 45, 1967.

Bouzaher, A., Lakshinarayan, C. R., Carriquiry, A., Gassman, P., and Shogren, J. F., Metamodels and nonpoint policy in agriculture, *Water Resour. Res.*, 29, 1579, 1993. California Regional Water Quality Control Board—Central Valley Region, Staff report on the modifications to beneficial uses and water quality objectives necessary for the regulation of agricultural subsurface drainage discharges in the San Joaquin Basin, Sacramento, CA, 1988.

Cohon, J. L. and Marks, D. H., A review and evaluation of multi-objective programming techniques, *Water Resour. Res.*, 10, 208, 1975.

Cummings, R. G., Optimum exploitation of ground-water reserves with saltwater intrusion, *Water Resour. Res.*, 7, 1415, 1971.

Cummings, R. G. and McFarland, J. W., Ground-water management and salinity control, *Water Resour. Res.*, 10, 909, 1974.

Danskin, W. R. and Freckleton, J. R., Ground-water flow modeling and optimization techniques applied to high ground-water problems in San Bernadino, California, in *Selected Papers in the Hydrologic Sciences, 1988–92*, U.S. Geological Survey Water Supply Paper 2340, S. Subitzky, Ed., 165, 1992.

Danskin, W. R. and Gorelick, S. M., A policy evaluation tool: multiaquifer management using controlled stream recharge, *Water Resour. Res.*, 21, 1731, 1985.

Daubert, J. T. and Young, R. A., Ground-water development in western river basins: large economic gains with unseen costs, *Ground Water*, 20, 80, 1982.

Denninger, R. A., Systems analysis of water supply systems, *Water Resour. Bull.*, 6, 573, 1970.

Feinerman, E. and Knapp, K. C., Benefits from ground water management: magnitude, sensitivity, and distribution, *Am. J. Agric. Econ.*, 65, 703, 1983.

Gardner, R. L. and Young, R. A., Assessing strategies for control of irrigation-induced salinity in the Upper Colorado River Basin, *Am. J. Agric. Econ.*, 70, 37, 1988.

Gisser, M., Ground water: focusing on the real issue, *J. Pol. Econ.*, 91, 1001, 1983.

Gisser, M. and Sanchez, D. A., Competition versus optimal control in ground water pumping, *Water Resour. Res.*, 16, 638, 1980.

Gordon, H. S., The economic theory of property rights, *Am. Econ. Rev., Papers Proc.*, 62, 124, 1954.

Gorelick, S. M., A review of distributed parameter ground water management modeling methods, *Water Resour. Res.*, 19, 305, 1983.

Gorelick, S. M., Large scale nonlinear deterministic and stochastic optimization: formulations involving simulation of subsurface contamination, *Math. Prog.*, North-Holland, 48, 19, 1990.

Gorelick, S. M., Remson, I., and Cottle, R. W., Management model of a ground water system with a transient pollutant source, *Water Resour. Res.*, 15, 1243, 1979.

Gorelick, S. M., Voss, C. I, Gill, P. E., Murray, W., Saunders, M. A., and Wright, M. H., Aquifer reclamation design: the use of contaminant transport simulation combined with nonlinear programming, *Water Resour. Res.*, 20, 415, 1984.

Gupta, R. S. and Goodman, A. S., Ground-water reservoir operation for drought management, *J. Water Res. Plan. Manage.*, 11, 303, 1986.

Heidari, M., Application of linear system's theory and linear programming to ground water management in Kansas, *Water Resour. Bull.*, 18, 1003, 1982.

Herfindahl, O. C. and Kneese, A. V., *Economic Theory of Natural Resources*, Charles E. Merrill, Columbus, OH, 1974.

Hippel, K. W., Multiple objective decision making in water resources, *Water Resour. Bull.*, 28, 3, 1992.

Howitt, R. E., Is overdraft always bad?, in *Proc. 12th Biennial Conf. Ground Water,* California Water Resources Center, University of California at Davis, Report No. 45, 50, 1979.

Illangasekare, T. H., Morel-Seytoux, H. J., and Verdin, K. L., A technique of reinitialization for efficient simulation of large aquifers using the discrete kernel approach, *Water Resour. Res.,* 20, 1733, 1984.

Kneese, A. V., and Schultze, C. L., *Pollution, Prices, and Public Policy,* The Brookings Institute, Washington, D.C., 1975.

Lall, U. and Lin, Y. C., A ground water management model for Salt Lake County, Utah with some water rights and quality considerations, *J. Hydrol.,* 123, 367, 1991.

Larson, S. P., Maddock, T., and Papadopulos, S., Optimization techniques applied to ground water development, *Mem. Int. Assoc. Hydrol.,* No. 13, E57–E67, 1977.

Lefkoff, L. J. and Gorelick, S. M., Simulating physical processes and economic behavior in saline, irrigated agriculture: model development, *Water Resour. Res.,* 26, 1359, 1990a.

Lefkoff, L. J. and Gorelick, S. M., Benefits of an irrigation water rental market in a saline stream-aquifer system, *Water Resour. Res.,* 26, 1371, 1990b.

Lemoine, P. H., Water Resources Management in the Salinas Valley: Integration of Economics and Hydrology in a Closed Control Model, Ph.D. dissertation, Stanford University, Stanford, CA, 1984, p. 208.

Li, C., Bahr, J. M., Reichard, E. G., Butler, J. J., and Remson, I., Optimal siting of artificial recharge: an analysis of objective functions, *Ground Water,* 25, 141, 1987.

Maddock, T., Algebraic technological function from a simulation model, *Water Resour. Res.,* 8, 129, 1972.

Maddock, T., Management model as a tool for studying the worth of data, *Water Resour. Res.,* 9, 270, 1973.

Maddock, T., The operation of stream-aquifer systems under stochastic demands, *Water Resour. Res.,* 10, 1, 1974.

Maknoon, R. and Burges, S. J., Conjunctive use of ground and surface water, *J. Am. Water Works Assoc.,* 10, 419, 1978.

Mitnick, B. M., *The Political Economy of Regulation: Creating, Designing and Removing Regulatory Forms,* Columbia University Press, New York, 1980.

Molz, F. J. and Bell, L. C., Head gradient control in aquifer used for fluid storage, *Water Resour. Res.,* 13, 795, 1977.

Morel-Seytoux, H. J. and Daly, C. J., A discrete kernel generator for stream-aquifer studies, *Water Resour. Res.,* 11, 253, 1975.

Nieswiadony, M., The demand for irrigation water in the High Plains of Texas, 1957–80, *Am. J. Agric. Econ.,* 67, 619, 1985.

Noel, J. E., Gardner, B. D., and Moore, C. V., Optimal regional conjunctive water management, *Am. J. Agric. Econ.,* 62, 489, 1980.

O'Mara, G. T. and Duloy, J. H., Modeling efficient water allocation in a conjunctive use regime: the Indus Basin of Pakistan, *Water Resour. Res.,* 20, 1489, 1984.

Peralta, R.C., Conjunctive use/sustained ground water yield design, *Proc. Special Conf. Comp. Appl. Water Res.,* ASCE/Buffalo, NY, 1391–1400, 1985.

Peralta, R. C. and Killian, P. J., Optimal regional potentiometric surface design: least-cost water supply/sustained ground water yield, *Trans. ASCE,* 28, 1098, 1985.

Randall, A., The problem of market failure, *Nat. Resour. J.*, 2, 131, 1983.

Reichard, E. G., The influences on the potential benefits of basinwide ground water management, *Water Resour. Res.*, 23, 77, 1987.

Remson, I. and Gorelick, S. M., Management models incorporating ground water variables, in operations research, in *Agriculture and Water Resources*, Yaron, D., and Tapiero, C. S., Eds., North Holland, 1980.

Rogers, P. P., Harington, J. J., and Fiering, M. B., New approaches in the use of mathematical programming for resource allocation, *Proc., World Bank Conf. External. Irrigated Agric.*, World Bank, 1983.

San Joaquin Valley Drainage Program, Developing Options, 28 p. 1987.

San Joaquin Valley Drainage Program, Preliminary Planning Alternatives, 1989.

Schwarz, J., Linear models for groundwater management, *J. Hydrol.*, 28, 377, 1976.

Shamir, U., Bear, J., and Gamliel, A., Optimal annual operation of a coastal aquifer, *Water Resour. Res.*, 20, 435, 1984.

Sharefkin, M., Shechter, M., and Kneese, A., Impacts, costs, and techniques for mitigation of contaminated ground water: a review, *Water Resour. Res.*, 16, 1771, 1984.

Tsur, Y., The stabilization role of ground water when surface water supplies are uncertain: the implications for ground water development, *Water Resour. Res.*, 22, 811, 1990.

Wanakule, N., Mays, L. W., and Lasdon, L. S., Optimal management of large-scale aquifers: methodology and applications, *Water Resour. Res.*, 22, 447, 1986.

White, A. F., Benson, S., Yee, A. Y., Wollenberg, H. A., Jr., and Flexser, S., Ground water contamination at the Kesterson Reservoir, California: 1. geochemical parameters influencing selenium mobility, *Water Resour. Res.*, 27, 1085, 1991.

Willis, R., A Planning model for the management of ground water quality, *Water Resour. Res.*, 15, 1305, 1979.

Willis, R. and Finney, B. A., Planning model for optimal control of saltwater intrusion, *J. Water Resour. Plan. Manage.*, 114, 163, 1988.

Willis, R. and Newman, B. A., Management model for ground water development, *J. Water Resour. Plan. Manage.*, ASCE, 103(WR1), 159, 1977.

Yazdanian, A. and Peralta, R. C., Maintaining target ground water levels using goal-programming: linear and quadratic models, *Trans. ASCE*, 29, 995, 1986.

Yeh, W. W-G., Systems analysis in ground-water planning and management, *J. Water Resour. Plan. Manage.*, 118, 224, 1992.

Young, R. A. and Bredehoeft, J. D., Digital computer simulation for solving management problems of conjunctive ground water and surface water systems, *Water Resour. Res.*, 3, 533, 1972.

Young, R. A., Daubert, J. T., and Morel-Seytoux, H. J., Economics of alternative institutions for managing interrelated stream-aquifer systems, unpublished manuscript, 1985, p. 34.

Yu, W. and Haimes, Y. Y., Multilevel optimization for conjunctive use of ground water and surface water, *Water Resour. Res.* 10, 625, 1974.

CHAPTER 8

Uncertainty Analysis of Subsurface Transport of Reactive Solutes Using Reliability Methods

Maged M. Hamed, Joel P. Conte, and Philip B. Bedient

INTRODUCTION

Physical parameters uncertainty in subsurface contaminant transport is manifested in the basic heterogeneity of the aquifer formation, the source term uncertainty, along with uncertainties related to the chemical, physical, and biological properties of the contaminant being transported. Such uncertainties greatly affect the predictive ability of deterministic groundwater flow and contaminant transport models and need careful consideration when making regulatory decisions based on such deterministic models.

Deterministic analytical screening models often provide a simple and easy-to-use tool for assessing the risk of contamination of groundwater supplies by organic chemicals. However, many of these models fail to account for parameter uncertainty prevalent in soil hydrology and chemical characteristics (DelVecchio and Haith, 1993).

In this paper, a simple and computationally efficient tool, which is based on first- and second-order reliability methods, is explored for the probabilistic assessment of the risk of groundwater contamination. The assessment procedure is demonstrated on a simple hypothetical case of transport of ethylbenzene in the subsurface.

UNCERTAINTY ANALYSIS METHOD

First- and second-order reliability methods (FORM and SORM, respectively) were originally developed to assess the safety of structural components and structural systems, and were recently applied to groundwater flow and transport problems (see, e.g., Sitar et al., 1987; Cawlfield and Sitar, 1988; Schanz and Salhotra, 1992; Cawlfield and Wu, 1993; Hamed et al., 1993). The problem of assessing the contamination risk is formulated from a component reliability view point, where situations with a single failure mode are analyzed. FORM and SORM are chosen because their many appealing features make them an excellent candidate for probabilistic groundwater modeling. These features include:

1. The ability of FORM and SORM to readily incorporate both analytical and numerical models. This overcomes the problem of having to make unrealistic limiting and restricting assumptions on problem geometry and boundary conditions;
2. the computational efficiency of FORM and SORM in the reliability analysis of problems that are characterized by low probability of occurrence as compared to the computationally prohibitive classical Monte Carlo simulation methods;
3. their provision for sensitivity measures at minimum extra computational cost. These sensitivity results are very valuable in recognizing the most important sources of uncertainty for the problem considered and for designing optimum future data collections at the site; and
4. their ability to address problems characterized by various degrees of statistical information: from the first two moments only, to the full joint probability density function of the input random variables.

Following is a brief review of the reliability methods, which is necessary to understand the formulation and approach described in the sequel. A full review of the reliability methods can be found in Madsen et al. (1986) and Melchers (1987).

THEORETICAL BACKGROUND

In this work the problem formulation is addressed from a *component reliability* viewpoint. This means that the interest is only in situations where the component studied has two possible states: the *failure state*, and the *survival state*. In this case, an n-dimensional function, termed the *limit-state function* of the input random variables, $g(X)$, is formulated with the following convention:

X	a vector of n basic random variables	
x	a realization of the basic random variables	
$g(x) < 0$	indicates the failure domain	(1)
$g(x) > 0$	indicates the safe (survival) domain	
$g(x) = 0$	indicates the limit-state surface	

The limit-state surface is, in general, an n-dimensional hypersurface which divides the performance space into a safe region and a failure region. The event of interest is usually the failure event; the probability of failure is given by the n-fold integral:

$$P_F = P[g(x) \leq 0] = \int_{g(x) \leq 0} f_X(x) dx \qquad (2)$$

where $f_X(x)$ is the joint probability density function of the basic random variables considered in the problem. The estimation of the aforementioned n-fold probability integral is complicated by the following factors:

1. The extreme difficulty and intractability of computing the above n-fold probability integral for large dimensional problems;

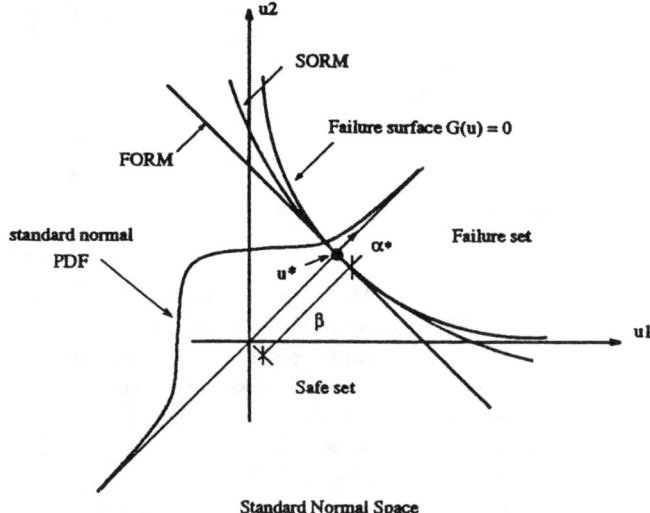

Figure 1. FORM and SORM approximations to the failure surface.

2. the model inaccuracies related to the limit-state function; and
3. uncertainties related to the joint probability density function.

FORM and SORM are analytical schemes to approximate the computation of the probability integral in Equation (2) and overcome the aforementioned problems. They consist of two main steps.

The first step involves the transformation of the random variables and the limit-state function, using a nonlinear one-to-one mapping to the standard normal space of uncorrelated normally distributed variates of zero mean and unit variance, such that the original density function $f_x(x)$ becomes the standard normal density function (Melchers, 1987).

In the second step, the transformed limit-state surface, $G(\boldsymbol{u}) = 0$, is approximated at a point on the surface, which is closest to the origin. This point is termed the *design point*, and it is the most likely failure point in the standard normal space.

FORM

In the first-order reliability method (FORM), the limit-state function is approximated by the tangent hyperplane at the design point. According to the first-order reliability method, the *reliability index* is given by the inner vector product

$$\beta_{\text{FORM}} = \alpha^\circ \boldsymbol{u}^\circ \tag{3}$$

where α° is the unit normal at the design point directed towards the failure region in the standard normal space and $\boldsymbol{u}^\circ$ is the design point. It follows that β represents the shortest distance from the origin to the limit-state surface in the standard normal space (Figure 1). The first-order approximation of the failure probability is obtained as

$$P_{F_{FORM}} = \Phi(-\beta) \qquad (4)$$

where $\Phi(.)$ is the standard normal cumulative distribution function.

SORM

The approximation given by FORM is accurate provided that the limit-state surface is nearly flat in the neighborhood of the design point. However, when the limit-state surface is curved, a second-order approximation of the limit-state surface at the design point is needed. This occurs when the limit-state function contains highly nonlinear terms, or when the input random variables have an accentuated non-normal character. SORM approximates the failure surface at the design point by a quadratic function (Fiessler et al., 1979). Various methods exist for such an approximation such as the curvature- and point-fitting method defined by Der Kiureghian et al. (1987). Figure 1 illustrates the approximations of the failure surface by FORM and SORM in a simplified two-dimensional case.

Determination of the Design Point

The major step in reliability analysis is to determine the design point in the standard normal space, especially for large reliability problems. The problem is formulated as a constrained nonlinear optimization as follows:

$$\text{Minimize } \frac{1}{2} u^T u \qquad (5)$$

$$\text{subject to } G(u) = 0 \qquad (6)$$

The above optimization problem consists of finding the point that lies on the limit-state surface and has a minimum distance from the origin in the standard normal space. Several algorithms have been developed to solve this problem, including the HL-RF method (Hasofer and Lind, 1974; Rackwitz and Fiessler, 1978), the gradient projection method, the sequential quadratic programming (SQP) method, and others (Liu and Der Kiureghian, 1991). For the purpose of this study, the SQP algorithm is used as the optimization procedure.

Sensitivity Measures

Other important quantities readily obtained in the FORM/SORM approach are the *uncertainty importance factors* (Hohenbichler and Rackwitz, 1986) and the parametric sensitivity factors or sensitivity of both the reliability index and estimate of the failure probability with respect to both probability distribution parameters and limit-state function parameters (Madsen, 1988).

For independent variates, the uncertainty importance factor is defined as the derivative of the first-order reliability index with respect to the corresponding variate in the standard normal space, and is given by:

$$\frac{\partial \beta}{\partial u_i}\bigg|_{u=u^\circ} = \alpha_i \qquad (7)$$

where α_i is the ith component of the unit normal vector to the limit-state surface at the design point.

In this work, only the uncertainty importance factors, expressed as $100\,\alpha_i^2$ are considered. For statistically independent variables, it was shown (Madsen, 1988) that omission sensitivity factors, defined as the relative error in the first-order reliability index when a basic variable X_i is replaced by a deterministic number equal to its median $X_{i,m}$, are given by:

$$\frac{\beta(X_i = X_{i,m})}{\beta} = \frac{1}{\sqrt{1 - \alpha_i^2}} \qquad (8)$$

It is therefore observed that the uncertainty importance factors, $100\alpha_i^2$, give a measure of the relative importance of modeling the uncertainty of a basic random variable X_i with respect to the final probability outcome.

The relative error in the first-order reliability index of representing a group of m mutually dependent variables, X_i, $i = 1, \ldots, m$ by their respective median is given by:

$$\frac{\beta(X_i = X_{i,m}, i = 1, \ldots, m)}{\beta} = \frac{1}{\sqrt{1 - \sum_{i=1}^{m} \alpha_i^2}} \qquad (9)$$

Therefore, the uncertainty importance factors associated to a group of mutually dependent variables can be expressed by the quantity $100\sum_{i=1}^{m} \alpha_i^2$.

Importance factors enable the identification of those random variables which have the least impact on the final reliability outcome. Each of these variables can then be replaced by a deterministic value (e.g., its median). Therefore, the importance factors are extremely helpful in reducing the number of basic random variables in large-size reliability models.

PROCEDURES

Description of Models Used

The proposed methodology is tested using the HPS model developed by Galya (1987), which is a semianalytical model that uses Green's function solutions, along with numerical integration to simulate uniform one-dimensional advective transport in the x direction with three-dimensional dispersion in the x, y, and z directions. It can incorporate retardation and first-order decay.

Input to the model includes the seepage velocity, dispersivities in the x, y, and z directions, soil porosity, aquifer thickness, first-order decay coefficient, soil bulk density, organic carbon content, receptor location, along with the source location, source area dimension, source concentration, and source infiltration rate. The model can accommodate any number of sources with varying concentration, and any number of receptor locations.

The transport model is extended to accommodate parameter uncertainty by linking it to the general purpose probability analysis program PROBAN (Veritas Research, 1992), which has the ability to handle various simulation and reliability analyses for both simple and complex problems formulated in either a component or a system reliability context. The built-in distributions library in PROBAN allows the assignment of a variety of marginal or joint probability density functions.

The interface between PROBAN and HPS is done using a FORTRAN subroutine. The coupled program (PROBAN-HPS) which is a probabilistic transport model is run on a SUN SPARCstation 2. This program works as follows: First, the user provides the input parameters to the model. These include the random variables, along with constant (deterministic) parameters. The definition of the input random variables is done by defining the probability density of each random variable along with the distribution parameters, such as the mean and the variance; or by just specifying the first two moments without specifying a certain probability distribution, in case only incomplete statistical information is available. Correlation between random variables, if any, can also be incorporated in the model. If the complete probability information is available, the input random variables are defined by their full joint density function which can be decomposed into a product of conditional distributions. The user also formulates the limit-state function that specifies the performance criteria for the component or system being studied. The model then proceeds to solve the reliability problem, following the steps discussed above, to estimate the probability of failure, the reliability index, and the sensitivity measures. During the FORM/SORM analysis, the reliability code calls, repetitively, the transport model for a specific realization of the random variables. Usually the method converges in less than 20 iterations. Figure 2 illustrates schematically the steps involved in the approach.

Problem Formulation

The problem is formulated as follows: Given a source of contamination, estimate the probability of having a contaminant concentration at a downgradient receptor well that is greater than a specific predetermined maximum permissible value, along with the sensitivity of this probability to the basic variability in the input random variables. The limit-state function is formulated as follows:

$$g(X) = C_t - C(X,t) \tag{10}$$

where C_t is the prespecified regulatory target concentration level at the receptor well, and $C(X,t)$ is the actual value of the contaminant concentration at the chosen well location. Thus, the event described by $[g(X) < 0]$ is equivalent to the event

UNCERTAINTY ANALYSIS OF SUBSURFACE TRANSPORT

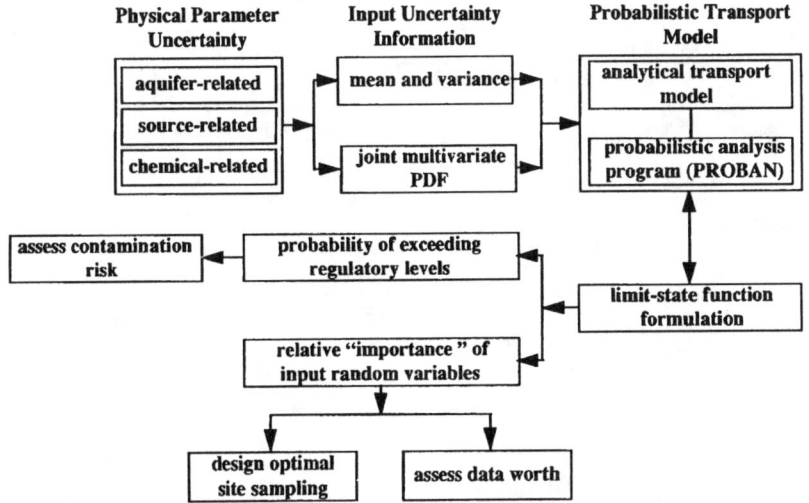

Figure 2. Flow chart of the probabilistic transport analysis model.

Table 1. Deterministic Input Parameters to the Case of a Reactive Solute Transport

Variable	Value
x-distance to well location (m)	100.0
y-distance to well location (m)	0.0
z-distance to well location (m)	1.0
(z-distance is measured from water table)	
Source type	continuous
Infiltration rate (m/year)	1.0
Simulation time (year)	20.0
Contaminant type	Ethylbenzene
K_{oc} (cm^3/g)	165.9

$\{C(X,t) > C_t\}$. In other words, the *failure state* in this case means failure to meet regulatory standards regarding the contaminant of interest at the well location within the time interval $(0,t)$. The reason for specifying a continuous source should be clear, since in that case the contaminant concentration at the receptor well increases with time in a monotonic fashion. Therefore, once the target concentration at the receptor well is exceeded (and *failure* occurs), the concentration at the well will always be greater than the target value as time progresses; the failure condition will persist.

Case Study: Input Parameters

Input to the model includes deterministic parameters, as well as random variables. A continuous source is assumed to leak ethylbenzene in an underlying aquifer, and the concentration at a well 100.0 m downgradient that is screened 1.0 m below the water table is studied. The deterministic parameters used in the case study are listed in Table 1.

The input random variables are categorized into aquifer-related, source-related, and chemical-related parameters. They are listed in Table 2. It should be noted that

Table 2. Input Random Variables to the Case of a Reactive Solute Transport

Variable	Distribution
Aquifer-Related Parameters	
Seepage velocity (m/year)	LN (126.7, 227.37)
Dispersivity (x-direction) (m)	SLN (10, 4, 0.01)
Dispersivity (y-direction) (m)	SLN (1, 0.4, 0.001)
Dispersivity (z-direction) (m)	SLN (0.1, 0.04, 0.0001)
Soil porosity	U (0.3, 0.5)
Soil bulk density (g/cm^3)	U (1.2, 1.8)
Fraction of organic carbon (% weight)	SLN (0.0031, 0.0003, 0.001)
Source-Related Parameters	
Source length (m)	U (50, 100)
Source width (m)	U (50, 100)
Chemical-Related Parameters	
1st-order decay coefficient (year^{-1})	U (1.14, 4.00)

Note: LN (mean, std. dev.): Lognormal; SLN (mean, std. dev., lower limit): Shifted Lognormal; U (lower and upper limits): Uniform.

the mean and the standard deviation are reported for the lognormal distributions. For the shifted lognormal distributions, however, the lower limit is also given. Both the upper and lower limits are given for the uniformly distributed parameters.

First-order kinetics has been widely used to describe processes like natural bio-attenuation, chemical reactions, and radioactive decay. Ethylbenzene here is assumed to undergo natural biodegradation by indigenous microorganisms following a first-order kinetics. Due to changing depths of groundwater elevation, and fluctuation in levels of nutrients, dissolved oxygen, and other electron acceptors, there is uncertainty in the value of the first-order decay coefficient for the contaminant. This is accounted for by assuming the decay coefficient to be random. The choice of the range of equally likely values of the first-order decay coefficient used in this work takes into account actual rates for natural bio-attenuation reported by Wilson et al. (1993) for ethylbenzene.

Results

The effect of changing the target concentration levels at the receptor well location on the probability of failure (or the probability of exceeding the target concentration level at the receptor well) is illustrated in Figure 3. Both FORM and SORM (curvature fitting) were used for the component reliability analysis. The decrease in failure probability with target concentration increase is intuitive, since it is less probable for ethylbenzene concentration to exceed a high value at the downgradient well location than a smaller value for a continuous waste source. It can be observed that for low target concentration values, FORM and SORM results are in good agreement, while they depart from each other for large target concentration values, indicating that the nonlinearity of the limit-state surface at the design point is appreciable. In this case, a second-order surface is expected to better approximate the failure surface at the design point. Although the results contained in Figure 3 are qualitatively intuitive, the quantitative aspects could not have been obtained without the probabilistic computations described in the above formulation.

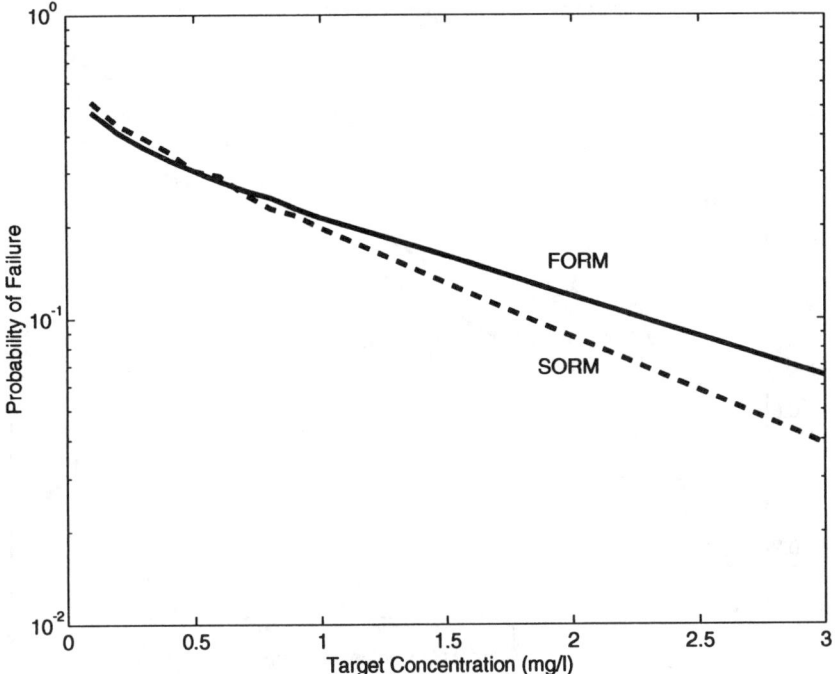

Figure 3. Effect of ethylbenzene target concentration levels on the probability of failure.

The effect of changing the target concentration levels at the receptor well location on the FORM and SORM reliability index is shown in Figure 4. Since there is a monotonic one-to-one relationship between the probability of failure and reliability index (see Equation 4), the same trend of agreement at low target concentration values and discrepancy at large target concentration values is observed for the FORM and SORM results. The reliability index is a measure of the component reliability in that it increases for decreasing probability of failure.

The importance factors for a range of ethylbenzene target concentration levels are shown in Figure 5. It is evident from the graph that over the range of target concentration selected, and for the prescribed probability distributions chosen for this case study, the probability of failure at the receptor well location is mostly contributed by the basic uncertainty in the seepage velocity, the first-order decay coefficient, and the source length (in the x-direction). Hence, although the impact of seepage velocity on the probabilistic outcome is evident, the significance of the chemical-related and source-related uncertainty should be recognized, and failure to account for these uncertainties would result in erroneous contamination risk assessment.

Figure 6 shows a comparison of the FORM, SORM, and Monte Carlo simulation (MCS) methods with regard to their estimate of the probability of failure at the given well location. For the case study presented here and in general, the computer run time for FORM and SORM is considerably less than that of the 5000 MCSs. The figure clearly indicates the power of SORM which provides an accurate estimate of

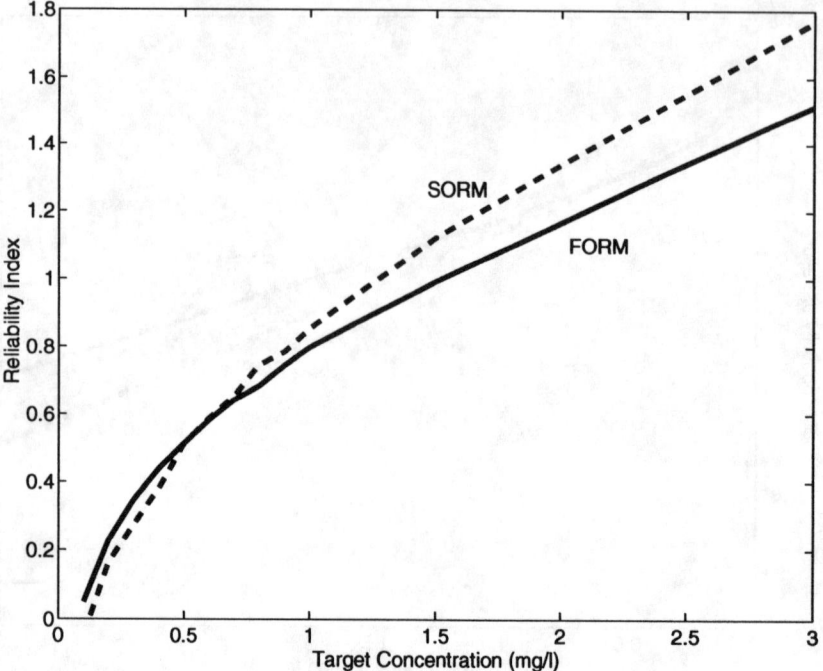

Figure 4. Effect of ethylbenzene target concentration levels on the reliability index.

the probability of failure at very low computational cost compared to MCS. The asymptotic convergence of the classic MCS is evident by the way the MCS estimate converges to the exact failure probability which is accurately predicted by SORM.

The consistency of the MCS estimate of the probability of failure is indicated by the reduction in the 90% confidence interval with an increasing number of simulations. In this particular case, the result obtained using FORM is obviously approximate as it departs from that of SORM and MCS, but it is produced at very low cost computationally. Once again, the issue of tradeoff between efficiency and accuracy is an important factor when selecting a reliability analysis method.

CONCLUSIONS

The recognition by environmental engineers and regulatory agencies of the impact of the ubiquitous heterogeneity of porous media and the importance of considering the effect of parameter uncertainty has led to the widespread use of probabilistic tools for modeling contaminant transport in the subsurface.

This study demonstrates the use of first- and second-order reliability methods (FORM and SORM, respectively) in the probabilistic analysis of groundwater reactive contaminant transport. Uncertainty importance factors were used to assess the relative impact of the variability of the basic random variables on the reliability index and failure probability. The general applicability of the method was illustrated on a simple

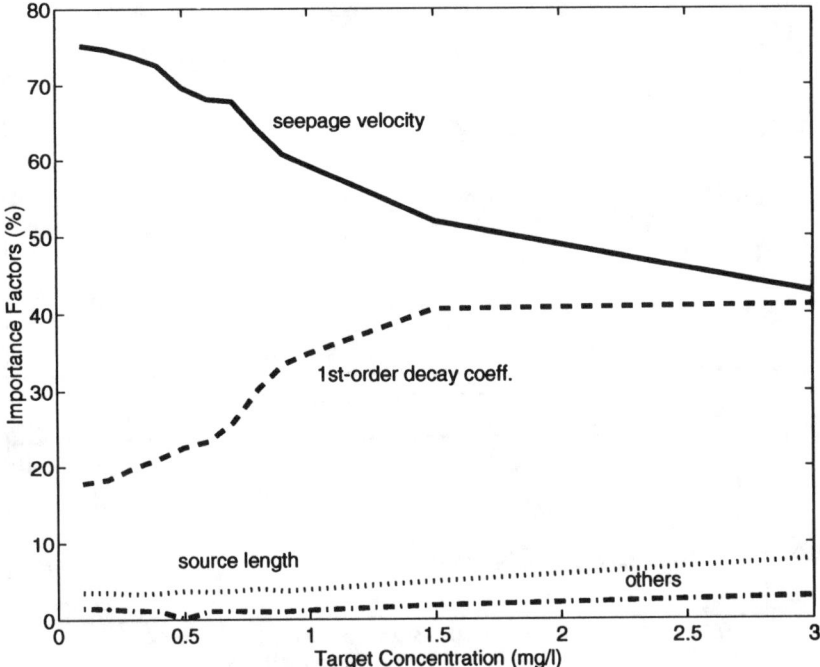

Figure 5. Effect of ethylbenzene target concentration on importance factors.

case of transport of ethylbenzene, and the probability of exceeding a prespecified concentration level at a downgradient well was analyzed.

FORM and SORM are efficient computational tools that can prove very useful for situations in which the failure event is of very low probability, and for which the classic MCS method is computationally prohibitive or even intractable. This means that the use of these approximate analytical reliability methods in conjunction with the MCS method can optimize the probabilistic treatment of groundwater problems. Although a large number of MCSs would be more accurate than FORM and SORM, the latter provide fast and fairly accurate results.

The difference between FORM and SORM was presented. SORM tends to produce more accurate results than FORM when the limit-state surface is significantly nonlinear at the design point in the standard normal space. However, SORM requires more computational effort than FORM. The selection of method should be based on problem dimensionality, available computer resources, and required level of accuracy. In other words, one should always conduct a careful tradeoff analysis between computational intensity and accuracy of results.

Chemical-related and source-related parameter uncertainty have been shown to be significant factors to consider in the probabilistic analysis of groundwater transport problems, and their importance should not be overshadowed by the aquifer-related parameter uncertainty. Although the probabilistic model used in this work is based on a semianalytical transport code, the methodology equally applies to more sophisticated and realistic numerical models as well. The integration of FORM and SORM

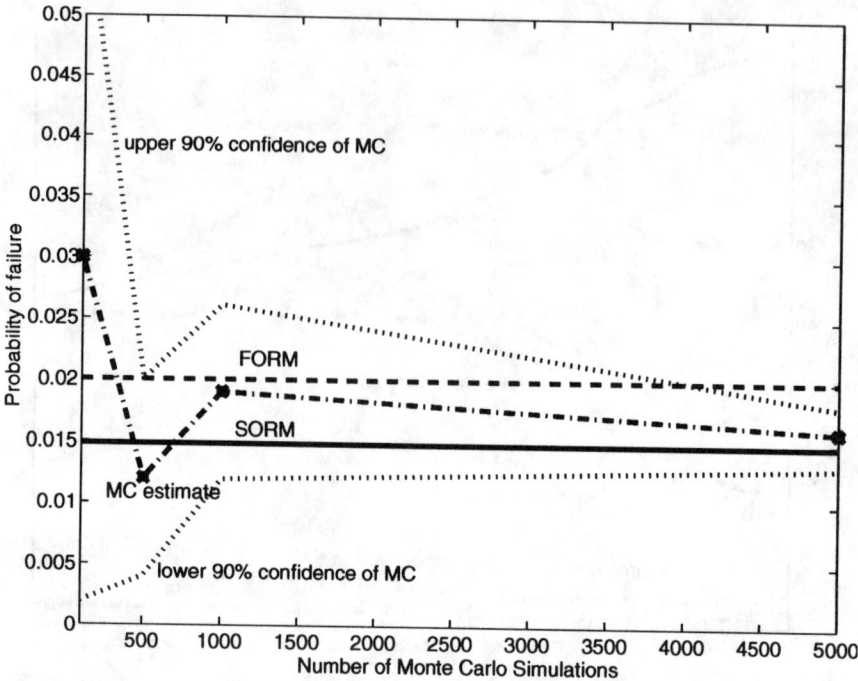

Figure 6. Comparison of FORM, SORM, and MCS estimate of the probability of failure.

with a numerical finite element transport code is the subject of ongoing research at Rice University.

REFERENCES

Cawlfield, J. D. and Sitar, N., Stochastic finite element analysis of groundwater flow using the first-order reliability method, in *Consequences of Spatial Variability in Aquifer Properties and Data Limitations for Groundwater Modelling Practice,* Peck, A., Gorelick, S., de Marsily, G., Foster, S., and Kovalevsky, V., Eds., IAHS Publication No. 175, International Association of Hydrological Sciences, Gentbrugge, Belgium, 1988, 191.

Cawlfield, J. D. and Wu, M.-C., Probabilistic sensitivity analysis for one-dimensional reactive transport in porous media, *Water Resour. Res.,* 29, 661, 1993.

DelVecchio, G. M. and Haith, D. A., Probabilistic screening of groundwater contaminants, *J. Env. Eng.,* ASCE, 119, 287, 1993.

Der Kiureghian, A., Lin, H.-Z., and Hwang, S.-J., Second order reliability approximations, *J. Eng. Mech.,* ASCE, 113, 1208, 1987.

Fiessler, B., Neumann, H.-J., and Rackwitz, R., Quadratic limit states in structural reliability, *J. Eng. Mech.,* ASCE, 105, 661, 1979.

Galya, D. P., A horizontal plane source model for groundwater transport, *Ground Water,* 25, 733, 1987.

Hamed, M. M., Bedient, P. B., and Conte, J. P., Reliability approach to the probabilistic modeling of groundwater flow and transport, in *Proc. Conf. Petroleum Hydrocarbons and Organic Chemicals in Ground Water: Prevention, Detection, and Restoration,* National Ground Water Association, Houston, TX, 1993, 317.

Hasofer, A. M. and Lind, N., An exact invariant first-order reliability format, *J. Eng. Mech.,* ASCE, 100, 111, 1974.

Hohenbichler, M. and Rackwitz, R., Sensitivity and importance measures in structural reliability, *Civ. Eng. Sys.,* 3, 203, 1986.

Liu, P. L. and Der Kiureghian, A., Optimization algorithms for structural reliability, *Struct. Saf.,* 9, 161, 1991.

Madsen, H. O., Omission sensitivity factors, *Struct. Saf.,* 5, 35, 1988.

Madsen, H. O., Krenk, S., and Lind, N. C., *Methods of Structural Safety,* Prentice Hall, Englewood Cliffs, NJ, 1986.

Melchers, R. E., *Structural Reliability, Analysis and Prediction,* Ellis Horwood Series in Civil Engineering-Structural Engineering Section, Ellis Horwood, 1987.

Rackwitz, R. and Fiessler, B., Structural reliability under combined load sequences, *Comp. Struct.,* 9, 489, 1978.

Schanz, R. W. and Salhotra, A., Evaluation of the Rackwitz-Fiessler uncertainty analysis method for environmental fate and transport models, *Water Resour. Res.,* 28, 1071, 1992.

Sitar, N., Cawlfield, J., and Der Kiureghian, A., First order reliability approach to stochastic analysis of subsurface flow and contaminant transport, *Water Resour. Res.,* 23, 794, 1987.

Veritas Research, *PROBAN: General Purpose Probabilistic Analysis Program,* Det norske veritas, Hovik, Norway, 1992.

Wilson, J. T., Kampbell, D. H., and Armstrong, J. M., Natural bioreclamation of alkylbenzenes (BTEX) from a gasoline spill in methanogenic ground water, in *Proc. 2nd Int. Symp. on In Situ and On-Site Bioreclamation,* Battelle, San Diego, CA, 1993.

CHAPTER 9

Groundwater Modeling and Litigation

Thomas A. Prickett and Wayne A. Pettyjohn

INTRODUCTION

One of the most nerve-racking projects that a groundwater hydrologist can be asked to undertake is to prepare for and to give testimony in a court of law. Preparing for and giving a presentation at a public hearing is usually uncomfortable as well. Perhaps nervousness begins to creep in as one conjures up images of past Perry Mason television episodes or unpleasant scenes from movies such as *The Verdict, Anatomy of a Murder,* or *The Firm.* With experience, however, some form of calm will begin to take shape as one finds that these proceedings have many predictable aspects to them and that the media portrayals are mostly overdone and not exactly faithful to the way things actually work in the real litigative world. As it turns out, most of this testimony is just plain hard work that is done under the guidance of a few rules set down by the law. Once you know some of the dominant rules, and have seen the way they are played out, most of the experiences repeat each other. This paper will begin with the author's background on this subject. Next, a reminder will be given concerning the groundwater modeling approach. Third, a discussion will be given about the legal approach including some of the details of interviews, case preparation, depositions, negotiations, and trial characteristics. Finally, a few philosophical statements will be given that might help bring this subject into proper perspective.

CASE BACKGROUND

As a background to this paper, the senior author of this paper has had court and hearing experiences in 23 cases. Statistics on these cases are eight hearings, five State-level courts, eight Federal-level courts, one International court, and one case that was before the U.S. Supreme Court. Seventeen cases were for the plaintiff and six were for the defendant. Fourteen of the cases involved mass transport models and nine involved only flow. The success rate was 83%.

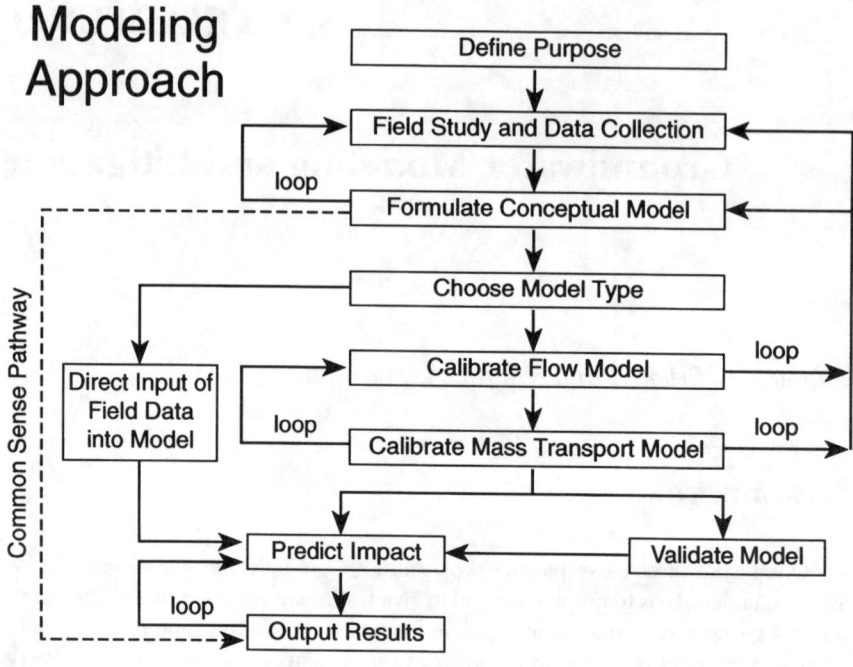

Figure 1. Modeling approach.

THE MODELING APPROACH

Figure 1 illustrates a flow chart of the groundwater modeling approach. The problem at hand dictates the purpose of the model. The project proceeds from the data collection phase on down to choosing some sort of model, calibrating it to the field data, and operating the model to obtain results that can be presented in court. The most direct route in the flow chart of Figure 1 is from top to bottom. The "looping" back pathways represent the rethinking processes that are always necessary when organizing and optimizing a comprehensive and consistent analysis of groundwater problems. Frequently, regulatory agencies and the seriousness of the legal action require extensive use of the looping pathways until uncertainty in the modeling results is minimized.

The "Common Sense Pathway" indicates that modeling "Output Results" must be understandable, despite the modeling jargon, high-level mathematics, and computer operations that comes between the conceptual model and the real world. Incidentally, modelers must pay particular attention to this pathway if they are to be successful in any problem-solving endeavor. If the answer is not clear or it does not make sense, the modeler will be dead in the water.

Although brief, the above description suffices to remind the reader of the general characteristics of modeling. Additional details of the modeling approach can be found in the latest modeling textbooks (e.g., Anderson and Woessner, 1992).

THE LITIGATIVE APPROACH

Although the litigative approach is more complex than what we are outlining now, a sufficient portion is discussed to give the reader an idea of the dominant parts. The five prominent components of the litigative approach include (1) an interview between the groundwater hydrologist and the attorney; (2) the case preparation; (3) depositions; (4) negotiations; and when all else fails; (5) the trial. Let us discuss each of these, one at a time.

Interview. Typically, the telephone rings and an attorney indicates an interest in hiring you as an expert witness in some sort of groundwater problem that his or her client has. If all looks good through the telephone conversation, there usually follows a face-to-face interview at the law firm or at the client's office. The attorney asks questions that are aimed at finding out if you are suitable for the task. Past experience and knowledge about your profession is mandatory. Becoming defensive in your answers to their questions is not looked upon favorably. When the attorneys are through, they usually will dismiss you and give you a call later. However, before you leave, you might indicate that you have some questions as well. Be careful here, as attorneys sometimes have big egos. The answers to whether the firm has ever been involved in a groundwater problem before, what the litigative experiences of the attorneys in court have been in groundwater problems, and who is going to be the lead attorney with regard to you can be very important. Some of the best lawyers are those that have dual degrees such as engineering, sciences, mathematics, and so forth. Find out about the attorneys and what their other interests are.

The senior author recalls an incident before his first jury trial where he admitted to a high-level New York-based attorney that he had no experience before a jury and was therefore asking for his advice. The attorney said not to worry and that he would do just fine. However, in almost the same breath, the attorney admitted that he had never tried a case in court before! As it turned out, the jury was sympathetic to the attorney's hesitations compared with the slick opposition lawyers and, although based upon a strong case anyway, the case was won. The moral to this story is to be aware of your leadership before you get too involved. Don't assume that your attorney knows everything that is going on. Be ready to "diplomatically" pass on your knowledge and experience.

Case Preparation. This is the most important aspect of the modeler's job. Ninety percent of your success in the court room is based upon how well you are prepared. If you are not prepared and feel uncomfortable about it, speak up. Remember that modeling usually comes at the end of a project. That's the part of the project when time and budgets are being strained.

The very first time giving testimony, the attorney said to keep it simple. The attorneys have what they call the KISS rule—Keep It Simple, Stupid. That is good advice. Furthermore, the modeler should remain a professional, be unbiased in the analysis, and not select data that fit only his or her opinion. One thing that will quickly ruin a case in court is to be exposed as an advocate of the client. If the client does not like what you have to say, then step aside.

On most occasions, the attorney suggests that you keep the amount of your analysis in written form to a minimum. Writing a report for a courtroom appearance is unusual. Reports are common, however, for hearings and permit proceedings. The general rule in the court situation is to not write it down unless you absolutely have to. Computer files included. The main reasoning seems to have to do with keeping strategy an unknown as long as possible. Finally, be prepared to work with many others in putting the case together. Most of the mass transport cases involve other expert witnesses such as chemists, biologists, engineers, historians, animators, and social scientists. Team theory has to be applied.

An approach the junior author follows in the preparation of a case is first to examine all of the data from the perspective of the client. The next step is to prepare a defense for the other party, noting the strengths and weaknesses of their arguments and data. The final step is a reevaluation of the client's position in order to overcome weak points, brought to light in step two, that are likely to be presented during the proceedings. The key to success is made of two ingredients—preparation and organization.

Depositions. To begin with, the interested reader is directed to *The Deposition Handbook* by Suplee and Donaldson (1992) referenced at the end of this paper. There are detailed guides on how to handle depositions. The deposition is the time when the opposing attorneys get their chance to question you on almost anything about what you have done, including your previous work and testimony, what your opinions are, what you are going to testify to, and what your plans are. The opposing attorneys want to see every shred of data, all reports that you are depending upon, all of the computer programs codes, printouts, and data files that you have produced or are using, and every handwritten note that you have, just as starters (Duces tecum—which means show us everything you've got).

The court neither appreciates nor allows surprises during trial, and the purpose of depositions is to overcome this Perry Mason approach. Through all of the legal proceeding, the expert must be both prepared and organized. All files must be available to opposing council, other than those documents that are privileged, which, no doubt, will be removed by your attorney. Data should be organized so that they can be quickly referenced during deposition, hearing, or trial. The senior author has given over 5000 pages of depositions. That experience has led to the conclusion that depositions are more challenging than trials. Depositions generally are less structured than a trial, are sometimes intrusive on your personal life, and there is no judge or jury around to observe abusive tactics of a cantankerous attorney. Experience leads one to believe that the real purpose of the deposition is to gain the advantage. Finding out what the opposition has done and what the opinions are appear to be secondary.

Depositions are challenging experiences. Most attorneys advise you along the following lines: Listen closely to the questions. Answer only the questions asked. Tell the truth. Don't volunteer anything. Don't teach during a deposition—wait for the trial to do that. Always leave an escape in your answer. Avoid absolutes. Proceed deliberately and slowly. Reconsider your phrasing or answers if your own attorney objects to a question. Be responsive to the questions. Beware of the use of "silence"—don't be tempted to fill in the blank. Remember that the record does not show the passage of time during silence. Pay attention at all times—if you begin to tire, ask for a break. Some attorneys use repetition and badgering techniques to try and eventually intimidate you into answering their questions in a fashion advantageous to them. Just remember what your studies indicated and stick with what you believe. Remain calm and don't argue with the opposing lawyer. Don't go beyond your own expertise. The words "Do you have an opinion about—???" carries a legal connotation which is different than something like: "What do you think about such and such?" You had better be sure of your opinions, while thinking about something denotes an indecisive stance. So, if one is unsure, don't say that you have an opinion on the subject. Just indicate that no opinions have been formed yet.

Depositions sometimes expose areas that one needs to do some additional work. Be aware of this, discuss this with your attorney, and do the necessary work to sufficiently defend your positions. Be also aware that your opposition will be working right up to the time of the trial, fortifying their weak points. You should be doing the same thing. If this is disturbing to opposition attorneys, there can be a deposition demanded during a trial.

Negotiations. Sometime after all parties have finished with depositions, a time of negotiation ensues to see if a settlement can be arranged. On occasion, cost sharing is discussed, likely consequences of going to trial are communicated, bluffing and accusations are delivered, and cleanup alternatives are talked about. The bottom line is that if negotiations fail, the case goes to trial. In our opinion, having to go to trial is a losing deal since each side is thus confident that they can win with their own set of expert witnesses. Neither side of experts is sufficiently convincing to break the impasse, at least on the groundwater issues.

An anecdotal story here is that some of these negotiations can be outrageous—particularly if groundwater decisions are being made without the benefit of a consultant. The case involved opposing attorneys negotiating the cleanup plan for a Superfund site. The first thing that they did was to negotiate the extent of a plume! The total data set of samples was scrutinized and any questionable lab results thrown out. For example a hit showing a questionable 10,000 µg/L value might be eliminated from the data set for one reason or another. The data point effectively became a zero value. The resulting plume extent got smaller. Subsequently, the uninformed groundwater consultants were asked to design a remediation plan based upon this "lawyer's plume". There is no need to finish the story other than to say that one should be aware of this sort of occasional problem and handle it accordingly. Incidentally, some lawyers negotiate deposition, hearing, continuation, and trial dates without consulting their experts. These attorneys are to be avoided.

The trial or hearing. There are many common occurrences prior to and during a trial or hearing. Trials are more formal; hearings can be more informal. The discussions below have applications more to trials than to hearings.

Getting ready for a trial has some similarities to staging an orchestra concert or some theatrical presentation: rehearsals, rearranging props and staging, organizing documents, exhibits, and equipment, and setting up nearby offices, secretaries, telephones, and copying machines. And of course, setting up your computer on the spot. Unlike concerts or theater, which have contained musical scores and fixed length scripts, trials are dynamic and reactive. The players must be ready for improvisation. If one likes competitive activities, has some showmanship capabilities, is well prepared, and thinks well on one's feet, then trials are exciting.

To begin with, read the book *Law for the Expert Witness* by Bronstein (1993). This gives ideas on how to handle yourself during one of these procedures. As instructed by most attorneys, the key is to be prepared. Further instructions are to keep it simple. Dress neatly and appropriately in what might be called the IBM look with a white shirt and dark suit. Watch your body language and eye contact. Talk to the attorney, the judge, and the jury, not to the floor.

Semantics is extremely important in the legal profession, and the meaning of selected words has a profound effect on evidence. For example, in civil actions a jury will be instructed to decide an issue on the basis of the "preponderance of evidence", which means something in excess of 50%. An expert is often asked if his or her opinion is true to a reasonable degree of scientific probability, which means greater than 50%. "Clear and convincing" evidence is somewhere in the neighborhood of 70 to 75%, and in criminal actions guilt must be proved "beyond a reasonable doubt", which might exceed 90%.

The above is background information. The order of business in the actual hearing or trial is to first present direct testimony, then to be cross-examined, and then, if necessary, to be involved in rebuttal testimony. The direct testimony should be well rehearsed and should contain no surprises. Most of the time, the direct testimony goes smoothly with the KISS rule driving most of the presentation. Use of live computer simulations in the courtroom should be discouraged because one can get deeply involved in running opposition parameters when it comes time for cross-examination. Any use of equipment, computers, slide projectors, and animation presentations during the courtroom activities opens one up to equipment failure possibilities. Have backups. Have professionals available to deal with problems if they occur. Attorneys and paralegals have virtually no facility with electronic devices.

The cross-examination is a difficult part of a trial or hearing. This is where the opposition attorneys get to question one about the details of the direct testimony. This is where one finds out how well one is prepared. Actually, if one knows what they are talking about, it is very difficult for an opposing lawyer to do anything critical. Be aware that what one said in deposition is going to come up in cross-examination. Be sure that you have studied the deposition record and are ready to answer any arguably conflicting statements. In addition, remember that what you have written in the literature may surface during this time. Any conflicting opinions therein may be discussed. Be ready.

One trick question that seems to be popular is to get you to admit that groundwater consultants can often differ in their opinions on the same subject. The correct response is to inform the attorney that he is correct but, given sufficient time and full consideration of all the facts, that these other competent consultants will eventually come around to the same opinion as yours. Further suggestions would be: Make sure that you understand the question. "I don't know" is an acceptable answer, particularly if you are asked about something beyond your expertise. One trick of some attorneys is to cut you off in the middle of your answer. You do have the right to complete your answer if you like, and let him know it.

Please refer to Figure 1 once more. This modeling approach illustration is a map that the opposition attorneys and their consultant also have and understand. It is a map that you can expect to be questioned about during cross-examination. There are critical questions that can be generated at every point along the pathways of that flow chart. Rest assured, whatever decisions that you made along the various pathways of Figure 1, the opposition will point out that they were the wrong ones. For example, a series of ill-conceived opposition criticism bent on seeking perfection in the analysis and incompleteness in whatever was done might be along the following lines:

> The purpose for modeling was wrong and incomplete. The field study and data collection program that you used were insufficient, incomplete, and inaccurate. The data base was not properly quality controlled. The conceptual model ignored important aspects of the opposition's conceptual model, whether it makes any difference or not. You chose the wrong model. If you chose a finite-difference model, it should have been a finite-element one. If you used grid intervals of 100 ft, they should have been 10 or 1000.
>
> If you chose to input field data directly into an analytical formula model, or into a numerical model without a calibration process, the computations are totally worthless since the real world is not homogeneous nor isotropic, and you undoubtedly broke the law of conservation of mass. (Actually, any numerical model breaks the law of conservation of mass when compared with real world data. The very process of adjusting aquifer parameters to match field heads or concentrations always produces differences compared with field data. The law of conservation of mass is in the real world. You are only approximating that in any of these models.)
>
> If you calibrated your model to a certain standard, it was not sufficiently accurate. You did not validate your model with an independent set of data. A calibrated model without validation is totally worthless. Your answers do not make any sense. And on and on.

Now, whether the cross-examination hits all of these or just some of them, the idea is to discredit your analysis and to position their groundwater consultant as the all-knowing expert of unquestioned accuracy. This opposition tactic of criticism can easily backfire when you expose the opposition expert along the same lines. As a matter of fact, the harder the opposition works on this type of criticism, the more they expose themselves. One experience comes to mind when an opposition expert included excruciating detail in his so-called superior model but was caught "calibrating the land surface elevation" values input to his model! This so-called expert was exposed as delivering self-serving smoke and mirrors.

There is a section in the book entitled *Chaos* (Gleick, 1987), wherein modeling is discussed as follows:

> Only the most naive scientist believes that the perfect model is the one that perfectly represents reality. Such a model would have the same drawbacks as a map as large and detailed as the city it represents, a map depicting every park, every street, every building, every tree, every pothole, every inhabitant, and every map. Were such a map possible, its specificity would defeat its purpose: to generalize and abstract . . . . Whatever their purpose, maps and models must simplify as much as they mimic the world.

Remember to keep your cool during criticism of this nature. And, if you do lose, the world won't come to an end.

The rebuttal testimony is prepared to counteract the opposition complaints about your testimony. It is impossible to plan much in this area until the complaints are made. In some cases, separate rebuttal witnesses are brought in. The ability to counter these witnesses has to be learned on-the-go since there is very little time to prepare your responses. Your rebuttal testimony must respond firmly to opposition accusations.

SUMMARY AND A FINAL REFLECTION

Hopefully this discussion will be of some assistance to a groundwater modeler who is facing a project leading to either a trial or a formal hearing. One needs all of the help that one can get. It is not a pleasant task, but one that is difficult to avoid at some time in a career. If we can leave a few thoughts here, they would be first to choose your case and attorney carefully. Second, prepare, prepare, prepare. That is the key to success in this arena. And finally, let us talk about keeping your cool in the line of fire. Every attorney tells you to do this but none of them tell you how. Here it is:

We will quote from the last pages of the book *The Terrible Truth About Lawyers* (McCormack, 1987) where they indicate how to maintain grace under pressure. They say effectively that

> . . . What allows lawyers—and can allow all of us—to function well under pressure is the ability to strike a balance between commitment and detachment. The effective professional should be wholly concentrated on and committed to the fulfillment of his professional role.
>
> All the while, however, the professional should be aware that he has a self beyond that role. Attack me in a negotiation, and you are not really attacking me; you are attacking the point of view that I am representing. Still, I am removed enough to be able to answer you calmly. The balance between commitment and detachment is what allows for the quality we call poise. Poise is something we instinctively admire: in athletes, in executives, in professionals. But poise is not really an end in itself. It is a stepping stone to a higher good: dignity—an effective dignity that lets us get things done not just with flair but with peace of mind—is really the ultimate object of the exercise.

Everyone should get a copy of that book and read it. It puts this litigation arena in its proper perspective. Good luck.

REFERENCES

Anderson, M. P. and Woessner, W. W., *Applied Groundwater Modeling*, Academic Press, San Diego, CA, 1992, 381 pp.

Aron, R., Duffy, K. T., and Rosner, J. L., *Cross-Examination of Witnesses*, Shepard's/McGraw-Hill, Colorado Springs, CO, 1989, 559 pp.

Aron, R., Duffy, K. T., and Rosner, J. L., *Impeachment of Witnesses: The Cross-Examiner's Art,* Shepard's/McGraw-Hill, Colorado Springs, CO, 1990, 440 pp.

Black, H. C., Ed., *Black's Law Dictionary,* West Publishing Company, St. Paul, MN, 1500 pp.

Bronstein, D. A., *Law for the Expert Witness,* Lewis Publishers, Boca Raton, FL, 1993, 223 pp.

Gleick, J., *Chaos,* Penguin Books USA, New York, 1987, 352 pp.

Suplee, D. R. and Donaldson, D. S., *The Deposition Handbook,* John Wiley & Sons, New York, 1992, 293 pp.

Machlowitz, D. S., *Legal Guide to Working with Environmental Consultants,* John Wiley & Sons, New York, 1992, 485 pp.

McCormack, M. H., *The Terrible Truth About Lawyers,* Avon Books, New York, 1987, 246 pp.

SECTION 3

On Unsaturated/Multiphase Flow and Transport Modeling

CHAPTER 10

On the Numerical Solutions of One-Dimensional Flow in the Unsaturated Zone

Aly I. El-Kadi

INTRODUCTION

Analysis and prediction of water flow patterns in the unsaturated zone are critical to many water resources and environmental problems. Examples of such problems include infiltration, which is an important part of the hydrologic cycle encompassing the associated movement and storage of subsurface water. Soil properties are a major factor in controlling rainwater partition between infiltration and runoff as well as in controlling moisture movement. Hence, an accurate estimation of infiltration and the factors affecting it is required to facilitate a reliable prediction of runoff and subsurface moisture distribution. Another example is related to agricultural management in which unsaturated water flow needs to be considered in decision making regarding irrigation. Finally, chemical transport in the subsurface environment is greatly influenced, under certain conditions, by flow in the unsaturated zone. Contamination may be caused by leakage from sanitary landfills or by recharge of sewage water under unsaturated flow conditions. Irrigation and rainwater dissolve and carry fertilizers, pesticides, and other chemicals under unsaturated conditions also. In most cases, understanding chemical transport and transformations in the unsaturated zone is essential for assessing the actual or potential contamination of groundwater aquifers.

Flow in the unsaturated zone is usually simulated by solving the Richards (1931) equation, which is derived by combining the mass conservation equation and Darcy's law. Recent studies have reported problems in solving such an equation within a numerical framework. This paper reviews the theory and various conventional numerical solutions pertinent to the problem. It also covers recent advances in numerical techniques that are mostly aimed at improving the efficiency of the solutions by optimizing the size of the spatial and temporal increments. Most of the attempts are related, in general, to equation transformation, solution iteration, and interblock parameter estimation. However, there is still room for improvement, because in some cases accuracy may require the use of small increments, on the order of a few centimeters and seconds for the spatial and temporal increments, respectively. Such a need is critical because of the burden involved in model applications for large-scale, multi-dimensional problems. Although only one-dimensional problems are addressed here, many of the issues involved can be readily extended to multidimensional cases.

MATHEMATICAL FORMULATION

Governing Equation

The Richards equation can be written in a pressure-based form, a water-content based form, or a mixed form (Celia et al., 1990). In the pressure-based form, the one-dimensional equation is

$$\frac{\partial}{\partial z}\left(K\frac{\partial h}{\partial z}\right) - \frac{\partial K}{\partial z} = C\frac{\partial h}{\partial t} \tag{1}$$

in which $K = K(h)$ is the hydraulic conductivity, h is the pressure head, t is time, and z is distance (positive downwards). The parameter C in Equation (1) is known as moisture capacity and is defined as

$$C(h) = \frac{d\theta}{dh} \tag{2}$$

The water-content based form is

$$\frac{\partial}{\partial z}\left(D\frac{\partial \theta}{\partial z}\right) - \frac{\partial K}{\partial z} = \frac{\partial \theta}{\partial t} \tag{3}$$

in which D is soil–water diffusivity and is defined as

$$D(h) = K\frac{dh}{d\theta} = \frac{K}{C} \tag{4}$$

The mixed form is

$$\frac{\partial}{\partial z}\left(K\frac{\partial h}{\partial z}\right) - \frac{\partial K}{\partial z} = \frac{\partial \theta}{\partial t} \tag{5}$$

Another form of the equation can be obtained by transforming the pressure-based form of the Richards equation via the Kirchhoff transformation (Carslaw and Jaeger, 1959). The transformed equation is

$$D(u)\frac{\partial^2 u}{\partial z^2} - v(u)\frac{\partial u}{\partial z} = \frac{\partial u}{\partial t} \tag{6}$$

in which u is defined as

$$u = u(h) = \int_{h_{max}}^{h} K(h)dh \tag{7}$$

and h_{max} is the upper bound of the pressure head in the domain. The functions $D(u)$ and $v(u)$ in Equation (6) are given as

$$D(u) = D[h(u)] \tag{8}$$

$$v(u) = \frac{1}{C[h(u)]}\frac{dK[h(u)]}{dh} = \frac{dK[h(u)]}{d\theta} \tag{9}$$

The parameter $v(h)$ is termed soil-moisture velocity by El-Kadi and Ling (1993). Apart from its nonlinearity, Equation (6) is similar in form to the solute transport equation.

Boundary and Initial Conditions

For infiltration under ponding, the boundary and initial conditions are

$$\begin{aligned} h &= h_i & \text{for all } z & \quad t = 0 \\ h &= h_s & z = 0 & \quad t > 0 \\ h &= h_L & z = L & \quad t > 0 \end{aligned} \tag{10}$$

where h_i is the initial pressure head, h_s and h_L are the prescribed values of h at $z = 0$ and $z = L$, respectively, and L is the column length. In the general case, h_i can vary with depth, and both h_s and h_L can vary with time. For specified flux at either the upper or lower boundary, the following condition applies:

$$\left(-K\frac{\partial h}{\partial z} + K\right)_{z=z_s} = q_s \tag{11}$$

in which z_s is the depth of the boundary below the soil surface and $q_s = q_s(t)$ is the specified flux. Equation (11) can be generalized to read

$$\left(\frac{\partial h}{\partial z}\right)_{z_s} = S_s \tag{12}$$

in which $S_s = S_s(t)$ is head gradient at the specified boundary. For the flux condition given by Equation (11), S_s is given by

$$S_s = 1 - \frac{q_s}{K} \tag{13}$$

For specified pressure-head boundary conditions on a soil column of length L, the transformed initial and boundary conditions are

$$\begin{aligned} u &= u(h_i) = u_i & \text{for all } z & & t = 0 \\ u &= u(h_s) = u_s & z = 0 & & t > 0 \\ u &= u(h_L) = u_L & z = L & & t > 0 \end{aligned} \quad (14)$$

in which u_i, u_s, and u_L are estimated from Equation (7), by substituting h with h_i, h_s, and h_L, respectively. For a specified flux (or gradient) condition at either boundary, Equation (12) can also be transformed to read

$$\left(\frac{\partial h}{\partial z}\right)_{z_s} = K(h)\big|_{z=z_s} S_s \quad (15)$$

which can be implemented in the numerical calculations.

Soil Hydraulic Properties

In addition to the boundary and initial conditions listed above, the solution requires specification of the functions $h(\theta)$ and $K(\theta)$. These hydraulic properties can be specified either in tabular or functional (analytical) forms. Examples of available functions include those by Brooks and Corey (1964), Brutsaert (1966), and van Genuchten (1978a). The latter functions are listed below:

$$S_e = \left[\frac{1}{1 + [\alpha|h|]^n}\right]^m \quad (16)$$

$$K(\theta) = K_s S_e^{(1/2)} [1 - (1 - S_e^{1/m})^m]^2 \quad (17)$$

$$S_e = \frac{\theta - \theta_r}{\theta_s - \theta_r} \quad (18)$$

In these equations, θ_s and θ_r are the saturated and residual water content, respectively; K_s is the saturated hydraulic conductivity; and α, n, and m are fitting parameters, with m related to n by

$$m = 1 - \frac{1}{n} \quad (19)$$

The parameters of the functions describing soil hydraulic properties can be estimated by fitting such functions to measured data of $h(\theta)$. Recent publications related to estimating the hydraulic properties of soils include the volume edited by van Genuchten et al. (1992). A routine to estimate the parameters of the van Genuchten

NUMERICAL SOLUTIONS OF ONE-DIMENSIONAL FLOW

functions from measured h versus θ is given in El-Kadi (1984). The routine has options for choosing other forms as well.

CONVENTIONAL NUMERICAL SOLUTIONS

Numerical solutions of the Richards equation via the finite-element and finite-difference techniques are readily available in the literature. Examples of related publications are Haverkamp et al. (1977) for the finite-difference technique and Istok (1989) for the finite-element technique. The formulations pertinent to the h-based form are described here. However, they can be extended to other forms as well.

Finite-Difference Technique

The Richards equation, given as Equation (1), can be approximated via the finite-difference approach and rearranged in the following form:

$$a_i h_{i+1}^{t+\Delta t} + b_i h_i^{t+\Delta t} + c_i h_{i-1}^{t+\Delta t} = d_i \qquad (20)$$

where i is the number of nodes (values from 1 to N) and N is the total number of nodes. The expression for the coefficients in this system of equations depends on the type of approximation adopted. In the implicit scheme, a weighted average of the derivatives at times t and $t + \Delta t$ is used to obtain an approximation at time $t + \Delta t/2$. The approximation yields

$$\begin{aligned}
a_i &= \frac{\varepsilon K_{i+1/2}}{\Delta z^2} \\
b_i &= -\frac{\varepsilon K_{i+1/2} + K_{i-1/2}}{\Delta z^2} - \frac{C_i}{\Delta t} \\
c_i &= -\frac{\varepsilon K_{i-1/2}}{\Delta z^2} \\
d_i &= \frac{\varepsilon K_{i+1/2} - K_{i-1/2}}{\Delta z} - \frac{C_i h_i^t}{\Delta t} - h_{i-1}^t \left[\frac{(1-\varepsilon)K_{i-1/2}}{\Delta z^2}\right] \\
&\quad - h_{i+1}^t \left[\frac{(1-\varepsilon)K_{i+1/2}}{\Delta z^2}\right] + h_i^t \left[\frac{(1-\varepsilon)(K_{i-1/2} + K_{i+1/2})}{\Delta z^2}\right]
\end{aligned} \qquad (21)$$

in which ε is a weighting factor ($0 < \varepsilon < 1$). Note that the well-known Crank-Nicolson scheme can be recovered by setting ε equal to one-half. The values $K_{i-1/2}$ and $K_{i+1/2}$ are known as the interblock hydraulic conductivities. They can be estimated conventionally by either the arithmetic or the geometric mean. For $K_{i-1/2}$, these means are respectively given by

$$K_{i-1/2} = 0.5(K_{i-1} + K_i) \qquad (22)$$

and

$$K_{i-1/2} = K_{i-1}^{1/2} + K_i^{1/2} \tag{23}$$

with similar expressions for $K_{i+1/2}$ as the mean of K_i and K_{i+1}. Haverkamp and Vauclin (1979) concluded that the geometric mean is the best estimate of interblock values.

A fully implicit scheme uses the derivatives at $t + \Delta t$ to obtain an approximation at time $t + \Delta t/2$, which can be achieved simply by setting ε in Equation (21) as equal to 1. The resulting coefficients in equation (20) are given by

$$\begin{aligned} a_i &= \frac{K_{i+1/2}}{\Delta z^2} \\ b_i &= -\frac{(K_{i+1/2} + K_{i-1/2})}{\Delta z^2} - \frac{C_i}{\Delta t} \\ c_i &= -\frac{K_{i-1/2}}{\Delta z^2} \\ d_i &= \frac{(K_{i+1/2} - K_{i-1/2})}{\Delta z} - \frac{C_i h_i'}{\Delta t} \end{aligned} \tag{24}$$

Equation (20) can be arranged in a system of equations in form

$$[E]\bar{h}_t + \Delta t = \bar{f} \tag{25}$$

which is solved using matrix solution techniques. The matrix $[E]$ and the vectors $\bar{h}$ and $\bar{f}$ are respectively defined as

$$[E] = \begin{bmatrix} b_1 & a_1 & & \\ c_2 & b_2 & a_2 & \\ & \cdots & & \\ & & c_N & b_N \end{bmatrix} \tag{26}$$

$$\bar{h} = \begin{pmatrix} h_1 \\ h_2 \\ \vdots \\ h_N \end{pmatrix} \tag{27}$$

$$\bar{f} = \begin{pmatrix} d_1 \\ d_2 \\ \vdots \\ d_N \end{pmatrix} \tag{28}$$

Equation (25) is nonlinear and can be solved using the Thomas algorithm within an iterative scheme. Such a scheme is necessary because the elements of the matrix $[E]$ and the vector $\bar{f}$ are not known at time $t + \Delta t$. Haverkamp et al. (1977) employed an explicit linearization scheme by estimating the coefficients in Equation (24) at time t to obtain a solution at time $t + \Delta t$. This scheme was implemented by Nofziger (1985). With such a linearization, no iterations are necessary, and the system of equation can be solved directly.

Finite-Element Technique

In the finite-element technique, the variable h of Equation (1) is approximated by

$$h(z,t) = \sum_{j=1}^{N} N_j(z) h_j(t) \tag{29}$$

where N_j are known as the basis functions, $h_j(t)$ are the nodal values of h, and n is the number of independent basis functions. Equation (1) can be written as

$$L'(h) = \frac{\partial}{\partial z}\left(K\frac{\partial h}{\partial z} - K\right) - C\frac{\partial h}{\partial t} = 0 \tag{30}$$

where L' is a quasi-linear differential operator defined in the solution domain $0 \le z \le L$.

When the approximate Equation (29) is substituted for h in Equation (30), its right-hand side will not be exactly zero but rather a small amount commonly referred to as the residual value. The value of such a residual may be minimized by requiring that $L'(h)$ be orthogonal to a set of mutually independent weighting functions, that is,

$$\int_0^L L'(h) W_i(z) dz = 0 \tag{31}$$

In the Galerkin method, the weighting functions are assumed to be equal to the basis functions; Equation (31) then reads

$$\int_0^L L'(h) N_i(z) dz = 0 \tag{32}$$

Substitution of Equation (30) into Equation (32) yields

$$\int_0^L \left[\frac{\partial}{\partial z}\left(K\frac{\partial h}{\partial z} - K\right) - C\frac{\partial h}{\partial t}\right] N_i dz = 0 \tag{33}$$

Integration by parts of Equation (33) yields

$$\int_0^L \left(K\frac{\partial h}{\partial z} - K\right)\frac{\partial N_i}{\partial z} dz - \int_0^L C\frac{\partial h}{\partial t} N_i dz = -q N_i\Big|_0^L \tag{34}$$

where

$$q = -\left(K\frac{\partial h}{\partial z} - K\right) \tag{35}$$

Combining Equations (29) and (34) results in

$$[A]\bar{h} + [B]\frac{d\bar{h}}{dt} = \bar{F} \tag{36}$$

where $[A]$ and $[B]$ are matrices that depend on soil properties and the shape functions, $\bar{h}$ is a vector that contains nodal point values, and $\bar{F}$ is a forcing vector that includes prescribed flux across the boundaries of the domain. The elements of the matrices $[A]$, $[B]$, and $[\bar{F}]$ are given by

$$\begin{aligned}
A_{ij} &= \int_0^L K\frac{dN_j}{dz}\frac{dN_i}{dz} dz \\
B_{ij} &= \int_0^L C\, N_j N_i dz \\
F_i &= -qN_i\Big|_0^L \int_0^L K\frac{dN_i}{dz} dz
\end{aligned} \tag{37}$$

Finally, the finite-difference method is used to discretize the system of Equation (36). After rearranging, the resulting system of equations can be expressed in the form given by Equation (25).

To perform the integration in Equation (37), the domain is divided into elements and specific shape functions are chosen. The integration can be performed analytically or numerically, depending on the type of these functions. The use of linear finite elements allows easy integration, yielding a nonlinear system of equations similar to Equation (20) with the coefficients given by

$$\begin{aligned}
a_i &= \frac{\varepsilon K_{i+1/2}}{\Delta z^2} + \frac{(C_{i+1} + C_i)}{12\Delta t} \\
b_i &= -\frac{\varepsilon K_{i+1/2} + K_{i-1/2}}{\Delta z^2} - \frac{(C_{i+1} + 6C_i + C_{i-1})}{12\Delta t} \\
c_i &= -\frac{\varepsilon K_{i-1/2}}{\Delta z^2} + \frac{(C_i + C_{i-1})}{12\Delta t} \\
d_i &= -\frac{(K_{i+1/2} - K_{i-1/2})}{\Delta z} - h_{i-1}^t\left[\frac{(1-\varepsilon)K_{i-1/2}}{\Delta z^2} + \frac{(C_i + C_{i-1})}{12\Delta t}\right] \\
&\quad - h_{i+1}^t\left[\frac{(1-\varepsilon)K_{i+1/2}}{\Delta z^2} + \frac{(C_{i+1} + C_i)}{12\Delta t}\right] \\
&\quad + h_i^t\left[\frac{(1-\varepsilon)(K_{i-1/2} + K_{i+1/2})}{\Delta z^2} + \frac{(C_{i+1} + 6C_i + C_{i-1})}{12\Delta t}\right]
\end{aligned} \tag{38}$$

Treatment of Boundary Conditions

As discussed earlier, boundary conditions may include a specified head or flux at either the upper or lower boundary. The system of Equation (20) should be modified to include the specified condition. For example, for a specified flux condition at $z = 0$, the coefficients of Equation (20) at node 1 for the fully implicit scheme read

$$a_1 = \frac{K_{1+1/2}}{\Delta z^2}$$
$$b_1 = -\frac{K_{1+1/2}}{\Delta z^2} - \frac{C_1}{\Delta t}$$
$$c_1 = 0 \tag{39}$$
$$d_1 = -\frac{(K_{1+1/2} - q_s)}{\Delta z} + \frac{(C_1 h_1^t)}{\Delta t}$$

and Equation (20) yields

$$a_1 h_2^{t+\Delta t} + b_1 h_1^{t+\Delta t} = d_1 \tag{40}$$

Similar equations can be written at $z = L$. The formulation for the specified head case can be recovered from Equations (39) and (40) by setting q_s equal to zero and replacing $h_1^{t+\Delta t}$ with the specified value for the head.

PROCEDURES FOR SOLVING THE SYSTEM OF NONLINEAR EQUATIONS

Approaches to solving the nonlinear system of Equation (25) include the Picard and Newton–Raphson methods (see, e.g., Istok, 1989), the modified Picard method (Celia et al., 1990), and the iterative conjugate gradient algorithm (Kirkland et al., 1992). In the Picard approach, an initial guess of the solution for pressure head is made at a given time step, which is generally taken as the values computed at the previous time step. In each iteration the values of $K(h)$ and $C(h)$ are estimated for each element or cell. The matrix $[E]$ and the $\bar{f}$ vector are then constructed, and a system of linear equations of the form of Equation (25) is solved. A vector of residual is then estimated as the difference between the computed and the guessed values at different nodes:

$$\bar{R}^k = \bar{f} - [E(h^{k-1})]\bar{h}^k \tag{41}$$

or

$$\bar{R}^k = \bar{h}^k - \bar{h}^{k-1} \tag{42}$$

where k is the iteration number. A test of convergence is then performed by comparing the maximum residual value against a specified error value. A solution is obtained if the convergence is reached or if

$$|\max r^k| \leq \delta \tag{43}$$

where δ is the acceptable tolerance. Otherwise, a new iteration is performed, and the whole procedure is repeated for all time steps.

The incremental Picard method uses a relaxation factor to speed-up the convergence process. The technique is similar to that outlined above, except that the residual vector is computed before the linear system of equations is solved. At the new iteration level, the residual vector is used as the right-hand side of a linear system of equations similar to Equation (25), with an unknown vector representing head increments:

$$[E(h^{k-1})]\,\overline{\Delta h} = \overline{R}^{k-1} \tag{44}$$

The increments are used to construct the updated solution via the expression

$$\overline{h}^k = \overline{h}^{k-1} + \omega \overline{\Delta h} \tag{45}$$

The optimal value for the relaxation factor ω is estimated by trial and error to minimize the number of iterations needed for convergence.

The Newton–Raphson method can be more efficient than the Picard method in some cases. It is similar to the incremental Picard method in that the residual vector is estimated before solving the linear system of equations. However, the increments are estimated by writing a Taylor series expansion for the residual vector and setting it equal to zero. The linear system of equations will thus contain the derivative of the matrix $[E]$ with respect to the vector $\overline{h}$:

$$\left[E(h^{k-1}) + \left.\frac{\partial[E(h)]}{\partial \overline{h}}\right|_{\overline{h}=\overline{h}^{k-1}}\right]\overline{\Delta h} = \overline{R}^{k-1} \tag{46}$$

Finally, Equation (45) is used to compute the solution at this iteration level.

The Thomas algorithm (von Rosenberg, 1969) is a direct matrix solution technique that uses triangular decomposition to solve a tridiagonal system of equations similar to that given by Equation (20). For such a system with N unknowns, the solution for h at node i is given by

$$h_N = g_N \tag{47}$$

$$h_i = g_i - \gamma_i h_{i+1}, \qquad i = N-1, N-2, \ldots, 1 \tag{48}$$

where g_i and γ_i are given by

$$g_1 = \frac{d_1}{b_1} \tag{49}$$

$$\gamma_1 = \frac{a_1}{b_1} \tag{50}$$

$$g_i = \frac{d_i - c_i g_{i-1}}{b_i - c_i \gamma_{i-1}} \qquad i = 2, 3, \ldots, N \tag{51}$$

$$\gamma_i = \frac{a_i}{b_i - c_i \gamma_{i-1}} \qquad i = 2, 3, \ldots, N \tag{52}$$

The technique thus consists of two steps. The first step is to create the arrays g_i and γ_i using the forward recursion relations in Equations (49) through (52), and the second is to determine the unknowns h_i using the backward recursion relations in Equations (47) and (48).

RECENT ADVANCES IN NUMERICAL SOLUTIONS

Studies that examined the accuracy of various numerical techniques pertinent to the one-dimensional unsaturated flow include those by Haverkamp et al. (1977) and van Genuchten (1982). Haverkamp et al. indicated that solutions obtained using implicit finite-difference schemes with implicit or explicit evaluation of hydraulic properties are the most accurate. van Genuchten compared a number of numerical solutions and concluded that Hermitian finite-element formulation with four- or five-point Lobatto integration scheme was the most accurate.

Recent studies have shown mass balance problems in the pressure-based form of the Richards equation (Celia et al., 1987; Milly, 1988). Milly (1985) modified the capacity term to force a global mass balance that overcomes the mass-conserving solution problem. Celia et al. (1990) indicated that the mixed form conserves mass; however, its solution is not always accurate. They also showed that treatment of the time-derivative is a critical factor in obtaining accurate results. They recommended using a lumped form of the time matrix in the finite-element formulation (see also Neuman, 1973; Cooley, 1983). The study by El-Kadi and Ling (1993) demonstrated, however, that the pressure-based form of the Richards equation, as well as other forms, can produce accurate and mass-conserving solutions, provided that care is taken in designing the spatial and temporal mesh. Numerical difficulties can also arise in simulating infiltration under heterogeneous conditions into initially dry soils or into coarser materials that are characterized by sharp wetting fronts. Kirkland et al. (1992) indicated that the water-content based form is more accurate for infiltration into dry soils; however, it is not suitable for unsaturated-saturated soils.

New advances to improve the efficiency of numerical solutions are related, generally, to pressure-head transformation, implicit-time iteration, and interblock parameter estimation (see, e.g., Ross and Bristow, 1990). For example, Zaidel and Russo (1992) introduced an estimate of finite-difference interblock conductivities that allowed the use of a relatively coarse spatial mesh. The interblock conductivity is estimated from

$$K_{i+1/2} = \int_{h_i}^{h_{i+1}} \frac{K(h)}{(h_{i+1} - h_i)} dh \tag{53}$$

Ross and Bristow (1990) applied the technique successfully to layered or gradational soils.

Gottardi and Venutelli (1992) implemented an accurate solution via a moving finite-element method in which grid points are moved along the wetting front and hence nodal coordinates change in time. For the transformed form of the Richards equation, given as Equation (6), the piecewise linear approximation of u is given by

$$u(z,t) \approx u'(z,t) = \sum_{j=1}^{N} [\alpha_j(z)u_j(t) + \beta_j(z)z_j(t)] \tag{54}$$

where N is the number of nodes and α_j and β_j are the bases functions which are given by

$$\alpha_j = \frac{\partial u'}{\partial u_j}$$
$$\beta_j = \frac{\partial u'}{\partial z_j} \tag{55}$$

The functions α_j and β_j can be related by assuming a linear approximation of u' within the element. Substituting u' for u in the governing equation defines a residual which, when minimized with respect to u_j and z_j, leads to a set of ordinary differential equations that can be solved to identify the vectors u and z. The authors applied the technique to a number of cases and showed that it can allow the use of a smaller number of nodes without sacrificing the accuracy. However, the method cannot be used for layered systems and under time-varying boundary conditions.

Kirkland et al. (1992) implemented a transformation that allows the use of the water-content based equation in saturated-unsaturated conditions. The transformation defines the generalized variable

$$\phi = \phi(z,t) = g[h(z,t), z] \tag{56}$$

with the water-retention relation written as

$$\theta = \theta(z,t) = f[h(z,t), z] \tag{57}$$

The transformed equation is given as

$$\frac{\partial}{\partial z}\left(\frac{K}{S}\frac{\partial \phi}{\partial z}\right) - \frac{\partial K}{\partial z} - \frac{\partial}{\partial z}\left(\frac{K}{S}\frac{\partial g}{\partial z}\right) = H\frac{\partial \phi}{\partial t} \tag{58}$$

in which

$$H = \frac{\partial f}{\partial h}\frac{\partial h}{\partial \phi} \tag{59}$$

and

$$S = \frac{\partial g}{\partial h} \tag{60}$$

The last term on the left-hand side in Equation (58) accounts for changes in soil properties owing to spatial variability in the water-retention parameters. The authors define ϕ as a linear function of the function f with constants that can be estimated based on the known retention data. A knowledge of ϕ and f allows a determination of the functions H and S as given by Equations (59) and (60).

Attempts to improve on the accuracy of numerical solutions also include solving the transformed form given by Equation (6) (Rubin, 1966; Raats and Gardner, 1974; Khanji et al., 1974; El-Kadi and Ling, 1993). Some studies indicated that such a transformation alleviates stability problems caused by the existence of very high pressure gradients near the wetting fronts in dry soils. El-Kadi and Ling (1993) showed that the solution within a finite-difference formulation, although allowing the use of a larger mesh size, suffers from the same difficulties associated with other forms regarding mass conservation and accuracy problems. They proposed using Peclet (Pe) and Courant (Co) number criteria for assessing the accuracy of numerical solutions of the Richards equation. Such a use is suggested by the similarity in shape between Equation (6) and the advective-dispersive equation, as well as the similarity in problems associated with sharp fronts. The Courant number reflects the importance of the convection process, whereas the Peclet number indicates the relative importance of the convection versus dispersion processes.

From Equation (6), the grid Peclet and Courant numbers can be respectively defined as

$$Pe(h) = \frac{v}{D}\Delta z = \frac{1}{K}\frac{dK}{dh}\Delta z \tag{61}$$

and

$$Co(h) = v\frac{\Delta t}{\Delta z} = \frac{dK}{d\theta}\frac{\Delta t}{\Delta z} \tag{62}$$

The expressions in Equations (61) and (62) can be evaluated regardless of the form of the Richards equation. The values of these parameters are generally higher for coarser materials and can vary by several orders of magnitude over the saturation range (El-Kadi and Ling, 1993). Smaller values for Δz and Δt should be used for large values of v and v/D order to reduce numerical dispersion or overshooting.

Dependency of the mesh size on the initial head-gradient suggests the need for a criterion that includes such an effect. El-Kadi and Ling (1993) defined a travel-distance form of the Courant number (Co°) that reads

$$Co° = \frac{Q\Delta t}{\Delta\theta\Delta z} = \frac{\left[K\frac{dh}{dz}\right]_{z=0}\Delta t}{\Delta\theta\Delta z} \tag{63}$$

Table 1. Mesh Sizes (Δz and Δt) and the Respective Values of the Courant Number (Co and $Co°$) and Peclet Number (Pe), as Estimated from the Listed References. Saturated Hydraulic Conductivity (K_s) and the Type of Boundary Condition at the Soil Surface are also Listed

Reference	K_s (cm/s)	Δz (cm)	Δt (s)	Co	$Co°$	Pe	Boundary Condition
El-Kadi and Ling (1993)	9.22×10^{-3}	2.5	20	0.011	0.202	0.13	Head
	8.7×10^{-4}	2.5	60	0.071	0.345	0.07	Head
	8.7×10^{-4}	2.5	210	0.045	1.208	0.057	Head
	9.5×10^{-6}	1.0	432	0.2	0.856	0.019	Head
Celia et al. (1990)	9.44×10^{-3}	1.0	10	1.532	1.799	0.137	Head
Kirkland et al. (1992)	6.0×10^{-3}	2.5	432	0.069	4.188	0.118	Flux
Haverkamp et al. (1977)	1.23×10^{-5}	1.0	40	0.104	0.290	0.019	Head
Paniconi et al. (1991)	1.11×10^{-5}	0.5	36	0.028	1.257	0.002	Head
	0.1389	1	9	1.136	1.954	0.013	Flux
	2.78×10^{-4}	0.5	9	0.01	—†	0.023	Flux
van Genuchten (1982)	1.05×10^{-2}	5	—‡	—‡	—‡	0.337	Head
Ross (1990)	9.0×10^{-3}	1.0	360	13.209	21.156	0.141	Head
	9.21×10^{-4}	1.0	360	1.563	4.095	0.089	Head
	1.23×10^{-5}	1.0	360	0.011	0.782	0.032	Head
Ross and Bristow (1990)	8.0×10^{-4}	2.5	288	1.284	1.453	0.345	Head
	3.0×10^{-4}	2.5	144	0.969	0.797	0.527	Head
	7.5×10^{-4}	2.5	50	0.179	0.152	2.135	Head
Cooley (1983)	8.68×10^{-4}	1.0	360	7.812	3.198	0.481	Head
Kabala and Milly (1990)	1.23×10^{-5}	4	2,000	1.295	0.520	0.076	Head
Warrick (1991)	1.52×10^{-4}	2.5	0.5	5.8×10^{-7}	0.608	0.013	Flux

† A mass-conserving solution was not possible.
‡ Value of Δt is not available.
(From El-Kadi, A. I. and Ling, G., *Water Resour. Res.*, 29, 3485, 1993.)

with $Q/\Delta\theta$ as an "effective" advective velocity. This number can be evaluated regardless of the type of boundary condition imposed at the soil surface.

Table 1 lists estimates of Pe, Co, and $Co°$ for the various case studies surveyed by El-Kadi and Ling (1993). The upper bound for the Peclet number is about 0.5, whereas that for $Co°$ falls below 4 in most cases (Figures 1 and 2). They respectively vary over two and one order of magnitude over the wide range of soil types covered. El-Kadi and Ling preferred the expression for Courant number as given by Equation (63) because of its narrower range of variability and its ability to consider the flow dynamics.

El-Kadi and Ling (1993) also proposed a scheme for time-step choice for cases in which the pressure-head gradients decline with increasing simulation time. A very small time step should be chosen first, on the order of seconds, to generally satisfy the Courant-number criterion. An expansion factor for the time step can be employed to increase the solution efficiency. Both the rate of convergence and the decline in the maximum value of the pressure-head gradient should be considered when estimating the size of the time step. The spatial mesh can also be expanded, as long as it satisfies the Courant and Peclet criteria where head gradients are high. Unfortunately, in cases where no decline in the pressure gradient occurs, accuracy may require the use of extremely small increments (seconds to minutes for the time step, and a few centimeters for spatial increment). Such ranges put severe limitations on model applications for large scale, multidimensional problems.

NUMERICAL SOLUTIONS OF ONE-DIMENSIONAL FLOW 163

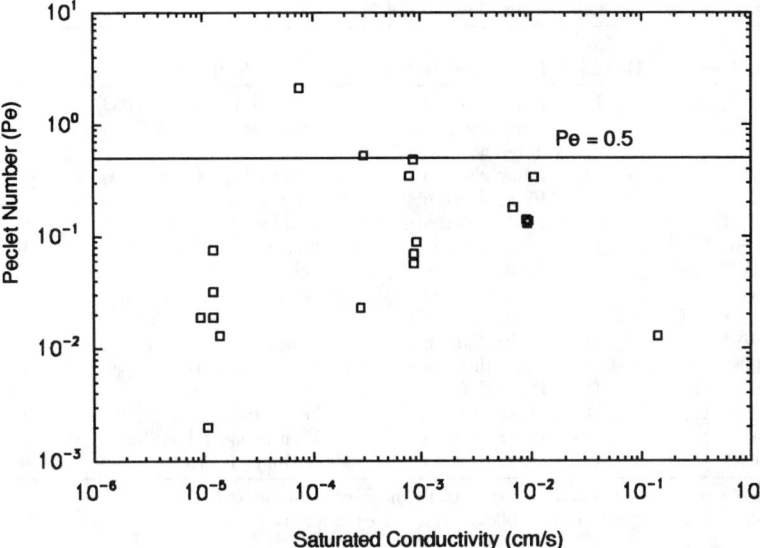

Figure 1. Values of the Peclet number relative to the value of the saturated hydraulic conductivity. (Redrawn from El-Kadi and Ling, 1993.)

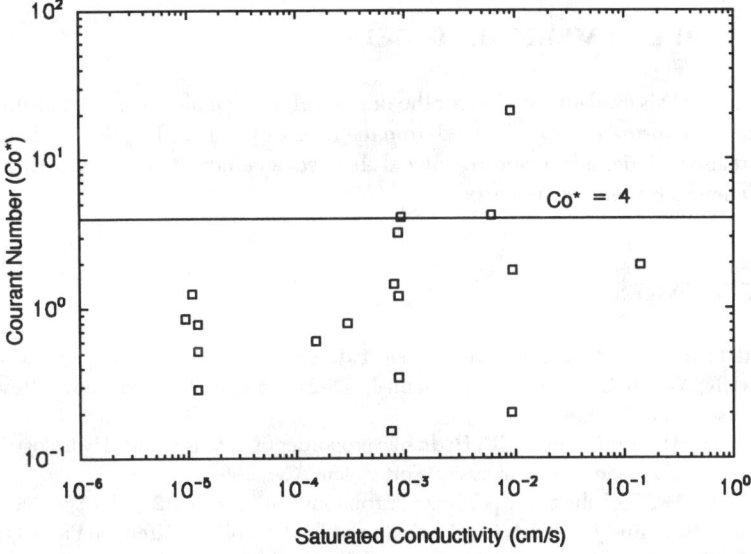

Figure 2. Values of the travel-distance form of the Courant number relative to the value of the saturated hydraulic conductivity. (Redrawn from El-Kadi and Ling, 1993.)

Table 2. Available Unsaturated Flow Models

Model Name	Flow Domain[a]	Numerical Method	Reference	Availability[b]
UNSAT2	2-D	Finite element	Davis and Neuman (1983)	P
TRUST	2-D	Integrated finite difference	Reisenauer et al. (1982)	P
FLUMP	2-D	Finite element	Narasimhan et al. (1978)	P
MUST	1-D	Finite difference	De Laat (1985)	R
UNSATID	1-D	Finite difference	Bond et al. (1984)	Pr
SWACROP	1-D	Finite difference	Wesseling et al. (1989)	R
FEMWATER	2-D	Finite element	Yeh and Ward (1980)	P
UNSAT-1	1-D	Finite element	van Genuchten (1978b)	P
UNSAT-H	1-D	Finite difference	Fayer and Gee (1985)	P
FLOWVEC	3-D	Finite difference	Li et al. (1983)	P
LANDFIL	1-D	Finite difference	Korfiatis (1984)	P
WATERFLO	1-D	Finite difference	Nofziger (1985)	P
3DFEMWATER	3-D	Finite element	Yeh (1987)	P
UNSAT	1-D	Finite element	Khaleel and Yeh (1985)	P
PC-SEEP	2-D	Finite element	Krahn et al. (1989)	Pr

[a] One dimension (1-D), two-dimensions (2-D), or three-dimensions (3-D).
[b] Public domain (P), restricted public domain (R), or proprietary (Pr).

The Peclet and Courant numbers can be useful for rough mesh design for complicated problems and as bases for comparing the efficiency of various numerical techniques. The efficiency of a numerical technique should not only be based on the mesh but also on the soil type and boundary conditions simulated. An efficient technique should allow for larger values than those suggested by El-Kadi and Ling (1993), namely 0.5 and 4, for the Peclet and Courant number, respectively.

AVAILABLE NUMERICAL MODELS

Table 2 lists available models for the numerical solution of flow in the unsaturated zone. The information was obtained from the report by van der Heijde (1992), whose report also includes information on model abstract, documentation status, verification/validation status, and code custodian.

REFERENCES

Bond, F. W., Cole, C. R., and Gutknecht, P. J., Unsaturated Groundwater Flow Model (UNSAT1D) Computer Code Manual, CS-2434-CCM, Electric Power Research Inst., Palo Alto, CA, 1984.

Brooks, R. H. and Corey, A. T., Hydraulic properties of porous media, Hydrology Paper No. 3, Colorado State University, Fort Collins, CO, 1964.

Brutsaert, W., Probability for pore-size distribution, *Soil Sci.* 101(2), 85, 1966.

Carslaw, H. S. and Jaeger, J. C., *Conduction of Heat in Solids*, Clarendon Press, Oxford, England, 1959.

Celia, M. A., Ahuja, L. R., and Pinder, G. F., Orthogonal collocation and alternating-direction procedures for unsaturated flow problems, *Adv. Water Res.*, 10, 178, 1987.

Celia, M. A., Bouloutas, E. T., and Zarba, R. L., A general mass-conservative numerical solution for the unsaturated flow equation, *Water Resour. Res.*, 26(7), 1483, 1990.

Cooley, R. L., Some new procedures for numerical solution of variably saturated flow problems, *Water Resour. Res.*, 19, 1271, 1983.

Davis, L. A. and Neuman, S. P., Documentation and User's Guide: UNSAT2–Variably Saturated Flow Model, NUREG/CR-3390, U.S. Nuclear Regulatory Commission, Washington, D.C., 1983.

De Laat, P. J. M., MUST, A Simulation Model for Unsaturated Flow, Report Series No. 16, International Institute for Hydraulic and Environmental Engineering Delft, The Netherlands, 1985.

El-Kadi, A. I., A computer program to estimate the parameters of soil hydraulic properties, International Ground Water Modeling Center, IGWMC-PLUTO 6330, Holcomb Research Institute, Butler University, Indianapolis, 1984.

El-Kadi, A. I. and Ling, G., The Courant and Peclet number criteria for the numerical solution of the Richards equation, *Water Resour. Res.*, 29, 3485, 1993.

Fayer, M. J. and Gee, G. W., UNSAT-H: An Unsaturated Soil Water Flow Code for Use at the Hanford Site: Code documentation, PNL-5585, Battelle Pacific Northwest Lab., Richland, WA, 1985.

Gottardi, G. and Venutelli, M., Moving finite element model for one-dimensional infiltration in unsaturated soil, *Water Resour. Res.*, 28, 3259, 1992.

Haverkamp, R. and Vauclin, M., A note on estimating finite difference interblock hydraulic conductivity values for transient unsaturated flow problems, *Water Resour. Res.*, 15, 181, 1979.

Haverkamp, R., Vauclin, M., Touma, J., Wierenga, P. J., and Vachaud, G., A comparison of numerical simulation models for one-dimensional infiltration, *Soil Sci. Soc. Am. J.*, 41, 285, 1977.

Istok, J., Groundwater modeling by the finite element method, Water Resources Monograph 13, American Geophysical Union, Washington, D.C., 1989.

Kabala, Z. J. and Milly, P. C. D., Sensitivity analysis of flow in unsaturated heterogeneous porous media: Theory, numerical model, and its verification, *Water Resour. Res.*, 26(4), 593, 1990.

Khaleel, R. and Yeh, T.-C., A Galerkin finite element program for simulating unsaturated flow in porous media, *Ground Water*, 23(1), 90, 1985.

Khanji, D., Vauclin, M., and Vachaud, G., Infiltration non permanente et bidimensionnelle dans une tranche de sol non saturee. Analyse numerique et resultats experimentaux, *C. R. Acad. Sci. (Paris)*, 278, 381, 1974.

Kirkland, M. R., Hills, R. G., and Wierenga, P. J., Algorithms for solving Richards' equation for variability saturated flow, *Water Resour. Res.*, 28(8), 2049, 1992.

Korfiatis, G. P., Modeling the Moisture Transport Through Solid Waste Landfills, Ph.D. thesis, Rutgers University, New Brunswick, NJ, 1984.

Krahn, J., Fredlund, D. G., Lam, L., and Barbour, S. L., PC-SEEP: A Finite Element Program for Modeling Seepage, Geo-Slope Programming, Calgary, Alberta, 1989.

Li, R.-M., Eggert, K. G., and Zachmann, K., Parallel processor algorithm for solving three-dimensional ground water flow equations, National Science Foundation, Washington, D.C., 1983.

Milly, P. C. D., A mass-conservative procedure for time-stepping in models of unsaturated flow, *Adv. Water Res.*, 8, 32, 1985.

Milly, P. C. D., Advances in the modeling of water in the unsaturated zone, *Trans. Porous Media*, 3, 491, 1988.

Narasimhan, T. N., Neuman, S. P., and Witherspoon, P. A., Finite element method for subsurface hydrology using a mixed explicit-implicit iterative scheme, *Water Resour. Res.*, 14(5), 863, 1978.

Neuman, S. P., Saturated-unsaturated seepage by finite elements, *J. Hydraul. Div.*, ASCE, 99(HY12), 2233, 1973.

Nofziger, D. L., Interactive simulation of one-dimensional water movement in soils: User's guide. Circular 675, Software in Soils Science, Florida Cooperative Extension Service, University of Florida, Gainesville, FL, 1985.

Paniconi, C., Aldama, A. A., and Wood, E. F., Numerical evaluation of iterative and noniterative methods for the solution of the nonlinear Richards equation, *Water Resour. Res.*, 27(6), 1147, 1991.

Raats, P. A. C. and Gardner, W. R., Movement of water in the unsaturated zone near a watertable, in *Drainage for agriculture: Agronomy, Schilfgaarde,* Jan Van, Ed., Am. Soc. Agron., Madison, WI, 17, 311, 1974.

Reisenauer, A. E., Key, K. T., Narasimhan, T. N., and Nelson, R. W., TRUST: A Computer Program for Variably Saturated Flow in Multidimensional, Deformable Media. NUREG/CR-2360, U.S. Nuclear Regulatory Commission, Washington, D.C., 1982.

Richards, L. A., Capillary conduction through porous mediums, *Physics*, 1, 313, 1931.

Ross, P. J., Efficient numerical methods for infiltration using Richards' equation, *Water Resour. Res.*, 26(2), 279, 1990.

Ross, P. J. and Bristow, K. L., Simulating water movement in layered and gradational soils using the Kirchhoff transform, *Soil Sci. Soc. Am. J.*, 54, 1519, 1990.

Rubin, J., Numerical analysis of bonded rainfall infiltration, in *Proc. Wageningen Symposium* IASH 82 Rijtema, P. E. Ed., UNESCO, Paris, 440, 1966.

van der Heijde, P. K. M., Identification and compilation of unsaturated/vadose zone models possible to setting soil remediation levels at superfund sites, GWMI 92–02, International Ground Water Modeling Center, Colorado School of Mines, Golden, CO, 1992.

van Genuchten, M. Th., Calculating the unsaturated hydraulic conductivity with a new closed form analytical model, Rep. 78-WR-08, Water Resources Program, Dept. of Civil Eng., Princeton University, Princeton, NJ, 1978a.

van Genuchten, M. Th., Numerical solutions of the one-dimensional saturated-unsaturated flow equation, Rep. 78-WR-11, Water Resources Program, Dept. of Civil Eng., Princeton University, NJ, 1978b.

van Genuchten, M. Th., A comparison of numerical solutions of one-dimensional unsaturated-saturated flow and mass transport equations, *Adv. Water Res.*, 5, 47, 1982.

van Genuchten, M. Th., Leij, F. J., and Lund, L. J., Eds., Indirect Methods for Estimating the Hydraulic Properties of Unsaturated Soils, University of California, Riverside, CA, 1992.

von Rosenberg, D. U., *Methods for the Numerical Solution of Partial Differential Equations,* Elsevier, New York, 1969.

Warrick, A. W., Numerical approximation of Darcian flow through unsaturated soil, *Water Resour. Res.*, 27(6), 1215, 1991.

Wesseling, J. G., Kabat, P., van den Broek, B. J., and Feddes, R. A., SWACROP: Simulating the dynamics of the unsaturated zone and water limited crop production, Winand Staring Centre, Department of Agrohydrology, Wageningen, The Netherlands, 1989.

Yeh, G. T., 3DFEMWATER: A Three-Dimensional Finite Element Model of Water Flow Through Saturated-Unsaturated Media, ORNL-6386, Oak Ridge National Laboratory, Oak Ridge, TN, 1987.

Yeh, G. T. and Ward, D. S., FEMWATER: A Finite-Element Model of Water Flow Through Saturated-Unsaturated Porous Media, ORNL-5567, Oak Ridge National Laboratory, Oak Ridge, TN, 1980.

Zaidel, J. and Russo, D., Estimation of finite difference interblock conductivities for simulation of infiltration into initially dry soils, *Water Resour. Res.*, 28(9), 2285, 1992.

CHAPTER 11

Consequences of Scale-Dependency on Application of Chemical Leaching Models: A Review of Approaches

R. J. Wagenet and J. L. Hutson

INTRODUCTION

The use of simulation models to address environmental problems is now a widespread practice (van der Heijde and Elnawawy, 1993). Yet, these models are limited by the assumptions inherent in their development, the availability and uncertainty of required input data, and the consequent reliability of output predictions. These issues are both compounded and confounded by the fact that spatial and temporal scales of variability dictate the availability and quality of data and therefore the appropriate modeling choices. As the constraints of available data at a particular scale are recognized, there must also follow a consideration of the degree of determinism and mechanism that can be invoked in the modeling process. One unfortunate consequence is a continuing use of simulation models derived for use at one scale to predict effects at larger scales. This process is a form of extrapolation of limited information on natural processes beyond the limits of scientific understanding. This type of misuse of models is important and widespread in vadose zone simulation modeling. Particularly important examples are the numerous pesticide leaching models which are now used to estimate pesticide transport and transformation at spatial scales from the soil profile to the regional, and for temporal scales from seconds to decades.

When errors in pesticide leaching estimations are made, they derive from errors in model selection as it relates to available data. These errors can be avoided by recognizing that proper use of these tools must also consider the purpose of the simulation and the relationship between a particular model and the ability to measure relevant input and validation data at each scale. Further discussion of these issues can be separated into modeling approaches, user groups, process representation, and scale. Example models will be used to illustrate the major issues.

MODELING APPROACHES

Pesticide leaching models are a type of solute transport model, and there has been much theoretical and experimental study of the latter, whether related to pesticides or other solutes (Nielsen et al., 1986; Wagenet and Rao, 1990). Generically, solute transport models (and therefore pesticide leaching models) have been categorized in several ways (Addiscott and Wagenet, 1985). The initial separation is between deterministic models that assume a system or process operates such that the occurrence of a given set of events leads to a uniquely definable outcome, or stochastic models which recognize that the natural system varies in space and time, leading to a model structure and output that represent the uncertainty that derives from the variability, usually presented in a statistical manner. To date, there are no purely stochastic pesticide leaching models that are widely used. In general, uncertainties and variabilities are ignored in the development of deterministic pesticide leaching models. Such models are thereby often criticized as being only one estimate of system behavior, and probably not a true estimate, if only one model execution is conducted to describe a variable space or any length of time. However, these models are an attractive starting point for describing pesticide leaching. First, they allow manipulation of basic input data and process representation so that realistic field cases can be considered. Second, they consist of a clear representation of their component processes, which allows independent studies of biological transformations, soil sorption reactions, and volatilization and internal soil hydraulic properties to be conducted as part of an integrated experimental/modeling program. Third, the development and use of a properly constructed pesticide leaching model is a learning experience for scientists who test their hypotheses, and for users who take time to understand the model, in the process gaining an appreciation for the way the soil-water-pesticide system behaves.

An important second level of categorization of deterministic pesticide leaching models is between those that are relatively mechanistic and those that are more functional. While these have been previously defined as representing distinct categories of leaching models (Addiscott and Wagenet, 1985), it is now wiser to regard them as extremes of a modeling spectrum which encompasses a number of hybrids of each type of model. A mechanistic model can be envisioned as an attempt to integrate, as best we know them, the fundamental physical, chemical, and biological processes related to leaching. These processes are also called rate models, because they use instantaneous rates of change of water potential or pesticide concentration to estimate the fluxes of water and pesticide. These models at their most comprehensive also include in rate terms such processes as sorption, biological degradation, and volatilization. In contrast, the purely functional leaching models simplify the representation of basic process, especially those pertaining to water flow and solute transport, to the point where no claim of fundamentality is made. These models are often known as capacity models, because they displace water and pesticide through the profile according to the capacity of a soil layer to retain water and the chemical dissolved in it. While these models are clearly different from mechanistic models, they have been presented in the form of mixed mechanistic/functional models, producing models that are intermediate in the spectrum (Addiscott, 1977; Hutson and Wagenet, 1993; Carsel et al., 1985). It is in these cases that modeling problems may arise, because the mixture of mechanism with an appropriate level of simplification is difficult if accuracy is to be retained. Such models can be deceptively attractive as they are presented to the nonscientific community. Properly constructed mixed deterministic/functional models usually require less soil characterization information as input data, and are able to execute faster, but they may provide less quantitative estimates of pesticide leaching than do the highly mechanistic models. Yet, these models are very useful when employed carefully, with full recognition of the assumptions used in their derivation. The ability to now exploit faster computers in repeated executions, using multiple ensembles of characterization sets in both mechanistic and functional models, will allow testing of the degree to which basic mechanism needs to be considered for accurate assessment of variability and uncertain leaching problems.

TYPES OF USERS

Pesticide leaching models are used as research tools, guides to regulation and management, and as educational aids. These models differ in not only their basic formulation, but also in their use, flexibility, and data requirements. As research tools, the models differ according to whether they are describing steady-state, laboratory conditions, or whether they are meant to describe nonsteady, or transient, field cases. Exact analytical mathematical solutions can be derived in the former cases, while more comprehensive numerical models are needed to describe field complexity. The deterministic, mechanistic, analytical research models are best used to understand basic transport processes (van Genuchten and Wagenet, 1989), derive transport parameters (Parker and van Genuchten, 1984), or to place pesticides into behavioral classes (Jury et al., 1983). These models are misused when applied to field cases. The deterministic, mechanistic, relatively comprehensive numerical models of pesticide leaching, for example LEACHM (Wagenet and Hutson, 1986; Wagenet et al., 1989; Hutson and Wagenet, 1992), which used to be the domain of the research community for the testing of new hypotheses, are now being extended into the management and regulatory communities. This use is leading to new methodologies for pesticide certification and registration, and also new formulations of the models to better serve such purposes.

Mixed mechanistic/functional models of pesticide leaching are also widely used (Pennel et al., 1990). The U.S. Environmental Protection Agency (EPA) model PRZM-2 (Mullins et al., 1993) and the USDA-ARS model GLEAMS (Knisel et al., 1989) are examples of models based on a mixture of mechanistic basic process and functional simplification. PRZM is widely used by EPA and GLEAMS is widely used by USDA as tools for evaluating possible environmental consequences of pesticide leaching. Another possible choice is a capacity-based option in LEACHM (Hutson and Wagenet, 1993). All of these models execute quite rapidly on most desktop computers, and serve to screen new chemicals quickly and in a straightforward fashion with a minimum of input data. In the case of such mixed models, issues of internal mass balancing and numerical dispersion become crucial as simplifications are made in the process representation of the model. For this reason, comparison of these mixed models with analytical solutions representing one well-defined case of the model simulation is essential as a minimum test of the quality of the numerical code.

Educational models of pesticide leaching provide an opportunity to present the basic principles of leaching to nonscientific audiences (Nofziger and Hornsby, 1986; Rao et al., 1976; Steenhuis et al., 1987). They are usually the simplest of functional models, simplifying basic process to the degree that it is usually improper to estimate pesticide concentrations in the profile or in deep drainage. Rather, educational models are useful for demonstrating relative leaching scenarios which result from changes in pesticides, soil properties, rainfall, or irrigation. The approximate depth of penetration of the applied mass, or the position in the profile of the center of mass is often presented. This is entirely appropriate for the broad comparison of scenarios that might result from alternative practices. However, attempts to take such results further and predict pesticide concentrations as a function of depth and time using superimposed analytical solutions are underway but to date have proved unreliable. These models are not intended for use outside broad educational purposes, as the nature of their simplifications are such that much of the subtle sensitivity to process interactions is lost.

Estimation of pesticide leaching is sometimes attempted using indices that are intended to represent the combination of processes that result in pesticide fate in soil profiles (Mackay et al., 1985; Rao et al., 1985; Loague et al., 1990; Wania et al., 1993). While they represent alternatives for estimating pesticide leaching, they are not derived with consideration of the dynamic processes in soil-water-air systems that determine water and pesticide fluxes. Nevertheless, these indices are occasionally used in large-scale groundwater contamination GIS assessments, even though there has been little examination of their accuracy and reliability. Comparison of indices with dynamic models of pesticide leaching and with field data is essential before these approaches are used further.

In all the above cases, the use of attractive color graphics and front-end user-friendly interfaces is growing. This interaction is both a positive and negative development, depending on the model, and the scale at which it is meant to be used. The user-friendly interface is often extolled as a way to bring the science of modeling into public use. However, this has both advantages and disadvantages, again depending upon the specific model and its intended scale of use. The nonscientific public, and even scientifically trained college graduates in natural sciences and engineering, can often be misled by a fancy color display that derives from a hidden modeling approach that has not been subjected to thorough scientific evaluation or that is being used at an improper spatial or temporal scale. Such cases are difficult to avoid, and it is perhaps best for model users to work cooperatively with model builders as applications for these packages are developed. Users should also ascertain whether or not the model has been published in refereed scientific literature, and used successfully at the scale of the present problem.

REPRESENTATION OF BASIC PROCESSES

A number of papers and reviews have been published recently on pesticide leaching models (e.g., Wagenet, 1993; Wagenet and Rao, 1990; Hern and Melancon, 1986). Tables 1 and 2 present the general concepts embodied in these models, as categorized above.

The basic equations used in mechanistic pesticide leaching models are the classic flux-gradient equation for water flow (Darcy and Richards equations), and the convection-dispersion equations that have been developed from studies of solute transport, summarized in Nielsen et al. (1986). Table 1 summaries these equations, along with the other representations of basic processes that are most often used in mechanistic pesticide leaching models. The nonsteady-state forms of both the water

Table 1. Process Representation in Mechanistic Pesticide Leaching Models

	Water Flow	Pesticide Transport
Steady-state equation	$q_w = -K\dfrac{dH}{dz}$	$q_s = -D\dfrac{dc}{dz} + q_w c$
Total mass	$T_w = \theta$	$T_s = \theta c_L + \rho c_s + \varepsilon c_G$
Equation of continuity	$\dfrac{\partial \theta}{\partial t} = -\dfrac{\partial q_w}{\partial z}$	$\dfrac{\partial T_s}{\partial t} = -\dfrac{\partial q_s}{\partial z}$
Nonsteady-state equation	$\dfrac{\partial \theta}{\partial t} = \dfrac{\partial}{\partial z}\left[K(\theta)\dfrac{\partial H}{\partial z}\right] \pm U$	$\dfrac{\partial}{\partial t}(\theta c_L + \rho c_S + \varepsilon c_G)$ $= \dfrac{\partial}{\partial z}\left[\theta D(\theta,q_w)\dfrac{\partial c_L}{\partial z} - q_w c_L + D_G(\varepsilon)\dfrac{\partial c_G}{\partial z}\right] \pm \phi$
Nonsteady-state flux	$q_w = \displaystyle\int_{z_1}^{z_2} \dfrac{\partial \theta}{\partial t} dz$	$q_s = \displaystyle\int_{z_1}^{z_2} \dfrac{\partial}{\partial t}(T_s) dz$
Degradation		$\dfrac{dc}{dt} = -\mu c^n$ (when $n = 1$, first-order kinetics)
Sorption		$c_s = K_d c_L$ (linear, equilibrium, Langmuir sorption) $c_s = K_F c_L^{1/n}$ (nonlinear, equilibrium, Freundlich sorption)
Volatilization		$c_G = (c_G^*/c_L^*) c_L$ (Henry's Law partitioning)

Note: θ = volume water content (L^3/L^3); qw = water flux (L/T); $K(\theta)$ = water content − dependent hydraulic conductivity (L/T); H = hydraulic potential (L); T_W = total water (L^3/L^3); q_s = pesticide flux (M/L^{2}T); T_S = total pesticide (M/L^3); c_L, c_S, c_G = pesticide concentration in liquid, sorbed and gas phases, respectively (M/L^3); ρ = soil bulk density (M/L^3); ε = air-filled porosity (L^3/L^3); $D(\theta,q_w)$ = apparent diffusion coefficient (L^2/T); $D_G(\varepsilon)$ = air-filled porosity-dependent gas phase diffusion coefficient (L^2/T); ϕ = source or sink of pesticide (M/L^{3}T); μ = pesticide degradation rate coefficient (when $n = 1$, units are reciprocal time); c_G^* = pesticide saturated vapor density (M/L^3); c_L^* = pesticide aqueous solubility (M/L^3); K_d = distribution coefficient (L^3/M); K_F = Freundlich coefficient.

Table 2. An Example Educational Pesticide Leaching Model

Depth of wetting front (d_{wf})	$d_{wf} = \dfrac{I}{\theta_{fc} - \theta_i}$
Depth of pesticide front (d_{pf}) Nonreactive chemical	$d_{pf} = \dfrac{I}{\theta_{fc} - \theta_i}$
Reactive chemical	$d_{pf} = \dfrac{I}{\theta_{fc}\left(1 + \dfrac{\rho k_d}{\theta_{fc}}\right)}$

Note: I = infiltrated water (L); ρ = soil bulk density (M/L^3); K_d = sorption distribution coefficient (L^3/M); θ_i = initial volumetric water content (L^3/L^3); θ_{fc} = field capacity water content (L^3/L^3).

Source: Rao et al., 1976.

flow and pesticide transport models are the most useful under field conditions, but they demand knowledge of relationships such as those between hydraulic conductivity, water content and matric potential, and between water flux density and dispersion. These issues and many others related to basic equations of pesticide leaching models can be found in Wagenet (1993), Wagenet and Rao (1990), and Jury and Valentine (1986).

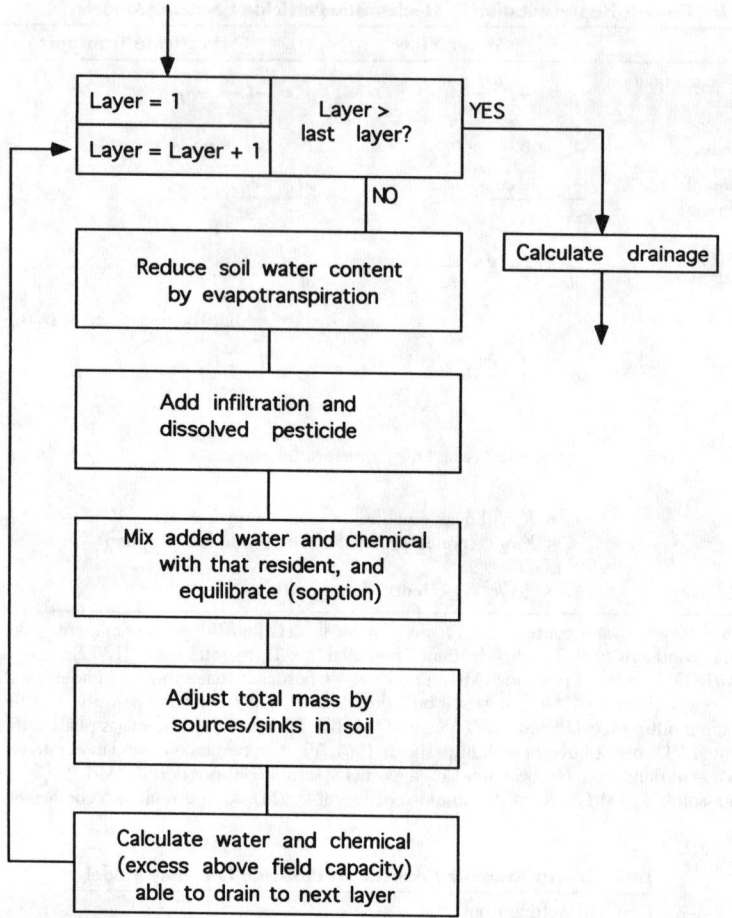

Figure 1. Flow chart representing the basic transport process often used in functional models.

It is much more difficult to present a mixed deterministic/functional model in terms of a common set of equations. Such models are to a much greater degree based upon subjective algorithms that represent fluxes, sources, and sinks of water and pesticide during leaching. A generalized flow chart (Figure 1) can be used as an example. The soil is usually separated into distinct layers, with a sequential passage of water and pesticide between layers. Water and pesticide that enter a layer are usually well mixed with resident water, followed by equilibration of pesticide in the layer according to a defined sorption isotherm. The transformation or degradation of the pesticide in that layer is calculated using a defined coefficient, most often a first-order rate coefficient. The passage to the next layer of water and dissolved pesticide is then estimated, usually from the difference between current water content and the layer's water-holding capacity. These calculations proceed until the last layer is reached, at which point there is either a return to the top layer for the next increment of added water and pesticide, or the model ends its calculations with final totals. There are a

Table 3. Basic Process Comparison of Deterministic Unsaturated Zone Pesticide Models

		———Mechanistic———		———Mixed———		———Functional———	
	|<——Research——>|	|<——Regulatory——>|	|<—Educational—>|
Process	Analytical Solutions	Complex Numerical	Complex Numerical	Analytical Solutions	Simple Numerical	Simple Numerical
Water Movement	|<-Steady->| State	|<-Richards->| Equation	|<-Water->| Balance	|<-Steady->| State	|<——Field Capacity/——>| wilting coefficient	
Solute Movement	|<————Convection-dispersion equation————>| |<——chromatographic——>|				|<-Front or Peak Location->|	
Sorption	|<————————————Linear Isotherms————————————>| |<-Nonlinear, kinetic——>|					
Degradation	|<————————————First-order kinetics————————————>| |<——Monod——>|					
Volatilization	|<————————Solubility / Vapor density————————>|					
Plant Uptake		|<——Simulation of Evapotranspiration——>|		|<——Specified ET——>|	

number of such models (Addiscott, 1977; Nicholls et al., 1982), and they are often quite useful for the management of pesticides (and fertilizers) and for field research in which leaching is less important to pesticide fate than other processes such as degradation. Mixed deterministic/functional models almost always assume linear equilibrium sorption and first-order decay, as do most mechanistic rate models. Key differences among these models arise mainly in the way that water flow and pesticide displacement are described. However, the order in which sorption, degradation, and flow are calculated depends on the assumptions of the model builder, and can lead to significant differences in calculated pesticide leaching.

Educational models of the pesticide leaching process are the simplest functional models. They are generally intended for nonscientific audiences and can be conceptually simple (Table 2). Such models are intended to demonstrate basic concepts, rather than actually use those concepts as in the mechanistic models. Educational models usually perform simple algebraic calculations that execute quite rapidly on most computers, with input requests and output usually presented with user-friendly interfaces and attractive color graphics displays. Users are provided with the opportunity to examine and contrast different scenarios of pesticide type and rate of application, soil properties, pesticide properties, rainfall, and irrigation. The prediction of pesticide concentrations in these cases is very tenuous and should be limited to calculation of the position of the pesticide center of mass in the profile, or the first arrival time of pesticide at a groundwater surface. Further refinement of these models by incorporating more basic process is now possible given the power of 386 and 486 operating systems. As this occurs, these models will blend with the larger group of mixed deterministic/functional models described above, while still retaining the simplicity which makes them valuable for educational purposes.

The above discussion of model categorization is summarized in Tables 3 and 4. The categories overlap somewhat, which is a reflection of the variety of pesticide leaching models that now exist. Example models are presented in Table 5.

Table 4. Features of Deterministic Unsaturated Zone Pesticide Leaching Models

| | Mechanistic | | Mixed | | Functional | |
	|<—— Research ——>|	|<—— Regulatory ——>|	|<— Educational —>|
Process	Analytical Solutions	Complex Numerical	Complex Numerical	Analytical Solutions	|<—— Simple ——>| Numerical	
Use	Soil Column Experiments	Field Studies/lysimeters	Field Scenarios	Screening Chemical and Soil Properties	Screening Field Conditions	Qualitative Guidance
Scale of Use	Laboratory <(i-2)	Field Plot (i-1)to(i+2)	Farm and greater	Screening (i-1)	Farm and greater	All (i-2)to(i+6)
Flexibility	Low	|<——— High ———>|		Limited	High	High
Predictions	|<——————— Quantitative ———————>|			|<——— Qualitative ———>|	
Data Needs	Small	|<——— Large ———>|		Small	|<——— Small ———>|	
CPU Time	Small	Large	Medium	|<——————— Small ———————>|		

Table 5. Example Pesticide Leaching Models

I. Deterministic models
  A. Mechanist
    1. Analytical° (van Genuchten and Wierenga, 1976; Jury et al., 1983; van Genuchten and Wagenet, 1989; Gamerdinger et al., 1990)
    2. Numerical° (Wagenet and Hutson, 1986; Boesten and Van der Linden, 1991; Hutson and Wagenet, 1992)
  B. Functional
    1. Partially analytical (Chen and Wagenet, 1992)
    2. Layer and other simple approaches (Addiscott, 1977; Nicholls et al., 1982; Carsel et al., 1985; Hutson and Wagenet, 1993)
II. Mixed deterministic-stochastic (Jury, 1982; Jury and Gruber, 1989; van der Zee and van Riemsdijk, 1987; van der Zee and Boesten, 1991)

° Refers to the solution of the flow equations.

SPATIAL AND TEMPORAL SCALE

The leaching of pesticides is manifested within and across the landscape at different spatial and temporal scales (Figure 2) (Hoosbeck and Bryant, 1990; Wagenet, 1993). One way to conceptualize the effects of scale on modeling choices is to understand the continuum of scales with which we deal. The diagonal line in Figure 2 describes those combinations of time and space scales upon which we have most often focused. Process representation progresses by conceptual steps from the molecular level through increasing spatial and temporal scales of soil aggregates, peds, horizons, pedons, catenas, and eventually to regional and continental scales. Modeling pesticide leaching has followed this progression of scale, with new models usually developed for a defined position on the diagonal. However, far less attention has been paid to the off-diagonal areas of Figure 2, which include the processes that occur at very short distances over very long times (below the diagonal), and those that occur at short times over long distances (above the diagonal). The consequences of these off-diagonal processes for accurate modeling of pesticide leaching has been almost ignored in the development of most pesticide leaching models.

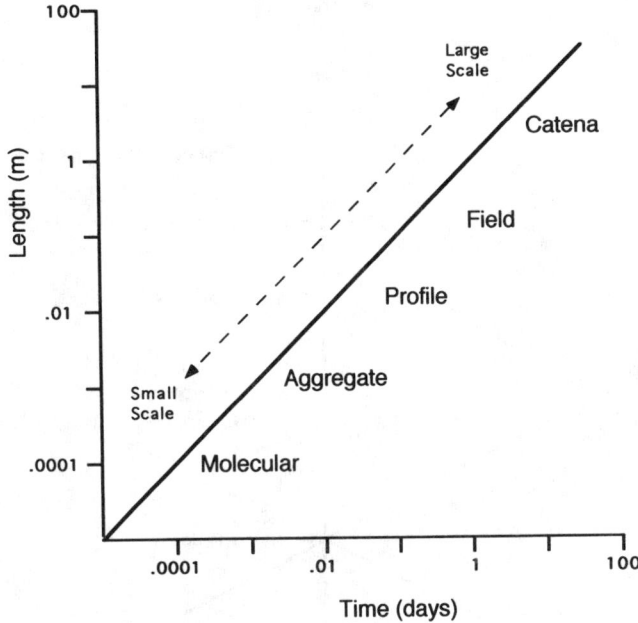

Figure 2. Spatial and temporal scales at which pesticide leaching and chemical fate processes may be measured and defined.

One of the principal issues in this debate is whether the models properly represent pesticide leaching with the processes currently modeled (Table 1). This is a valid question, and it relates to the off-diagonal areas of Figure 2. It is clear that we have spent insufficient research time on processes occurring in these off-diagonal areas. For example, pesticide sorption is now recognized as being a combination of equilibrium and kinetic processes for a number of pesticides. While an equilibrium process is appropriate for modeling pesticide-soil interaction on the diagonal, the kinetic process occurs at a short scale of space, but over a very long time. This produces pesticide leaching of very small concentrations over very long time periods, a pattern observed in some field monitoring programs. However, there are currently no widely used pesticide leaching models that include this process, which is sometimes termed chemical nonequilibrium. Similarly, physical nonequilibrium exists when physical processes occurring at very short time scales vary over extended distances. Preferential flow of water and pesticides, sometimes termed macropore flow, is an example. Most mechanistic pesticide leaching models do not yet include this process, due to the inability to accurately measure and predict the complexity of flow paths that determine the rapid transport event. Some recent models, for example, the USDA-ARS Root Zone Water Quality Model (Ahuja et al., 1991), consider preferential flow.

Other key issues related to modeling of pesticide leaching that occur in off-diagonal areas include the activity of microorganisms at very low, and very local, pesticide concentrations over extended time periods, and the effect of climate change on both short-term and long-term hydrology, at all spatial scales. Other, more conventional questions on representation of basic process related to modeling pesticide leaching exist (e.g., effects of tillage on pesticide mobility, bound residues, pesticide

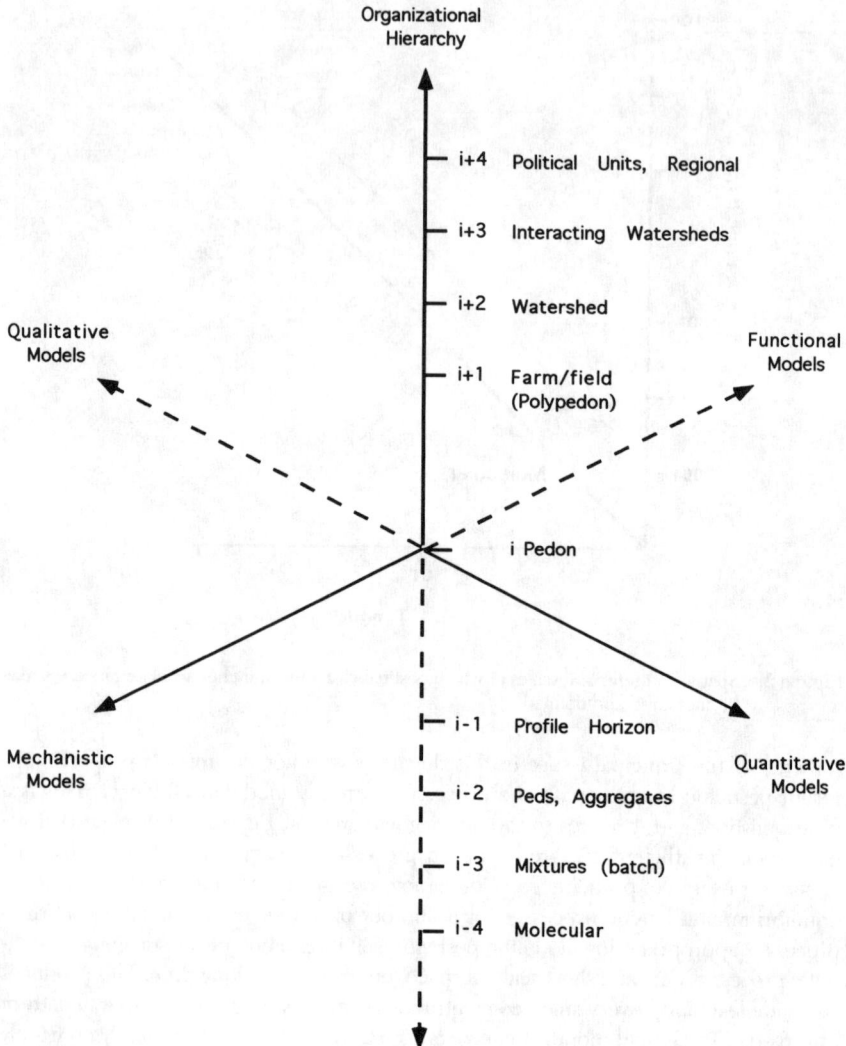

Figure 3. Spectrum of scales and modeling approaches used for estimating pesticide leaching. (Adapted from Hoosbeck and Bryant, 1990.)

uptake by plants). It is now important for the modelers to incorporate new processes in their models, if only in an exploratory fashion, to determine the possible effect of considering the off-diagonal processes.

Application of hierarchy theory to establish scales of soil processes and the modeling and experiments appropriate to describe pesticide fate have been presented elsewhere (Wagenet, 1993). In short, example scales can be identified (Wagenet et al., 1994) that represent progressively larger units of space and time at which soil and pesticide processes can be considered (Figure 3). At any particular scale, a plane created by the intersection of two lines related to leaching models can be imagined. One line represents the continuum of mechanistic to functional models. The second line

Table 6. Pesticide Models at Different Spatial and Temporal Scales

Scale	System	Example Processes Considered	Possible Modeling Approaches
i+4	County, state, province	Pesticide contamination of groundwater	Mass balance functional models of fluxes combined with GIS [valid (i+1) to (i+4)]
i+3	Soil regions (interacting watershed)	Ecosystem response to pesticide	Statistical models; compartment flux and pool models (not leaching models)
i+2	Watershed	Delivery to surface water Groundwater recharge, erosion	Distributed (functional) or statistical hydrologic models: Lumped parameter models. Functional management models
i+1	Polypedon (field/farm)	Runoff, erosion, lateral flow atmospheric dispersal	Two- or three-dimensional; lateral flow; Adhoc stochastic use of deterministic models; Functional management models
i	Pedon	(i−1), (i−2) processes combined with spatial variability Chemical fluxes to groundwater and air	Comprehensive deterministic models; (i−1) models applied with knowledge of variability of water flow, sorption, degradation, volatilization
i−1	Profile horizon	Nonequilibrium physical and chemical transport; volatilization, degradation	One- or two-dimensional deterministic, mechanistic leaching with consideration of mediating processes, soil horizonation
i−2	Soil peds, aggregates	Chemical-soil interactions in flowing systems Intraaggregate diffusion	Miscible displacement theory applied to laboratory lysimeters, soil columns. Macropore, preferential flow modeling
i−3	Soil-water-chemical mixtures	Sorption relationships; degradation kinetics	Langmuir, Freundlich isotherms Michaelis-Menten, first-order degradation
i−4	Molecular (pore/particle)	Phase partitioning; polarity; hydrophobicity	Chemical reactions in solution; Molecular structure, solubility

is the continuum of quantitative to qualitative models. Pesticide leaching models are generally developed for scales of the soil peds (i−2) to the farm/field (i+1), although some of these models are now applied to the larger scales of watersheds and political regions, such as counties or states.

Contemporary mechanistic pesticide leaching models have generally been constructed from knowledge gained from studies at lower i-scales (Table 6). Basic understanding of such processes as phase partitioning (i−4), sorption relationships (i−3), microbiological degradation (i−3), and chemical diffusion-hydrodynamic dispersion (i−2) are combined as a mechanistic leaching model is developed. The most comprehensive mechanistic models are properly used at the (i−2) to (i+1) scales. As larger areas and longer times are considered at scales beyond (i+1), it is necessary to simplify process representation owing to an inability to satisfy the large data demands of leaching models originally intended for a lower scale where more frequent and precise measurement of system behavior is possible. As a result, mixed deterministic/functional pesticide leaching models are more easily applied at scales of (i+2) to (i+4) than are mechanistic models. It is possible to use them as a component of spatially distributed hydrological models (i+2), as a dynamic tool in geographic information system (GIS) studies (i+2,3,4), and perhaps in ecosystem studies where pesticide leaching is only one of the long-term processes considered (i+3). With appropriate consideration of the simplifying assumptions, mixed models of pesticide leaching models are very useful in such cases.

The relative importance of processes defined at different scales must be considered. Small-scale processes may or may not be masked by larger-scale processes. Sometimes precise definition of small-scale process is unnecessary at larger scales, but in other situations leaching may be controlled by very variable small-scale processes. We need to appreciate the circumstances under which representation of short-term processes can be simplified in long-term applications without biasing the outcome of simulations. The need for field measurements to test and refine leaching models has resulted in extensive experimental activity in field soils. However, complicating factors such as spatial and temporal variability of processes and the parameters that describe them has led to a diversity of opinion on several issues, such as appropriate measurement or estimation methods at successively larger scales. These issues cannot be discussed fully here; readers are referred to Hern and Melancon (1986), Lauren et al. (1988), Roth et al. (1990), Wagenet et al. (1991), and Wagenet et al. (1994).

FINAL POINTS

A wide variety of pesticide leaching models exist, and those that are properly constructed, that are used by the intended user in a professional manner, and that are not stretched beyond their limits, are very good tools. However, the proper construction or use of these tools requires some background in soil science (both in the laboratory and in the field), computer programming, and a modicum of common sense. It is important that potential model builders, as well as potential users, appreciate the capabilities and limitations of existing pesticide leaching models, or know someone who does. It is easy to misapply pesticide leaching models, which has resulted in erroneous estimations of pesticide transport to groundwater.

Several other issues need to be recognized. First, it is possible to categorize pesticide leaching models in several ways. The assumptions regarding the degree of representation of basic process leads to an initial distinction between leaching models that are mechanistic and those that are functional. A further separation can be made according to the intended user group, such as research scientists, regulators, farmers, planners, and educators.

Second, current pesticide leaching models are used to address problems at spatial and temporal scales ranging from the one-dimensional soil profile of a farmer's field to a region that might include interacting watersheds or be defined by political boundaries. It is important to use measurement and monitoring methods that are relevant at the spatial and temporal domains being modeled. For example, pesticide leaching is very dependent on water flow, and the measurement of water flow at the scale of a farmer's field involves different methods than does measurement of water flow on a watershed basis. Different models would be used to describe the dynamics of leaching at these two scales, and different measurements are also required. This concept is often overlooked as the predictions of a pesticide leaching model are evaluated against the "truth" of whatever field data happens to exist.

Third, it is clear that greater attention has to be given to the nonequilibrium chemical and physical processes that greatly control water flow, pesticide partitioning, and mobility. Other important issues related to pesticide leaching remain, such as irreversibly bound residues, degradation kinetics, metabolite properties, and the effect of pesticide formulation on subsequent behavior. In the presence of such studies there remains the need to focus on nonequilibrium processes, as results to date indicate that they have significant influence on pesticide leaching. Without further understanding of such basic processes, there will be little improvement in current pesticide leaching models.

Finally, there should be recognition that modeling of pesticide leaching, and also the leaching of other solutes, is a valuable component of environmental assessments. The models can certainly be improved, and resources should be identified for that purpose. We must gain enough confidence in using leaching models that we use them on a prospective, rather than always a retrospective, basis. This method seems the only economical way to screen new chemicals, or to evaluate new management practices, either in agricultural or other environmental cases where pesticide is applied. Limited resources will always prevent field studies in every possible case of interest. Adoption of an integrated, team approach to modeling is always the best choice, and it is particularly wise in the case of modeling pesticide leaching.

SUMMARY

Leaching of soil-applied chemicals, particularly pesticides, represents a significant threat to soil and groundwater quality and an economic loss to the farmer. The ability to estimate this threat is crucial to scientists who are engaged in environmental fate research, to regulators who must wisely manage agrochemical applications, and to both indigenous ecosystems and consumers that are subjected to the consequences of chemical introduction into the environment. Pesticide leaching models have been developed in response to the concerns of each of these groups. Yet, models are imperfect, and if they are to be used as effective tools in environmental management, the ability to discriminate models according to the type of model and its proper use is vital. Recognition that models describe only a proportion of the real world leads to appreciation that there are scales, both temporally and spatially, that are different in their natural processes. The user group defines the spatial and temporal scale at which modeling of chemical leaching is required to address user-specific questions. Pesticide leaching models, when properly used, will be appropriate at distinct and different scales. User choices need to be made considering how scale-dependency influences the complexity of the model, the process representation and consequent demand for input data, and the resolution and uncertainty in space and time of the results. Additionally, much work remains in testing and refining models through an iterative process of field experimentation and model revision.

REFERENCES

Addiscott, T. M., A simple computer model for leaching in structured soils, *J. Soil Sci.*, 28, 554, 1977.

Addiscott, T. M. and Wagenet, R. J., Concepts of solute leaching in soils: a review of modelling approaches, *J. Soil Sci.*, 36, 411, 1985.

Ahuja, L. R., DeCoursey, D. G., Barnes, B. B., and Rojas, K. W., Characteristics and importance of preferential macropore transport studied with the ARS Root Zone Water Quality Model, in *Proc. Nat. Symp. Preferential Flow*, Gish, T. J. and Shirmohammadi, A., Eds., *Amer. Soc. Agric. Engr.*, St. Joseph, MI, 1991, 32.

Boesten, J. J. T. I. and Van der Linden, A. M. A., Modeling the influence of sorption and transformation pesticide leaching and persistence, *J. Environ. Qual.*, 20, 425, 1991.

Carsel, R. F., Mulkey, L. A., Lorber, M. N., and Baskin, L. B., The pesticide root zone model (PRZM): A procedure for evaluating pesticide leaching threats to groundwater, *Ecol. Model.*, 30, 49, 1985.

Chen, C. and Wagenet, R. J., Simulation of water and chemicals in macropore soils, Part I. Representation of the equivalent macropore influence and its effect on soilwater flow, *J. Hydrol.*, 130, 105, 1992.

Gamerdinger, A. P., Wagenet, R. J., and van Genuchten, M. Th., Application of two-site/two-region models for studying simultaneous nonequilibrium transport and degradation of pesticides, *Soil Sci. Soc. Am. J.*, 54, 957, 1990.

Hern, S. C. and Melancon, S. M., *Vadose Zone Modeling of Organic Pollutants*, Lewis Publishers, Chelsea, MI, 1986.

Hoosbeck, M. R. and Bryant, R. B., Towards the quantitative modeling of pedogenesis—A review, *Geoderma*, 55, 183, 1993.

Hutson, J. L. and Wagenet, R. J., A pragmatic field-scale approach for modeling pesticides, *J. Environ. Qual.*, 22, 494, 1993.

Hutson, J. L. and Wagenet, R. J., LEACHM: Leaching Estimation and Chemistry Model—a process based model of water and solute movement, transformations, plant uptake and chemical reactions in the unsaturated zone, Research Series 92–3, Department of Soil, Crop and Atmospheric Sciences, Cornell University, Ithaca, NY, 1992.

Jury, W. A., Simulation of solute transport using a transfer function model, *Water Resour. Res.*, 18, 363, 1982.

Jury, W. A., Grover, R., Spencer, W. F., and Farmer, W. J., Behavior assessment model for trace organics in soil. I. Model description, *J. Environ. Qual.*, 12, 558, 1983.

Jury, W. A. and Gruber, J., A stochastic analysis of the influence of soil and climatic variability on the estimate of pesticide groundwater pollution potential, *Water Resour. Res.*, 25, 2465, 1989.

Jury, W. A. and Valentine, R. L., Transport mechanisms and loss pathways for chemicals in soils, in *Vadose Zone Modeling of Organic Pollutants,* Hern, S. C. and Melancon, S. M., Eds., Lewis Publishers, Chelsea, MI, 1986, 2.

Knisel, W. G., Leonard R. A., and Davis, F. M., *GLEAMS User Manual,* Southeast Watershed Research Laboratory, Tifton, GA, 1989.

Lauren, J. G., Wagenet, R. J., Bouma, J., and Wosten, J. H. M., Variability of saturated hydraulic conductivity in a glossaquic hapludalf with macropores, *Soil Sci.,* 145, 20, 1988.

Loague, K., Green, R. E., Giambelluca, T. W., Liang, T. C., and Yost, R. S., Impact of uncertainty in soil, climatic, and chemical information in a pesticide leaching assessment, *J. Contam. Hydrol.*, 5, 171, 1990.

Mackay, D., Paterson, S., Cheung, B., and Neely, W. B., Evaluating the environmental behavior of chemicals with a Level III fugacity model, *Chemosphere,* 14, 335, 1985.

Mullins, J. A., Carsel, R. F., Scarbrough, J. E., and Ivery, A. M., PRZM-2, A model for predicting pesticide fate in the crop root and unsaturated soil zones: Users manual for release 2.0, EPA/600/R-93/046, Environmental Research Laboratory, U.S. Environmental Protection Agency, Athens, GA, 1993.

Nicholls, P. H., Walker, A., and Baker, R. J., Measurement and simulation of the movement and degradation of atrazine and metribuzin in a fallow soil, *Pestic. Sci.* 13, 484, 1982.

Nielsen, D. R., van Genuchten, M. Th., and Biggar, J. W., Water flow and transport processes in the unsaturated zone, *Water Resour. Res.* 22, 895, 1986.

Nofziger, D. L. and Hornsby, A. G., A microcomputer-based management tool for chemical movement in soil, *Appl. Agric. Res.,* 1, 50, 1986.

Parker, J. C. and van Genuchten, M. Th., 1984. Determining transport parameters from laboratory and field tracer experiments, Bull. 84–3, *Va. Agric. Exp. Stn.*, Blacksburg, VA, 1984.

Pennel, K. D., Hornsby, A. G., Jessup, R. E., and Rao, P. S. C., Evaluation of five simulation models for predicting aldicarb and bromide behavior under field conditions, *Water Resour. Res.*, 26, 2679, 1990.

Rao, P. S. C., Davidson, J. M., and Hammond, L. C., Estimation of nonreactive solute front locations in soils, in *Proc. of the Hazardous Waste Res. Symp.* EPA-608/9-76-015, July 1976, US/EPA Cincinnati, OH, 1976.

Rao, P. S. C., Hornsby, A. G., and Jessup, R. E., Indices for ranking the potential for pesticide contamination of groundwater, in *Proc. 44th Annual Meeting of the Soil and Crop Science Society of Florida,* Vol. 44, 1985.

Roth, K., Flühler, H., Jury, W. A., and Parker, J. C., Eds., *Field-Scale Water and Solute Flux in Soils, Proc. of the Centro Stefano Franscini Ascona* Birkhäuser Verlag, Basel, 1990.

Steenhuis, T. S., Pacenka, S., and Porter, K. S., MOUSE: A management model for evaluating groundwater contamination from diffuse surface sources aided by computer graphics, *Appl. Agric. Res.,* 2, 277, 1987.

van der Heijde, P. K. M. and Elnawawy, O. A, *Compilation of groundwater models,* EPA/600/R-93/118, Robert S. Kerr Environmental Research Laboratory, U.S. Environmental Protection Agency, Ada, OK, 1993.

van der Zee, S. E. A. T. M. and Boesten, J. J. T. I., Effects of soil heterogeneity on pesticide leaching to groundwater, *Water Resour. Res.*, 27, 3051, 1991.

van der Zee, S. E. A. T. M. and van Riemsdijk, W. H., Transport of reactive solute in spatially variable soil systems, *Water Resour. Res.*, 23, 2059, 1987.

van Genuchten, M. Th. and Wierenga, P. J., Mass transfer studies in porous sorbing media. I. analytical solutions, *Soil Sci. Soc. Am. J.*, 40, 473, 1976.

van Genuchten, M. Th. and Wagenet, R. J., Two-site/two-region models for pesticide transport and degradation: theoretical development and analytical solutions, *Soil Sci. Soc. Am. J.*, 53, 1303, 1989.

Wagenet, R. J., A review of pesticide leaching models and their application to the field and laboratory data, in *Proc. IXth Symp. Pesticide Chemistry: Degradation and Mobility of Xenobiotics*, Instituto di Chimica Agraria ed Ambientale, Piacenza, Italy, 1993.

Wagenet, R. J., Bouma, J., and Grossman, R. B., Minimum data sets for use of soil survey information in soil interpretive models, in *Spatial Variabilities of Soils and Landforms*, Mausbach, M. J. and Wilding, L. P., Eds., SSSA Special Publication No. 28, Soil Science Society of America, Madison, WI, 1991, chap.10.

Wagenet, R. J., Bouma, J., and Hutson, J. L., Modeling water and chemical fluxes as driving forces in pedogenesis, in *Quantitative Modeling of Soil-Forming Processes*, Soil Science Society of America, Madison, WI, 1994.

Wagenet, R. J. and Hutson, J. L., Predicting the fate of non-volatile pesticides in the unsaturated zone, *J. Environ. Qual.*, 15, 315, 1986.

Wagenet, R. J., Hutson, J. L., and Biggar, J. W., Simulating the fate of a volatile pesticide in unsaturated soil: a case study with DBCP, *J. Environ. Qual.*, 18, 78, 1989.

Wagenet, R. J. and Rao, P. S. C., Modeling Pesticide Fate in Soil, in *Pesticides in the Soil Environment*, Cheng, H. H., Ed., SSSA Book Series, No.2, Soil Science Society of America, Madison, WI, 1990, chap.10.

Wania, F., Mackay, D., Paterson, S., Di Guardo, A., and Mackay, N., Compartmental models in environmental science, in *Proc. IXth Symp. Pesticide Chemistry: Degradation and Mobility of Xenobiotics*, Instituto di Chimica Agraria ed Ambientale, Piacenza, Italy, 1993.

CHAPTER 12

Stochastic Modeling of Water Flow and Solute Transport in the Vadose Zone

T.-C. Jim Yeh

INTRODUCTION

It is difficult to predict any natural process. The complexity of natural systems often prohibits our understanding of governing principles of processes in the systems. For some natural processes such as water flow and solute transport in aquifers and the vadose zone (the geological medium, i.e., soils and rocks above the regional groundwater table), we seem to understand the governing principle of the processes (i.e., many laboratory-scale experiments have manifested the validity of Darcy's law for flow and solute transport through saturated and unsaturated porous media). The principle, however, is often limited to a narrow range of scales. For example, when applying the principle to field-scale problems, we encounter the problem of extrapolating the principle to much larger scale systems which involve spatial and temporal variabilities of the system characteristics at many different scales.

The spatial variability of hydrologic properties of geologic media has been long recognized. It is generally agreed that accurate predictions of water flow and solute transport in the vadose zone require a detailed characterization of the spatial distribution of hydrologic properties. Physical, time, and economic constraints often prohibit such a detailed characterization of the vadose zone. Consequently, for most analyses of water flow and solute transport in the vadose zone one has to rely on a homogeneous assumption or to portray the vadose zone with a limited amount of samples. Such simplified approaches always lead to the question of the accuracy of our prediction and to the assessment of the uncertainty associated with our predictions.

While the accuracy and uncertainty of our predictions have frequently been addressed in the simulation of flow and solute transport through saturated aquifers, they are seldom visited in the analysis of vadose zone hydrological processes due to the complexity of the processes. The purpose of this paper is to explore these important issues and to provide an overview of some stochastic approaches which have been developed recently in an attempt to tackle these issues.

CONCEPTS OF WATER FLOW AND SOLUTE TRANSPORT MODELING IN THE VADOSE ZONE

Scales of Heterogeneity, REV, Dispersion and Measurement Scale

Strictly speaking, water movement in the vadose zone is directly associated with the movement of air. Analysis of water flow and solute transport in the vadose zone should also consider the flow of air. However, in many cases the movement of air can be ignored to simplify the analysis, and the Richards equation is generally used:

$$\frac{\partial}{\partial x_i}\left(K_{ij}(\psi)\frac{\partial(\psi + x_1)}{\partial x_j}\right) = (C(\psi) + \beta S_s)\frac{\partial \psi}{\partial t} \qquad (1)$$

where x_i and x_j are the spatial coordinates ($i, j = 1, 2,$ and 3); x_1 corresponds to the vertical direction; $K_{ij}(\psi)$ is the hydraulic conductivity tensor which is a function of the soil water pressure head, ψ. The pressure head is positive if the porous medium is fully saturated and it is negative when the medium is unsaturated. The moisture capacity term, $C(\psi)$, represents the amount of change in moisture content per unit change in negative pressure when the geological medium is partially saturated. It corresponds to the slope of the water release curve or moisture-pressure head relationship, $\theta(\psi)$, of a given soil at different soil-water pressure head values. When the soil is fully saturated, the change in water storage due to change in the positive pressure is denoted by the specific storage term, S_s, which is related to the compressibility of the porous medium and water. In Equation (1), β is a saturation index of the porous medium for the sake of convenience in mathematics. It is zero when the medium is unsaturated but is equal to one if the medium is fully saturated.

Hydrologic properties, such as $K(\psi)$, $C(\psi)$, and $\theta(\psi)$, of geological media vary with the degree of saturation or moisture content and pressure head. Their dependence on soil-water pressure or moisture content, is generally described by a mathematical formula over a full range of soil-water pressure and moisture content although a tabulated relationship between the properties and pressure is sometimes used. One formula frequently used to depict the unsaturated hydraulic conductivity, and moisture release curves is the exponential model (Gardner, 1958),

$$K(\psi) = K_s \exp(\beta\psi) \qquad (2)$$

$$\theta(\psi) = (\theta_s - \theta_r)\exp(\beta\psi) + \theta_r$$

where K_s is the saturated hydraulic conductivity, β is the pore-size distribution parameter, representing the rate of reduction in conductivity as the soil desaturates, θ_s is the saturated moisture content, and θ_r is the residual moisture content. The exponential model has been very popular owing to its simplicity and convenience in mathematical analysis. However, it fits the observed $K(\psi)$ or $\theta(\psi)$ data over a limited range of pressure head values. Other well-liked models for $K(\psi)$, and $\theta(\psi)$ are those by Mualem (1976) and van Genuchten (1980):

$$K(\psi) = K_s \frac{(1 - (\alpha\psi)^{n-1}[1 + (\alpha\psi)^n]^{-m})^2}{[1 + (\alpha\psi)^n]^{m/2}} \qquad (3)$$

$$\theta(\psi) = (\theta_s - \theta_r)[1 + (\alpha\psi)^n]^{-m} + \theta_r$$

in which α, n, and m are soil parameters and $m = 1 - 1/n$. These models are valid over a broader range of pressure values than the exponential model (van Genuchten and Nielson, 1985). Because of the use of these mathematical models for the functional relationship between conductivity, pressure head, and moisture content, soils can often be categorized by the parameters such as α, β, n, θ_s, θ_r, and K_s. For example, coarse textured soils are reported to have large values of β and K_s, and fine-textured soils to have small values. However, values of these parameters are not necessarily unique for a given geological medium due to hysteretic behavior in $K(\psi)$ and $\theta(\psi)$ relationships. These parameter values are often different during the wetting and drying histories of the medium.

For solute transport in the vadose zone, a convection-dispersion equation is used to account for mixing. The equation can be expressed as:

$$\frac{\partial}{\partial x_i}\left(D_{ij}\frac{\partial c}{\partial x_j}\right) - q_i\frac{\partial c}{\partial x_i} = \theta\frac{\partial c}{\partial t} \qquad (4)$$

where c is the concentration of the solute, D_{ij} is the dispersion coefficient tensor, and q_i is the specific discharge components which can be derived from the analysis of Equation (1). The dispersion coefficient is generally defined as:

$$D_{ij} = (\alpha_L - \alpha_T)\frac{q_i q_j}{q} + \alpha_T q \delta_{ij} + D_m \qquad (5)$$

in which α_L and α_T are the longitudinal and transverse dispersivities, respectively, $q = (q_i q_i)^{1/2}$ and is the magnitude of the specific discharge, δ_{ij} is the Kronecker delta ($\delta_{ij}=1$, if $i=j$ and 0 otherwise), and D_m is the molecular diffusion which is generally small and can be omitted. This linear relationship between dispersivity and specific flux is an extension of the relationship for solute transport in saturated porous media. Very few laboratory experiments have attempted to verify this relationship for unsaturated porous media because of difficulties in the design of the experiment. Nevertheless, this relationship is generally adopted in vadose zone hydrology.

As a result, predicting water flow and solute transport at a given field site requires the solution of Equation (1) with specified initial and boundary conditions to obtain soil-water pressure head and moisture content distributions. Subsequently, the specific discharge is determined from using Darcy's law along with the soil-water pressure head distribution. With the specific discharge information, Equation (4) is then solved for concentration distributions at different times. Solutions to Equations (1) and (4), however, require the specification of physical properties of the site such as hydraulic conductivity, moisture capacity, storage coefficient, and dispersivity values. Many field studies show that these physical properties always exhibit a high degree of spatial variability. As an example, Figure 1 shows the three-dimensional saturated hydraulic

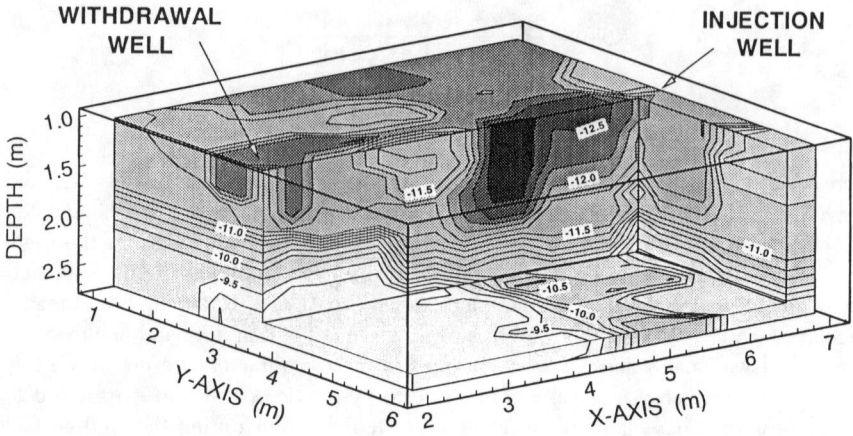

Figure 1. A view of the three-dimensional distribution of lnK_s in a coastal sandy aquifer at Georgetown, South Carolina.

conductivity distribution over a 5m × 5m × 3m portion of a sandy aquifer at the Georgetown site in South Carolina. This three-dimensional view, resulting from an intensive sampling effort involving a total of 330 hydraulic conductivity measurements (30 in the horizontal and 11 in the vertical), reveals layered structures and complex variability within each layer (Mas-Pla et al., 1994). Had not such a detailed site characterization been conducted, such a small portion of the aquifer would have been treated as a homogeneous unit with a uniform hydraulic conductivity, and effects of these complex structures on flow and solute transport, thus, would have been lost. Similarly, Figure 2 illustrates the complex spatial distribution of the parameter values of the natural log of saturated hydraulic conductivity (lnK_s), the natural log of pore-size distribution parameter ($ln\alpha$), n, θ_s, and θ_r of the moisture release curves (50 in the horizontal × 9 in the vertical) of a vertical soil profile at the trench site in Las Cruces, New Mexico. Obviously, to delineate such complex features and to predict their effects would require an intensive sampling effort and a high-resolution numerical simulation.

Multiscale heterogeneity is common sense. Its importance in hydrological analyses of large-scale geological formations has been reemphasized by many researchers (e.g., Dagan, 1986; Gelhar, 1986). For example, the size of the heterogeneity within a core sample is related to variation in pore size and its geometry. Such variability is denoted as laboratory-scale heterogeneity. On the other hand, heterogeneities due to geologic stratification or layering in a formation is classified as field-scale heterogeneity. Regional-scale heterogeneity represents the variation of geologic formations or facies. Variations among sedimentary basins are, then, classified as the global-scale heterogeneity, and so on. Therefore, heterogeneity exists at all scales of observation. This scale-dependent heterogeneity further complicates the analysis of flow and solute transport in aquifers and the vadose zone.

To resolve problems of heterogeneities at the laboratory-scale, hydrologists rely on the concept of representative elementary volume (REV). For example, in a saturated core sample, flow takes place through a complex network of interconnected pores or openings. Obviously, it is practically impossible to describe in any exact mathematical manner the intricate pore structure which controls the flow through porous

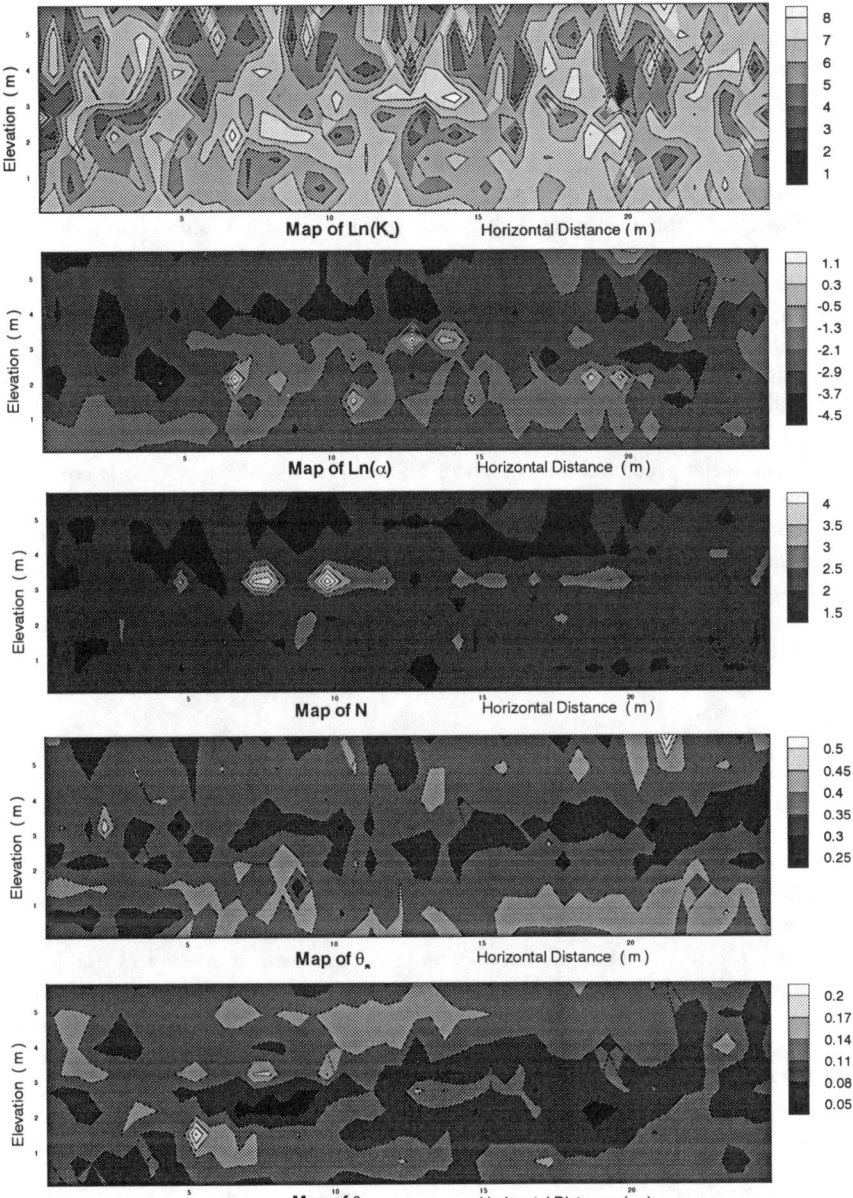

Figure 2. Cross-sectional views of *lnKs*, *ln*α, *n*, θ$_s$, and θ$_r$ at the La Cruces Trench, New Mexico.

media. As a result, one has to abandon the basic equations governing fluid flow (such as the Navier-Stokes equations) at the pore-scale level. Similar to the continuum hypothesis in fluid mechanics, groundwater hydrologists have to overlook the microscopic or pore-scale flow patterns inside individual pores and consider some average flow over a certain volume of porous media. This volume over which the flow is averaged is defined as an REV (Bear, 1979). Using the REV concept, we essentially bypass both the microscopic level at which we consider what happens to each fluid particle and the pore level at which we consider the flow pattern within each pore and between pores. We, then, move to the macroscopic level at which only average phenomena over the REV are considered. As a result, the properties we defined represent averaged values and the medium can be considered as a continuum upon which our differential calculus applies. The same approach has also been adopted by soil physicists and hydrologists to deal with flow through unsaturated porous media. For example, the moisture content of a core sample represents an average moisture content over the volume of the core sample and does not represents the moisture content at each individual pore. In fact, some of the pores in the core sample may be fully saturated, some are unsaturated, but on the average they are unsaturated. Similarly, the REV assumption embedded in the Richards equation ignores discontinuous water films in pores and assumes they are continuous on the average over the entire soil core.

Because the predicted flow behavior based upon the classic Richards equation represents an average over the REV, flow behaviors deviating from the average due to the heterogeneity at scales smaller than the size of the REV are overlooked. Generally, neglecting the effect of small-scale variations has little impact on the assessment of the quantity of water flow in the vadose zone (e.g., average infiltration, groundwater recharge, etc.). Nevertheless, such small-scale variations can have profound impacts on solute transport in porous media, because the small-scale variations represent fast or slow flow channels where solutes are likely to travel and they can cause the spread of solute arrival.

To include effects of small-scale variations in the analysis of solute transport in porous media, the concept of hydrodynamic dispersion is generally used. That is, concentration fluxes from the fast and slow flow channels are included in a transport equation via a dispersive flux, and Fick's law is used to represent this flux (dispersive flux is linearly proportional to the concentration gradient or it is the product of dispersion coefficient and the concentration gradient). Therefore, the classical solute transport Equation (4) includes a convective term ($q \cdot \nabla C$) and a dispersive term ($\nabla D \cdot \nabla C$), representing mass fluxes resulting from the average flow velocity and flow velocities deviating from the average, respectively. This dispersion approach is similar to the molecular diffusion concept in chemistry, and hydrodynamic dispersion in surface water hydrology. However, the mechanism causing the velocity variation in porous media under steady-state uniform flow conditions is mainly attributed to the heterogeneity of porous media, whereas the random collision of molecules causes chemical diffusion and channel roughness and shear effects result in hydrodynamic dispersion in rivers.

Many laboratory experiments of transport in fully saturated and partially saturated porous media have demonstrated that the dispersion coefficient varies with the magnitude of the average water flow velocity and is approximately linearly proportional to the average velocity. The constant of proportionality is, then, defined as dispersivity, representing heterogeneities at scales smaller than the size of REV (i.e., the average grain size in a uniformly packed soil column; Bear, 1972). The dispersivity is considered one of the properties of porous media.

The validity of the REV approach has been tested and substantiated by numerous laboratory experiments (Bear, 1972). Thus, the REV concept, becomes the foundation of many principles of groundwater hydrology. However, when we apply these principles to large-scale geological formations, we somehow forget the basic assumption of the REV approach. That is, the flow behavior predicted by a continuum-based REV model represents an average behavior of the flow over the REV, and does not necessarily depict phenomena measured or observed at scales much smaller than the size of the REV. For example, the moisture content observed by neutron probe located at a local clay lens embedded in sand will be different than that predicted by any mathematical model (e.g., Phillips infiltration solution), which assumes that the soil is homogeneous. The moisture content at this observation point will reflect the response of the clay lens to the stress caused by infiltration and will not necessarily represent the response of other parts of the soil, unless the entire soil consists of the same material.

Similarly, the concentration calculated from Equation (4) depicts the average concentration of a chemical species over the REV only. This implies that the predicted concentration based on Equation (4) may not be equivalent to that measured at a volume much smaller than the size of the REV. Again, to be consistent with the theory, the measurement scale for the concentration must be at least the same as the size of REV. This requirement is likely to be met in a laboratory tracer experiment where a soil column is packed with a relatively uniform sand. In such a soil column, the variation in grain size and its geometry are the source of the heterogeneity. Additionally, the concentration of a tracer collected at the end of the column is the average concentration over numerous pores and meets the requirement of REV. Consequently, success in predicting tracer movement in soil columns has been widely reported.

On the other hand, a field-scale vadose zone consists of heterogeneities of many different scales, such as variations in pore-geometry and size, macropores (i.e., cracks, root cavities, worm holes, etc.), fractures, layers, facies, and sedimentary structures. Treating such a vadose zone as a homogeneous one is tantamount to employing a large-size REV, which must be many times larger than the largest heterogeneity. However, our sampling devices (such as the length of porous cup of a tensiometer, lysimeter, or the volume of soil that a neutron probe affects) are generally much smaller than the size of the REV. Therefore, predictions based upon an assumption of homogeneity will likely deviate from our observations unless the observations are large enough to incorporate many heterogeneities.

APPROACHES FOR THE INVESTIGATION OF SPATIAL VARIABILITY

There are many different approaches for predicting water flow and solute transport in the vadose zone. These include the classical approach which relies on Richards and convection-dispersion equations, and a system approach which treats the vadose zone as a black box (such as mass transfer function model; Jury et al., 1986). In the system approach, the governing principle of the process in the vadose zone is determined by the relationship between the input and output of the vadose zone. The system approach is not new to hydrologists and had been widely utilized to predict discharge from watersheds, storm-runoff relationships, etc. However, this system approach is often criticized for its empiricism and the lack of physical principles. This approach is also limited to predicting the integrated behavior of a system (such as discharge at the downstream outlet of the watershed where streamflow, runoff, and groundwater flow from all parts of the watershed are concentrated, or the average concentration breakthrough of nitrate at the water table beneath an entire farm, etc.). Worst of all, the system approach must rely on the knowledge of input and output histories and model calibrations. This requirement implies that one must carry out tracer experiments at a given site prior to prediction. A calibrated model for one section of the vadose zone cannot necessarily be applied to another, as illustrated by many recent field tracer experiments (e.g., Butters et al., 1989; Butters and Jury, 1989; Roth et al., 1991). As a result, the usefulness of this type of model is questionable, especially for thick vadose zones. Because of these drawbacks and limitations of the system approach, our discussion of approaches for spatial variability will focus on the classical approach.

Deterministic Approach

For reasons of different interests, soil physicists tend to focus on water flow and solute transport in the shallow depth of the vadose zone under irrigated farm lands. Groundwater hydrologists tend to concentrate on the relationship between the flow and solute transport processes within the entire vadose zone and regional aquifers. Regardless of their difference in interests, for decades both soil physicists and hydrologists have relied on deterministic approaches to predict flow and solute transport in many highly heterogeneous vadose zones. The deterministic approach implies that parameter values in the mathematical model [Equations (1) and (4)] are perfectly known and specified at all points in the solution domain. In general, the deterministic approach can be categorized into an equivalent homogeneous and a heterogeneous approach. The equivalent homogeneous approach assumes that a heterogeneous aquifer can be treated as an equivalent homogeneous one whose hydraulic properties are constant in space. Such constant properties are called effective properties. The effective hydraulic properties are often obtained by conducting large-scale infiltration tests and then applying inverse approaches to identify the parameter value. Alternatively, one may average the values from many small-scale tests (i.e., the arithmetic, geometric, or harmonic mean of conductivity values obtained from laboratory column tests). These effective parameters are then used as input to mathematical models to predict water flow or transport of contaminants in the vadose zone. The heterogeneous approach, on the other hand, utilizes all available field data to depict the heterogeneous nature of the vadose zone. This approach is intended to describe the behavior of water flow or transport of contaminants in the vadose zone at higher resolutions than the homogeneous approach.

Regardless of whether treating the vadose zone as a homogeneous or heterogeneous medium, the deterministic approaches suffer from many drawbacks. First, there are no conclusive means to obtain effective parameters of the equivalent homogeneous vadose zone using data from large-scale hydraulic tests. This problem largely stems from difficulties in solving the inverse problem with the nonlinear Richards equation and difficulties in monitoring processes in the deep vadose zone. On the other hand, if one relies on small-scale hydraulic tests, the hydraulic parameter values measured at various parts of the vadose zone are likely very different. Then, how are the data to be averaged to obtain the effective hydraulic properties for the equivalent homogeneous vadose zone? And provided that such an effective hydraulic conductivity can be defined, how can the predicted results be related to our observations? The heterogeneous approach is not immune from problems either: Can we predict flow and transport in heterogeneous vadose zones using only a limited amount of data collected from small-scale tests? What is the magnitude of uncertainty in our predictions if only a limited number of data are available? To answer these questions, a probabilistic approach is necessary, and a stochastic approach seems most appropriate.

Stochastic Approach

The theories based on stochastic approaches to tackle the spatial variability problems have been developed only in the past decade. Many recent field experiments in both saturated zones (e.g., Freyberg, 1986; Sudicky, 1986; Garabedian et al., 1991) and unsaturated zones (Yeh et al., 1986; Greenholtz et al., 1988; McCord et al., 1991, Wierenga et al., 1989; Herkelrath et al., 1991) have already indicated that the stochastic theories are promising at least qualitatively, regardless of many simplifying assumptions used in their development. In the following sections, the stochastic approach will be introduced by first discussing the concept of statistical representation of heterogeneity. Then, we will discuss recently developed stochastic theories of flow and transport in the vadose zone.

Statistical Representation of Heterogeneity

Vadose zones are inherently heterogeneous at various observation scales. Characterizing the heterogeneity at a scale of our interest generally requires information of hydrologic properties at every point in the vadose zone. To delineate such a detailed hydraulic property distribution of the vadose zone at a site of tens of kilometers in width and tens and hundreds of meters in depth obviously requires numerous measurements, considerable time, and great expense, and is virtually impossible based on our current technology. The alternative is to obtain a small number of representative samples of the site and to estimate the variability of properties at the site in a statistical framework. That is, the spatial variation of a property is described by its mean, standard deviation, and probability distribution. This statistical approach has been frequently used in the past. For example, field data show that the distribution of porosity in a given aquifer or saturated moisture content is usually normal (Law, 1944; Bennion and Griffiths, 1969; Anderson and Cassel, 1986) and recent work by Hoeksema and Kitanidis (1985) suggests that the storage coefficient may be log-normally distributed. Saturated hydraulic conductivity data are usually reported to be approximately log-normal (Law, 1944; Bulness, 1946; Warren et al., 1961; Bakr, 1976; Freeze, 1975; Sudicky, 1986; Anderson and Cassel, 1986). There are exceptions, however, as shown by Jensen et al. (1987).

Generally speaking, data sets for unsaturated hydraulic properties of soils are not as abundant as those of saturated materials. White and Sully (1992) and Russo (1993)

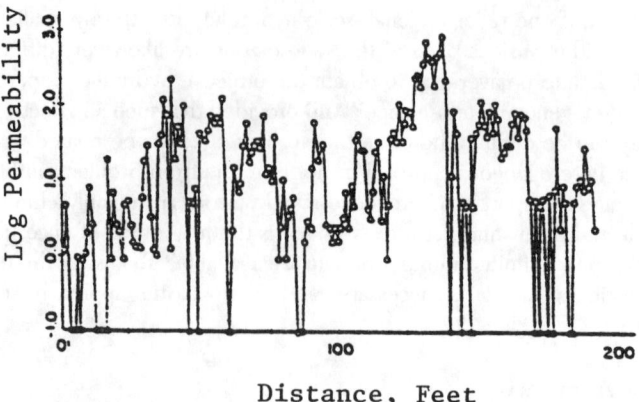

Figure 3. Log saturated conductivity values of Mt. Simon Sandstone, IL (From Bakr, A. A., Ph.D. Dissertation, New Mexico Institute of Mining and Technology, Socorro, NM, 1976.)

reported that the size distribution parameter, β, tends to have log-normal distribution. Russo showed that saturated water content, $θ_s$, initial water content, parameters for Mualem and van Genutchen models, α, and n are log-normally distributed at a field site in Israel. Greminger et al. (1985) reported that α, n, and m for moisture release data sets in a Yolo loam exhibit log-normal distributions. Soil-water pressure head distribution (ranging from 0 cm to 100 cm) at a field site was reported to have a normal distribution after a rainfall or flooding (Saddiq et al., 1985).

Early analyses of flow through heterogeneous aquifers assumed hydraulic conductivity to be a random variable (e.g., Freeze, 1975). Recent studies of hydraulic conductivity data (Bakr, 1976; Byers and Stephens, 1983; Hoeksema and Kitanidis, 1985, etc.) showed that although the hydraulic conductivity values vary significantly in space, the variation is not entirely random but correlated in space. Similarly, Russo (1993) reported that unsaturated soil parameters such as, α, β, n, and K_s are also correlated spatially. Evidence of significant short-range autocorrelation, usually ≤2 m, was found for hydraulic conductivity, soil-water characteristic, texture, and bulk density in the coastal plain of North Carolina (Anderson and Cassel, 1986). Such a correlated nature implies that the parameter values are not purely random (i.e., statistically independent) in space and each of them must be treated as a stochastic process, instead of a single random variable.

To explain the stochastic conceptualization of spatial variability of hydrologic parameters, the saturated hydraulic conductivity data measured along a vertical bore hole in a sandstone in Illinois (Bakr, 1976) are used as an example (see Figure 3). The value of hydraulic conductivity at a point, x_0, along the bore hole can be conceptualized as one of many possible geological materials that may have been deposited at that given point. Thus, the hydraulic conductivity at that point is a random variable, $K(x_0, ω)$. The ω indicates that there are many possible values of K at x_0. Similarly, the hydraulic conductivity values at other locations along the bore hole are random variables. As a result, hydraulic conductivity values of the entire depth of the bore hole may be considered as a collection of many random variables in space. Namely, if conductivity is observed at locations $x_1, x_2, x_3 \ldots x_n$, then $K(x_1, ω)$ is a random variable,

$K(x_2, \omega)$ another random variable, and so on out to $K(x_n, \omega)$. Each has a probability distribution and furthermore the probability distributions may be interrelated. The chance of finding a particular sequence of hydraulic conductivity values along the bore hole, $K(x, \omega_1)$, depends not only on the probability distribution of the hydraulic conductivity at one location but also on those at other locations. This implies that actual hydraulic conductivity values, along the bore hole are one possible sequence of $K(x, \omega_1)$ out of all the possible sequences, $K(x, \omega)$. In the vocabulary of stochastic processes, the probability of finding the sequence is then defined as the joint probability distribution or joint distribution. All of these possible sequences are called an ensemble, and a realization refers to one of the possible sequences.

In order to determine the probability of occurrence of a particular sequence of random variables, a joint distribution of these random variables must be known. This joint distribution is completely defined only if the probabilities associated with all possible sequences of $K(x, \omega)$ values along a bore hole are known. Obviously, the joint distribution is not available in real-life situations because hydraulic conductivity values sampled along a bore hole represent only one realization out of the ensemble of the hydraulic conductivity values along the bore hole. Therefore, one must resort to simplified assumptions, i.e., stationarity and ergodicity.

Stationarity (or strict stationarity) implies that any statistical property (joint distribution, mean, and variance) of a stochastic process remains stationary or constant in space. Ergodicity means that by observing the spatial variation of a single realization of a stochastic process, it is possible to determine the statistical property of the process for all realizations. Since in reality one always deals with a specific geologic formation (i.e., one realization) rather than an ensemble of such formations, one has no choice but to adopt the assumption of ergodicity as a working hypothesis for the stochastic approach. With this assumption in mind, the ensemble parameter, ω, will be dropped from our notations in subsequent discussions for convenience.

Because stationarity is a very stringent assumption, and because in many cases important properties of a stochastic process can be assessed by moments, the mean and covariance function (the first and the second moment, respectively), an assumption of weak or second-order stationarity is often invoked. The first moment (mean) of $K(x)$ is defined as:

$$\mu = E[K] = \int_0^\infty K f(K) \, dK \tag{6}$$

where $E[\ ]$ stands for the expected value, i.e., the average over the entire ensemble, and $f(K)$ is the joint density distribution of K. The covariance function is defined as:

$$c(\xi) = \text{cov}\,[K(x+\xi), K(x)] = E[(K(x+\xi)-\mu)(K(x)-\mu)] \tag{7}$$

Second-order stationarity implies that the mean is a constant and the covariance function depends on the separation distance, ξ, only, which is the distance that separates any two samples in the calculation of the covariance function. This assumption allows us to characterize the stochastic process by using only its mean and covariance function. If the separation distance is set to zero, the covariance function becomes the variance. An autocorrelation function is simply defined as the ratio of covariance function to its variance, i.e.,

$$\rho(\xi) = \frac{c(\xi)}{\sigma^2} \qquad (8)$$

The autocorrelation function represents the persistence of the value of a property in space.

Generally, the autocorrelation function value of hydraulic conductivity data tends to drop rapidly as the separation distance increases. The decline of the correlation can be represented by many different autocorrelation models. The one commonly used is an exponential decay model (Bakr et al., 1978; Gelhar and Axness, 1983; Yeh et al., 1985a,b,c):

$$\rho(\xi) = \exp\left\{-\left[\left(\frac{\xi_1}{\lambda_1}\right)^2 + \left(\frac{\xi_2}{\lambda_2}\right)^2 + \left(\frac{\xi_3}{\lambda_3}\right)^2\right]\right\} \qquad (9)$$

where ρ is the autocorrelation function, ξ is the separation vector, and the integral scales (or correlation scales) in the x, y, and z directions are λ_1, λ_2, and λ_3, respectively. The integral scale is defined as the area under an autocorrelation function if the area is a positive and nonzero value (Lumley and Panofsky, 1964). For the exponential model the integral scale is a separation distance at which the correlation drops to the $\exp(-1)$ level. At this level, the correlation between data points is considered insignificant. That is, data points separated by distances larger than the correlation scale are only weakly associated with each other.

On an intuitive basis, the correlation scale may be interpreted as the average length of clay lenses or sedimentary structures (e.g., cross-bedding, stratification, etc.). Hydraulic property values of samples taken within the clay lens tend to be similar; correlation between samples is near unity. However, sample values are quite different if one sample is taken within the clay lens and the other outside the lens; the resulting correlation will be small. Thus, the autocorrelation function is a statistical measure of spatial structure of hydrogeologic parameters.

Using the stochastic representation, spatial variability of hydrogeologic properties thus can be characterized by the means, and covariance functions of the properties. This approach does not provide information about the values of soil properties at any location in the vadose zone but does provide a way to quantify the spatial variability of the properties. That is, we only know where the mean value of the properties lies; how widely the property values spread around the mean value; and how these values are correlated over space in a statistical sense.

Stochastic Modeling of Flow and Solute Transport in the Vadose Zone

Based on the stochastic conceptualization of field heterogeneity, many stochastic methods for predicting water flow and solute transport in heterogeneous vadose zone have been developed during the past decade. Similar to the deterministic approach, the stochastic modeling of flow and solute transport in the vadose zone can also be classified as an effective parameter (equivalent homogeneous) and a heterogeneous approach.

Effective Parameter Approach

The principle of this approach is identical to the equivalent homogeneity concept in the deterministic approach. As discussed earlier, one of the major problems facing the effective approach is how to extrapolate small-scale measurements to the large-scale effective parameters. Another critical issue is the assessment of the discrepancy between the behavior observed at scales smaller than the REV and that predicted by the effective parameter models. The stochastic perturbation-spectral method to be discussed in the following paragraphs provides a way to address these issues although debate on the validity of the effective parameter concept continues (see Smith and Freeze, 1979; Anderson, 1989).

The perturbation-spectral analysis is an analytical approach which has been used extensively by Gelhar and his co-workers (Gelhar, 1976; Bakr et al., 1978; Gutjahr et al., 1978; Mizell et al., 1982; Gelhar et al., 1979; Gelhar and Axness, 1983; Yeh et al., 1985a,b,c; Mantoglou and Gelhar, 1987a,b,c). To illustrate the approach, let us consider steady-state vertical flow in a heterogeneous but locally isotropic vadose zone (i.e., the hydraulic conductivity at the scale of core samples is isotropic) with infinite lateral extent. The governing equation is:

$$\frac{\partial}{\partial x_i}\left[K(\psi, x) \frac{\partial(\psi + x_1)}{\partial x_i}\right] = 0, \ i=1,2,3 \tag{10}$$

where K is the local-scale hydraulic conductivity which is assumed isotropic and a function of the spatial coordinates, and the water pressure head, ψ. Its dependence of ψ is assumed to be described by the exponential model (2). Equation (10) uses the Einstein summation convention (i.e., repeated indices imply summing over the range of the indices). If $K \neq 0$, Equation (10) can be rewritten as:

$$\frac{\partial^2 \psi}{\partial x_i^2} + \frac{\partial lnK}{\partial x_1} + \frac{\partial lnK}{\partial x_i}\frac{\partial \psi}{\partial x_i} = 0 \tag{11}$$

If the natural log of hydraulic conductivity, lnK, and the hydraulic head, ψ, are assumed to be second-order stationary stochastic processes, Equation (11) becomes a stochastic partial differential equation. All the variables in Equation (11) can, then, be decomposed into means and perturbations, i.e.,

$$\begin{aligned}\psi &= H + h, \ E[\psi] = H, \text{ and } E[h] = 0. \\ \alpha &= A + a, \ E[\alpha] = A, \text{ and } E[a] = 0. \\ lnK_s &= F + f, \ E[lnK_s] = F, \text{ and } E[f] = 0.\end{aligned} \tag{12}$$

where H, A, and F are the means of ψ, α, and lnK_s, respectively, and the perturbations are h, a, and f, correspondingly. By substituting Equation (12) into Equation (2), the unsaturated hydraulic conductivity can be expressed as

$$lnK = F + f + AH + Ah + aH + ah \tag{13}$$

and the mean of the unsaturated hydraulic conductivity is then approximated as

$$lnK_m = E[lnK] \approx F + AH \tag{14}$$

where the expected value of the product of ah is neglected. Substituting Equations (12) and (13) into Equation (11), the approximate mean flow equation is

$$\frac{\partial^2 H}{\partial x_i^2} + \frac{\partial(F + AH)}{\partial x_i} + \frac{\partial(F + AH)}{\partial x_i} + \frac{\partial F}{\partial x_i}\frac{\partial H}{\partial x_i} + \frac{\partial E[ah]}{\partial x_1}$$
$$+ E\left[\frac{\partial(f + Ah + aH + ah)}{\partial x_i}\frac{\partial h}{\partial x_i}\right] \approx 0 \tag{15}$$

Subtracting the mean Equation (15) from Equation (11) and dropping products of perturbation terms gives:

$$\frac{\partial^2 h}{\partial x_i^2} + A(2J_i - \delta_{1i})\frac{\partial h}{\partial x_i} - J_i(J_i - \delta_{1i})a + J_i\left(\frac{\partial f}{\partial x_i} - H\frac{\partial a}{\partial x_i}\right) \approx 0 \tag{16}$$

This is an approximate equation describing the relationship between the perturbations in lnK_s, α, and ψ in the steady-state flow with mean gradients, $J_1 = \partial H/\partial x_1 + 1$, $J_2 = \partial H/\partial x_2$, and $J_3 = \partial H/\partial x_3$.

The above mathematical procedures are equivalent to visualizing the heterogeneous vadose zone as a collection of finite elements, and flow in each element is described by the governing flow equation with a given hydraulic conductivity value. A collection of an infinite number of elements whose hydraulic property parameter values (f and a) are spatially correlated is then equivalent to an ensemble in the stochastic sense. Taking the expected value (ensemble average) of Equation (11) with stochastic parameters is tantamount to homogenizing the heterogeneous vadose zone and to ignoring the details of the flow behavior in each element but examining the average behavior of the flow in all the interconnected elements. The perturbation equation thus depicts the deviation of the flow from the mean flow. However, the equivalence between the spatial homogenization and the ensemble average may depend on the validity of the ergodicity assumption.

In order to solve the perturbation equation, Fourier-Stieljes integral representations for the perturbation terms are used, i.e.,

$$h(x) = \int_{-\infty}^{+\infty} e^{ik \cdot x} dZ_h(k)$$
$$a(x) = \int_{-\infty}^{+\infty} e^{ik \cdot x} dZ_a(k) \tag{17}$$
$$f(x) = \int_{-\infty}^{+\infty} e^{ik \cdot x} dZ_f(k)$$

These representations lead to an expression relating the complex Fourier amplitude of h to those of a, and f fluctuations (i.e., dZ_h, dZ_a, and dZ_f):

$$dZ_h = \frac{[iJ_n K_n(dZ_f - HdZ_a) - (J_n J_n - J_1)dZ_a]}{[k^2 + iA(2J_n k_n - k_1)]} \quad (18)$$

The dZs have the following properties:

$$E[dZ(k)dZ^*(k')] = 0 \quad k \neq k' \quad (19)$$
$$E[dZ(k)dZ^*(k')] = S(k)dk \quad k = k'$$

where S is the spectrum of the stochastic process, and the asterisk denotes the complex conjugate (Lumley and Panofsky, 1964). The relationship between spectra of these stochastic processes can, then, be derived. The spectrum can further be related to the covariance function in the spatial domain using the following relationship:

$$R(\xi) = \int_{-\infty}^{\infty} e^{ik\xi_*} S(k)dk \quad (20)$$

Note that the covariance or cross-covariance function becomes the variance and cross-variance if the separation distance is set to zero, which represents the spatial variability of the processes. In other words, the variance represents a statistical measure of the deviation of the head value at a point from the ensemble average head value.

Based on Equation (18), Yeh et al. (1985a,b) analyzed the effects of spatial variability on steady-state vertical infiltration in the vadose zone under unit mean gradient conditions. Closed forms of head variance expressions were derived for one- and three-dimensional flow fields in which unsaturated hydraulic conductivity is a spatially varying stochastic process. The head variance based on the result by Yeh et al. (1985) can be generalized as

$$\sigma_h^2(H) = J_1^2 \sigma_f^2 \lambda_{f1}^2 \rho_f^2 \, G(\rho_f, A\lambda_{f1}, J_1) + J_1^2 \sigma_a^2 \lambda_{a1}^2 H^2 \rho_a^2 \, G(\rho_a, A\lambda_{a1}, J_1) \quad (21)$$

where ρ_f and ρ_a are the aspect ratios (the ratio of the horizontal to the vertical correlation scale) for lnK_s and α, respectively, λ_{f1} and λ_{a1} are the correlation scale of lnK_s and α in the vertical direction, and the function, G, is the integral described in Equation (5) by Yeh et al. (1985b). This formation assumes that lnK_s and α are uncorrelated.

Equation (21) predicts that the head variance decreases or increases with the mean head value, H. This result implies that the spatial variability of soil-water pressure head in the field grows as the soil is dry and becomes small when the soil is near saturation. It is qualitatively consistent with the results of several field observations of soil-water pressures (e.g., Yeh et al., 1986; Greenholtz et al., 1988; Herkelrath et al., 1991). Equation (21) also shows that as the soil becomes dry (large negative values of H), the influence of the variability of α grows. As a result, the correlation structure of the soil-water pressure varies with H. At small values of H (wet conditions), the correlation scale is dominated by that of lnK_s and at large values of H (dry conditions) the correlation scale is controlled by that of α. These findings are quite different from those in the saturated flow analysis (Bakr et al., 1978). Similarity between the results of the unsaturated flow and those of the saturated flow cases exist. Head values are

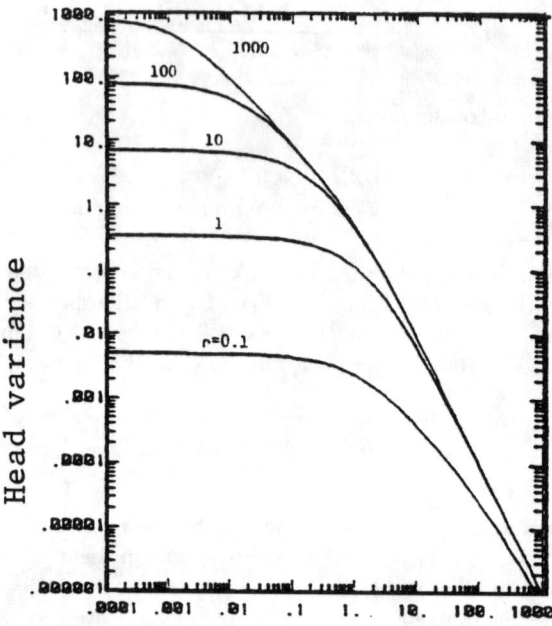

Figure 4. Dimensionless head variance as a function of the dimensionless mean pore-size distribution parameter, $A\lambda_1$, with different aspect ratios, ρ.

anisotropic (i.e., correlated over a longer distance in the direction perpendicular to the mean flow than the direction parallel to the flow), even if the hydraulic conductivity field is statistically isotropic. In addition, the head variation predicted by a three-dimensional model is much smaller than that by a one-dimensional model. This means that a three-dimensional model would be much more appropriate for analyzing flow and solute transport in field problems. It is also manifested from Equation (21) that the correlation scale is an important factor in the calculation of head variance or uncertainty in head prediction.

Figure 4 depicts the behavior of normalized head variance as a function of $A\lambda_1$ with different aspect ratios assuming that lnK_s and α have the same correlation structure. As the aspect ratio increases (the soil formation is close to perfectly stratified), the head variance is large but the head variance decreases as the aspect ratio becomes small (analogous to flow parallel to bedding). For flow into the vadose zone to aquifers, the aspect ratio is likely much greater than one and, as such, the variance in soil-water pressure head during infiltration is likely to be large. Such a large variability in soil-water pressure can induce great lateral movement of water or contaminants in the vadose zone, indicating a complex multidimensional flow phenomenon in the vadose zone.

Yeh et al. (1985) also defined the effective unsaturated hydraulic conductivity for heterogeneous vadose zone. Their study shows that the effective hydraulic conductivity in one-dimensional flow, perpendicular to layering, is

$$K_e(H) = K_g(H) \exp[-\sigma^2_{lnK}(H)/2] \tag{22}$$

where K_g is the geometric mean of $K(H)$, and $\sigma^2_{lnK}(H)$ is the variance of unsaturated hydraulic conductivity which is a function of the mean pressure head, H, and for flow parallel to bedding is:

$$K_e(H) = K_g(H) \exp[\sigma^2_{lnK}(H)/2] \tag{23}$$

and the anisotropy ratio for the effective unsaturated hydraulic conductivity is

$$\frac{K_{he}(H)}{K_{ve}(H)} = \exp[\sigma^2_{lnK}(H)] \tag{24}$$

More specifically,

$$\frac{K_{he}(H)}{K_{ve}(H)} = \exp\left[\frac{\sigma^2_f}{(1 + A\lambda_{f1})} + \frac{\sigma^2_a H^2}{(1 + A\lambda_{f1})}\right] \tag{25}$$

Similar expressions were derived by Mantoglou and Gelhar (1987) for transient flow situations. The anisotropy ratio of the effective unsaturated hydraulic conductivity shown in Equation (25) demonstrates that the hydraulic anisotropy depends not only on the property of porous media but also on the degree of mean saturation (or H) of the media. The horizontal effective conductivity tends to become greater than vertical as the mean soil-water pressure head, H, becomes more negative. Such a moisture-dependent anisotropy concept appears explaining large lateral migration of water and contaminants observed in the vadose zone (e.g., McCord et al., 1991).

Nevertheless, these effective hydraulic conductivity formulae provide a practical means for relating small-scale measurements of unsaturated hydraulic properties to the effective property of a large-scale vadose zone. For instance, at a given contaminated field site where only a limited number of unsaturated hydraulic conductivity measurements is available, one can use the conductivity data set to determine the variances and means of $lnKs$ and β, and their correlation scales. Then, the effective unsaturated hydraulic conductivity can be estimated using Equations 22 and 23 and thus, the mean flow behavior can be determined. Once the mean flow is determined, the head variance can be evaluated, which can be used as a measure of the error in the prediction by the effective parameter model as the result of unmodeled vadose zone heterogeneity. Polmann et al. (1991) demonstrated this application to predicting wetting front movements at a field site. The theoretical head variance can also be used as a model calibration target for defining a detailed hydraulic conductivity distribution in the model area, provided that other sources of error are also considered.

The estimated effective hydraulic conductivity can also be used to predict the mean migration path of contaminant plumes which result from the average flow. Since flow predictions based on the effective hydraulic conductivity ignore the fast and/or slow moving solute particles due to subscale velocity variations which can cause the spread of the plume, the hydrodynamic dispersion concept must be employed. Generally speaking, the REV involved in this case is large. Thus, the dispersion due to the variation in hydraulic conductivity values at scales smaller than this REV is generally called macrodispersion and the dispersivity is called the macrodispersivity. The concept of macrodispersion is an extension of G. I. Taylor's shear flow dispersion concept (see Fisher et al., 1979 for detailed discussion). The concept states that the distribution of a tracer plume, after being displaced for a large distance and experienced "enough" velocity variation, can be described by the classical convection-dispersion equation, and the dispersion coefficient can be related to the velocity variation. Application of this concept to solute transport in porous media, however, requires the knowledge of the relationship between the macrodispersivities and the spatial variability of hydraulic conductivity measurements, since it is difficult to measure water velocity in porous media. To facilitate this requirement, Gelhar and Axness (1983) developed a formula relating the macrodispersivity to the variability of hydraulic conductivity in saturated aquifers. Later, Mantoglou and Gelhar (1985) derived expressions for the macrodispersivities in the unsaturated zone by the perturbation-spectral technique. Based on their study, the longitudinal macrodispersivity is given as:

$$A_{11} = \sigma^2_{lnK}(H) \, \lambda/\gamma^2 \tag{26}$$

where A_{11} is the longitudinal macrodispersivity at large time, $\sigma^2_{lnK}(H)$ is the variance of the unsaturated hydraulic conductivity, λ is the correlation scale, and γ is a flow correlation factor which depends on the direction of the mean flow and the orientation of the heterogeneity. Detailed discussions on the components of the macrodispersivity tensors are given in Gelhar and Axness (1983) and Mantoglou and Gelhar (1985). One important point about the macrodispersivity is that the macrodispersivity is an asymptotic parameter. It exists only when the tracer plume has been displaced for a large distance in the geological formation and has encountered many different heterogeneities in the formation.

From a practitioner's point of view, Equation (26) implies that the macrodispersivity values can be estimated from the knowledge of the variation of local-scale unsaturated hydraulic conductivity values without conducting a large-scale field tracer experiment. Large-scale field tracer experiments may be the most appropriate means for determining solute movement in the vadose zone, but such experiments are often impractical in terms of time and expenses, and are infested with sampling difficulties. On the other hand, local-scale unsaturated hydraulic conductivity values can be obtained from core samples without difficulties. These conductivity data may be limited and insufficient for any detailed numerical simulations, but may be suitable for estimating the statistical parameters required for the estimation of macrodispersivity. For this reason, the macrodispersivity approach is a practical tool for predicting many pollution problems in the vadose zone without resorting to extensive site characterization.

It should be pointed out that the macrodispersivity is an effective parameter (i.e., an ensemble averaged parameter). It represents the subscale velocity variation averaged over many possible vadose zones of similar heterogeneities, or the variation

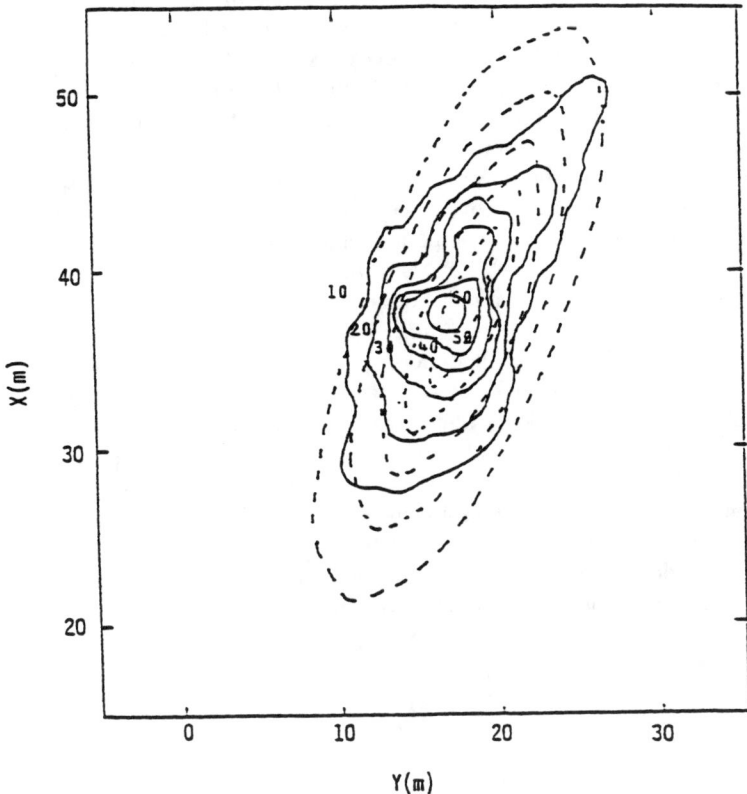

Figure 5. Schematic illustration of an observed depth-averaged concentration profile (solid) and a mean profile (dash), calculated from the macrodispersion equation. (Modified from Sudicky, 1986.)

averaged over many parts of the vadose zone. Thus, the macrodispersivity approach produces the mean concentration distribution which is likely different from the actual concentration distribution observed in a given field site. To demonstrate the difference, Figure 5 shows the observed concentration distribution in the Borden site aquifer and the corresponding mean concentration distributions calculated from the classic convection-dispersion equation with macrodispersivity values. Due to this difference a measure of the deviation of the real concentration distributions (concentration variance) from the mean becomes necessary. For solute transport in aquifers, an expression for the concentration variance was developed by Vomvoris and Gelhar (1990) who found that the concentration variance is proportional to the mean concentration gradient and to the variance and correlation scales of log-hydraulic conductivity and is inversely proportional to local dispersivity values. They concluded that the concentration variance could be large, depending on the magnitude of the mean concentration gradient. However, after the plume has been displaced for a large distance, when the mean concentration gradient is small, the macrodispersion approach will produce satisfactory results. Although no such expression has been developed for solute transport in the vadose zone, it is expected that the variance of concentration in the modeling of solute transport in the vadose zone will behave similarly to that in the saturated condition, if a gravity drainage scenario ($J_1 = 1$, and $J_2 = J_3 = 0$) is considered. The major difference will lie in the fact that the concentration variance will grow with the mean soil-water pressure. This result implies that in the dry soil the mean concentration distribution will be significantly different from that observed in the field site. Thus, the uncertainty in the prediction of solute transport in the vadose zone based on the effective macrodispersivity approach will be large but the approach provides a convenient and practical tool for many purposes.

Another effective approach for solute transport in porous media focuses on the evolution of the second spatial moment of a contaminant plume, instead of the asymptotic macrodispersivity at large time. Extending Taylor's (1922) theorem of diffusion, Dagan (1987) used first-order perturbation analysis of the stochastic steady-state flow equation and derived the groundwater velocity covariance function for spatially correlated random hydraulic conductivity fields. Assuming tracer particles move along with the velocity fields, Dagan obtained mathematical expressions for the second spatial moments of the tracer particles at any given time. The spatial second moment represents the spatial displacement variance of particle position around the mean position, and in turn, the "size" of a tracer plume. If the hydraulic conductivity field is considered to be statistically isotropic, for early time periods (i.e., $t \ll \lambda/v$), the spatial displacement variances are given by:

$$\sigma_{11}^2(t) = \frac{8}{15}\sigma_y^2 v^2 t^2 + 2D_L t \tag{27}$$

and

$$\sigma_{22}^2(t) = \sigma_{33}^2(t) = \frac{1}{15}\sigma_y^2 v^2 t^2 + 2D_T t \tag{28}$$

for three-dimensional flow, where σ_{11}^2, σ_{22}^2, and σ_{33}^2 are the displacement variances in the direction of flow, and lateral directions, respectively. Local-scale dispersion coefficients in the longitudinal and the transverse directions are D_L and D_T, respectively, v is the mean velocity, t is time, and λ is the correlation scale of the hydraulic conductivity field. The parameter σ_y^2 represents the variance of the natural log of hydraulic conductivity in three-dimensional flow and the variance of the natural log of transmissivity in two-dimensional flow. The results of the two-dimensional flow analysis are:

$$\sigma_{11}^2(t) = \frac{3}{8}\sigma_f^2 v^2 t^2 \tag{29}$$

$$\sigma_{22}^2(t) = \frac{1}{8}\sigma_y^2 v^2 t^2 \tag{30}$$

For large time periods (i.e., $t \gg \lambda/v$), the three-dimensional results are:

$$\sigma_{11}^2(t) \approx 2\sigma_y^2 v\lambda t + 2D_L t \tag{31}$$

$$\sigma_{22}^2(t) = \sigma_{33}^2(t) \approx \frac{2}{3}\sigma_y^2 \lambda^2 t \tag{32}$$

and the two-dimensional results,

$$\sigma_{11}^2(t) \approx 2\sigma_y^2 \lambda v t \left[1 - \frac{3\ln(vt/\lambda)}{2vt/\lambda}\right] \tag{33}$$

$$\sigma_{22}^2(t) \approx \sigma_y^2 \lambda^2 [\ln(vt/\lambda) - 0.933] \tag{34}$$

Note that the displacement variance is the spatial variance of a tracer plume, representing the size of the plume at a relative concentration equal to the $\exp(-1)$ level if the plume is assumed to have a normal distribution. Macrodispersion coefficients can be determined by taking the time derivative of the displacement variances. Since the spatial displacement variances [Equations (27) through (30)] for early time periods depend on t^2, the macrodispersion coefficient grows with time or mean travel distance (this is the so-called scale-dependent dispersion). For late time periods, the concentration variances [Equations (31) through (34)] are a function of t only, and the macrodispersion coefficients or marcodispersivities are constant over time. Note that Equations (27) through (34) evaluate the spatial displacement variance (or the second moment of a plume) and do not predict the shape of the concentration plume but its "size". Thus, the classical convection-dispersion equation which assumes the validity of Fick's law is avoided. Similar expressions for the displacement variances in statistically anisotropic media were reported by Dagan (1988), Neuman and Zhang (1990), and Zhang and Neuman (1990). Dagan (1990) and Rajaram and Gelhar (1993a,b) developed similar formulae for correcting the contribution of variability in the mean plume position to the overall spatial displacement variance. This subject is well explained in the textbooks by Fisher et al. (1979) and Csanady (1973).

It should be easy to see that the moments Equations (27) through (34) can be adopted for unsaturated porous media, if the unsaturated flow is steady and under gravity drainage condition (or unit mean gradient conditions) and the correlation structures of $\ln Ks$ and β are assumed identical. Under these circumstances, the variance of unsaturated hydraulic conductivity will be the controlling factor for the macrodispersion. Therefore, the moment equations for solute transport in the unsaturated zone can be described by Equations (27) through (34), with replacing σ_f^2 with $\sigma_{\ln K}^2(H)$. Explicit formulation of the equations is given in Russo (1993).

In general, the results based on Equations (27) through (34) seem to compare favorably with those obtained from a field tracer experiment (Freyberg, 1986; Sudicky, 1986) conducted in a sand aquifer in Canada. A recent field tracer experiment by Garabedian (1987) in a glacial outwash aquifer also indicated that the stochastic results (i.e., Gelhar and Axness, 1983; Dagan, 1987) are robust, regardless of many assumptions used in the development. However, such a statement may be premature at this time since many disputes on the use of Dagan's two-dimensional model for the Borden site continue (see Kemblowski, 1988; White, 1988). Moreover, Naff et al.

(1988) developed a three-dimensional macrodispersion model for perfectly stratified aquifers and attempted to reproduce the field experimental results at the site. However, they found that the three-dimensional model, which is more realistic than the two-dimensional model developed by Dagan (1987), does not reproduce the tracer concentration distribution as well as the two-dimensional model. Naff et al. (1989) attributed the discrepancy to temporal variation in flow patterns which is not considered in all stochastic models. Dagan (1989) defended the two-dimensional approach with the conjecture that there exists an invisible clay lenses of large lateral extent, prohibiting vertical spreading of the tracer plume. This conjecture, however, was never verified. Clearly, the robustness of the spatial moment approach for saturated porous media still remains to be tested.

Overall, the major advantage of the analytical approaches by Dagan (1987) and others is that they provide an explicit expression relating the variability of local hydraulic conductivity measurements to the macrodispersivity, and the spatial variance of the mean concentration distribution. Thus, one can easily estimate the evolution of the size of a solute plume in the aquifer if the statistical parameters characterizing the variability of the small-scale hydraulic conductivity values are known. The major mathematical drawback of the method is that the solution (or the formula) may be valid only for small values of variations in hydraulic conductivity because of the omission of product terms in the analysis. Several studies have shown the results to be valid for the variance of $f = 1$. For cases where a large variance in f is expected [such as in fractured rocks where the variance of lnK has been reported to be about 8.7 (Neuman, 1987)], the accuracy of the first-order approach is questionable. One should also keep in mind that the spatial moment approach predicts the "size" of the mean concentration distribution. The "size" is given in an ensemble average sense and again it will deviate from that of the real plume even if the model is technically flawless for large variance of f or other parameters.

Further, the perturbation approach assumes stationarity. In a basin-scale aquifer, the hydraulic conductivity field is likely to be nonstationary because of changes in deposition environments which may contain heterogeneities of a variety of scales (e.g., Gelhar, 1986; Anderson, 1989). The effect of multiscale heterogeneity on solute transport is well illustrated in Figure 6 which shows the distributions of chloride tracer at different times during a two-well forced gradient experiment in the aquifer at the Georgetown site. The tracer was injected uniformly through the injection well but the plume split into two parts as time progressed; a small portion of the plume moved slowly at the upper portion of the aquifer and the major portion moved rapidly along the bottom of the aquifer. A three-dimensional numerical simulation by Yeh et al. (1994), using a detailed hydraulic conductivity data set (Figure 1) demonstrated that the stratification and the low permeable inclusion in the middle part of the aquifer (see Figure 1) are responsible for the split of the plume. Without including these large-scale features, the effective parameter approach is unable to mimic this type of behavior unless a strong density effect is considered, which is not the case in reality. Thus, the inclusion of multiscale heterogeneity in the macrodispersion approach remains to be explored. Additionally, the traditional second moment analysis based on unit mode probability distribution is inadequate for depicting the split of the plume. It is impossible to distinguish a plume with one peak concentration from a plume with multiple peaks based on the second moment. The need for new statistical measures is clear.

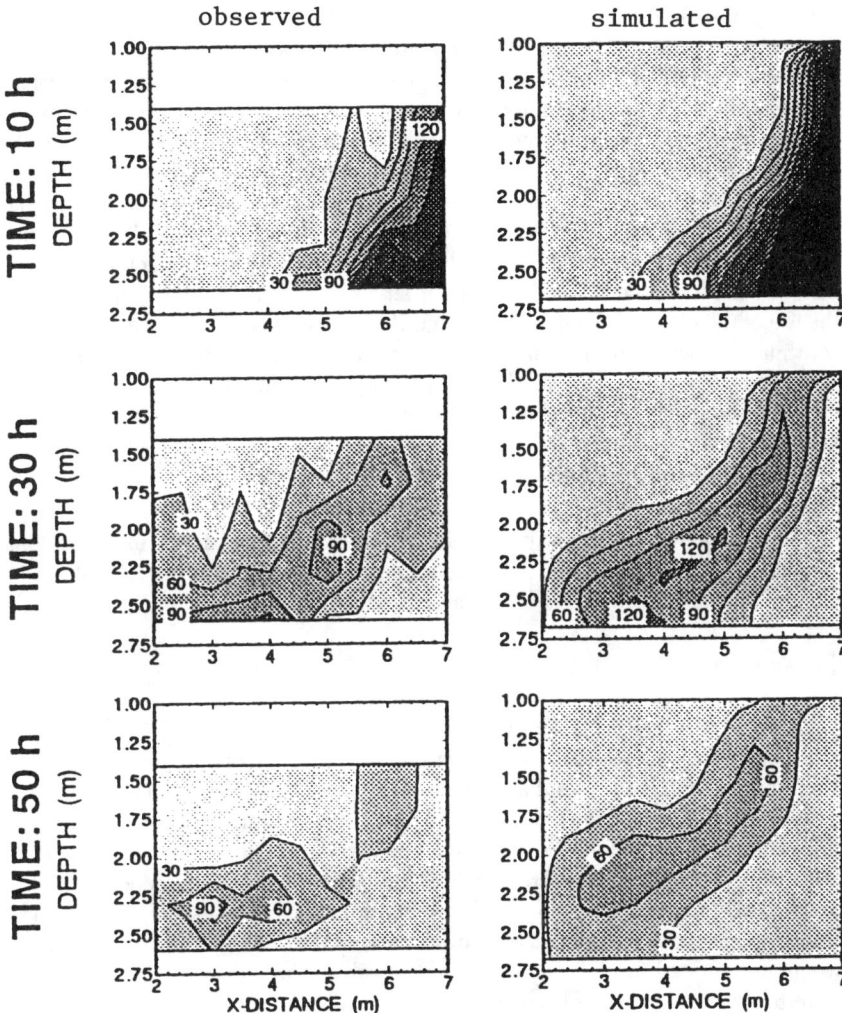

Figure 6. Comparisons between the observed and simulated chloride plume distributions along a vertical profile at various times during the pumping-withdrawal tracer experiment in the sandy aquifer, Georgetown, South Carolina.

Although few data sets are available for testing the macrodispersion concept for the vadose zone, it should be obvious that the macrodispersion concept suffers from the same or more severe difficulties in unsaturated porous media than in saturated media. First of all, when the soil dries, the perturbation grows and thus the first-order approximation will fail. Consequently, the spatial variances based on the first-order analysis may be erroneous. More importantly, as the soil becomes highly heterogeneous, many preferential channels develop and the contaminant plume will likely split into many smaller plumes. The conventional moment analysis based on unit-mode assumption is, again, no longer meaningful. Secondarily, the thickness of the vadose zone is generally much smaller than the distance required for the development of asymptotic macrodispersivity partially due to the large vertical downward gradient in the case of vertical infiltration. The mean plume determined by the macrodispersivity concept is likely very different from the reality. Also, the steady-state flow assumption embedded in the theory does not reflect the reality in most of cases, except for the flow field behind wetting front resulting from constant infiltration. Although constantly irrigated farm lands may be one case where the macrodispersion theory may apply, the depth to the water table is generally too shallow for the ergodic assumption embedded in the theory. The flow field beneath a tailing pond in a thick vadose zone may fit the requirement but the effect of larger scale heterogeneities may prohibit the use of the effective parameter concept. Regardless, it is clear that there are many constraints attached to such an effective approach due to its analytical nature. As a result, this approach is not general even though it advances our knowledge of the effect of heterogeneity on the movement of solutes in the vadose zone.

Heterogeneous Approach

Several methods including geostatistics, Monte Carlo simulations, and conditional simulations can be considered as the heterogeneous approach in the stochastic modeling of flow and solute transport in the vadose zone.

Geostatistics

To illustrate the theory and utility of geostatistics in predicting flow and solute transport in the vadose zone, we will consider the problem of parameter estimation. For example, if one assumes that local-scale dispersion is negligible, and movements of solutes are mainly controlled by the major flow pattern, one may use a two-dimensional, finite difference, flow and solute transport model to simulate solute movements in an unsaturated region. For accuracy or other reasons, one may design a 2000 node finite element mesh for the entire basin. However, only 50 measured unsaturated conductivity data sets scattered around the entire region are available. To assign unsaturated conductivity curves objectively to the remaining nodes, one may have to resort to the use of mathematical tools. Geostatistics is one of the possible mathematical tools.

Geostatistics is not new and has been widely used in the mining industry to estimate ore grades (Journel and Huijbregts, 1978) during the past few decades. Recently, this technique has been applied to groundwater hydrology to address spatial variability problems. In principle, geostatistical concepts are similar to stochastic concepts but some terminology differs. For example, the term "random function" is used in geostatistics to define a collection of correlated random variables. That is, at a point x_1, the function, $F(x_1)$ is a random variable and the random variables at x_1 and $x_2 + \xi$ are not independent but correlated. According to this definition, it is clear that the random function is equivalent to the stochastic process defined earlier. Similarly, a "regionalized variable" is used in geostatistics to define a function $f(x)$ which takes a value at every point x of coordinates (x_1, x_2, x_3) in three-dimensional space (Journel and Huijbregts, 1978). In other words, a regionalized variable is simply a particular realization of a certain random function or stochastic process.

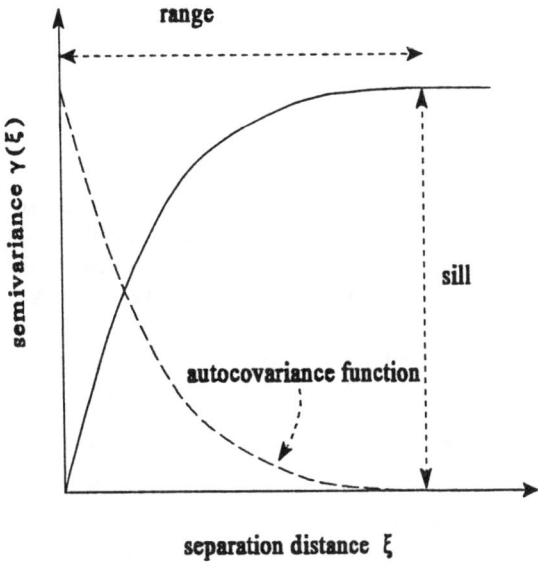

Figure 7. General behavior of a variogram and autocovariance function.

Two important parts of geostatistics are (1) identification of the spatial structure of the variable (variogram estimation, trend estimation, etc.) and (2) interpolation or estimation of the value of a spatially distributed variable from neighboring values taking into account the spatial structure of the variable (kriging, cokriging, etc.; de Marsily, 1986).

Like the autocorrelation function described previously, the variogram is another means of defining the spatial structure of a random field but it is based on the intrinsic hypothesis which is less stringent than the second-order stationarity assumption. Recall that second-order stationarity requires that the stochastic process has a constant mean and that the covariance function of the process depends on the separation distance only. However, the intrinsic hypothesis requires only that the mean of the differences between two data points is constant or depends on the separation distance, ξ, and the variance of the difference depends on the separation distance. For example, if we consider the hydraulic conductivity, $K(x)$, as a random field under the intrinsic hypothesis, it has to satisfy the following two conditions:

$$E[K(x + \xi) - K(x)] = m(\xi) \qquad (35)$$

$$1/2 \, var[K(x + \xi) - K(x)] = \gamma(\xi)$$

where m and γ denote mean and variogram, respectively, and they are functions of ξ, and not of x. Figure 7 shows the general behavior of a variogram. As indicated in the figure, when the variance of the variable is finite, the variogram approaches an asymptotic value equal to this variance, $var[K(x)]$. This is called the sill of the variogram, and the distance at which the variogram reaches its asymptotic value is called the range, λ. The range is analogous to the correlation scale discussed earlier: beyond the range, the regionalized variables $K(x)$ and $K(x+\xi)$ are no longer correlated. In fact, the variogram is a mirror image of the covariance function if the data are a second-order stationary process.

The second part of geostatistics is kriging which is an estimation technique. To illustrate the principle behind kriging, let us suppose that we are studying the saturated hydraulic conductivity parameter distribution in the vadose zone, $T(x)$, and having measurements of its values at a number of locations, $x_1, x_2, \ldots x_n$, we wish to predict its values at the location x_0. Intuitively, we would use the conductivity values measured at sample locations, $x_1, x_2, \ldots x_n$, in predicting the unknown conductivity value $T(x_0)$. In fact, at least conceptually, this is identical to how a contour map of conductivity is manually drawn by hydrogeologists. If we express this concept in a mathematical formula, we would write

$$\overline{T}(x_0) = F(T(x_1), T(x_2), \ldots, T(x_n)) \tag{36}$$

In other words, the unknown $\overline{T}(x_0)$ is a function of the known conductivity values. The question now is how do we choose the function F? To answer this we must first decide on the "criterion" which we will use to measure the accuracy of $\overline{T}(x_0)$ as a predictor of $T(x_0)$. The simplest and most widely used measure of accuracy is the "mean square error" (MSE),

$$MSE = E[\overline{T}(x_0) - T(x_0)]^2 \tag{37}$$

and if we adopt this as our criterion, then the problem is to find that form of F which minimizes MSE. If we do not restrict the form of F in any way, then the solution is the conditional expectation of $T(x_0)$, given $T(x_1), T(x_2), \ldots, T(x_n)$, i.e.,

$$\overline{T}(x_0) = E[T(x_0) | T(x_1), T(x_2), \ldots, T(x_n)] \tag{38}$$

The intuitive interpretation of the conditional expectation, as explained by Priestley (1981), may be given as follows. Suppose we consider all the possible realizations of the process $T(x)$ as shown in Figure 8. Within this ensemble, we would expect that there must be a subset of all the realizations which consist of the values of Ts at our sample locations, $x_1, x_2, \ldots x_n$. The condition imposed on the expectation simply means that we will only consider this subset and discard the realizations which do not agree with the measured T values at $x_1, x_2, \ldots x_n$ when we take the expectation. Since the subset will have different values for T at the unsampled location, x_0, the way to make our "best" prediction of the value at x_0 is to take the expected value or the average of the different values of T in the subset at location x_0. The average value of T at x_0 over this subset is precisely what we mean when we refer to the conditional expectation of $T(x_0)$, given $T(x_1), T(x_2), \ldots, T(x_n)$.

However, the conditional expectation requires the joint distribution of $T(x_1), T(x_2), \ldots, T(x_n)$, which we hardly ever know. The nearest we can approach this problem is to argue that in many cases we would expect such joint distributions to be approximately multivariate normal. If the joint distributions were normal, then the

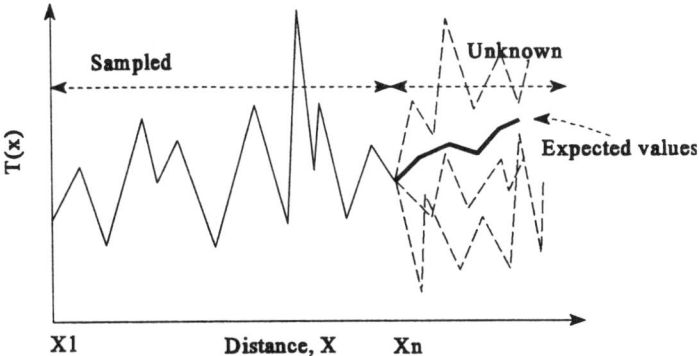

Figure 8. Schematic illustration of the conditional expectation concept.

conditional expectation is a linear function of $T(x_1)$, $T(x_2)$, ..., $T(x_n)$ in which case we can write the predictor $T(x_0)$ more explicitly as

$$\overline{T}(x_0) = a_{01}T(x_1) + a_{02}T(x_2) + \ldots + a_{0n}T(x_n) \quad (39)$$

where $a_{01}, a_{02}, \ldots, a_{0n}$ are constants; a_{01} denotes the weight associated with the measurement at the location x_1 and the estimate at location x_0, and so on up to a_{0n}. The remaining step is to determine the values of the coefficients, $a_{01}, a_{02}, \ldots, a_{0n}$, which minimize the MSE. Since the MSE is a quadratic function of the $T(x)$, the values of these coefficients may be determined from a knowledge only of the autocovariance function (or variogram in the case of intrinsic random fields) of the process.

We can, of course, decide to consider only linear predictors, whether or not the process is multivariate normal, the argument being that if the process is jointly normal then the linear predictor is optimal, whereas if the process is not jointly normal then, in general, we would be unable to evaluate the conditional expectation and so we might as well seek the best linear predictor.

Now, we will briefly examine the theory of kriging which is essentially a conditional expectation as discussed previously. The most general form of kriging is termed "Universal Kriging", in which the values of points may be estimated from irregularly distributed samples in the presence of trends (or nonstationary field). If the data set is stationary, a less involved operation can be used to estimate the values of points. This technique is referred to simply as "Ordinary Kriging" or "Kriging" (see Journel and Huijbregts, 1978). In this introductory review, we will focus on the theory of kriging. A review of universal kriging is available in de Marsily (1986).

To explain the development of kriging, we will again use the estimation of conductivity values at unsampled locations in between sampled locations as an example. To find the saturated hydraulic conductivity estimate $\overline{T}(x_0)$ of the unknown quantity $T(x_0)$, kriging uses a weighted linear sum of all the measured transmissivity values, $T(x_i)$:

$$\overline{T}(x_0) = \sum_{i=1}^{n} a_{0i}T(x_i) \quad (40)$$

where a_{0i} are the kriging weights. In order to limit the choice of the kriging weights, we will impose two conditions. The first condition is that the expected value of the estimate of $T(x_0)$ should be the same as the expected value of $T(x_0)$, i.e.,

$$E[\overline{T}(x_0)] = E[T(x_0)] \tag{41}$$

This condition implies that the kriging estimator has to be unbiased. It then follows that in the constant mean case the sum of the weighting factors has to be unity.

$$\sum_{i=1}^{n} a_{0i} = a_{01} + a_{02} + \ldots + a_{0n} = 1 \tag{42}$$

The second condition is that the error of estimation should be minimal:

$$E[(\overline{T}(x_0) - T(x_0)] = minimum \tag{43}$$

This latter condition, along with Equation (11) leads to a system of equations of the form:

$$\sum_{i=1}^{n} a_{0j}\, \gamma(x_i - x_j) + \varepsilon = \gamma(x_i - x_0) \tag{44}$$

where $\gamma(x_i - x_j)$ represents the variogram corresponding to a separation distance, ξ, equal to the distance between points x_i and x_j. Similarly, $\gamma(x_i - x_0)$ represents the variogram over a distance equal to that between the point x_0 to be estimated and the point x_j. ε is a Lagrange multiplier. Combining the constraint that the kriging weights must sum to one, the system of equations can be solved to obtain optimal a_{0i} values which can then be input to Equation (31) to obtain the estimate of $T(x_0)$. The estimator, $\overline{T}(x_0)$, is a linear combination of the n data values. The n weights a_{0i} are calculated to ensure that the estimator is unbiased and that the estimation variance is minimal. Thus, kriging is a best linear unbiased estimator (BLUE). If the random field is multivariate normal, then kriging is equivalent to the conditional expectation which is an optimal estimator (in the mean square error sense).

The kriged estimate $\overline{T}(x_0)$ is based on samples located a distance away from the estimated point. The values at these distant points are only partially related to the value at the kriged point, the degree of relationship being expressed by the variogram. Therefore, we do not expect our estimate, $\overline{T}(x_0)$, to be exact. This point should be clear if we reexamine the conditional expectation discussed previously. That is, in the case where the T is a multivariate normal random process, the estimate, $\overline{T}(x_0)$, is simply the average of T values at the location x_0 from a subset of $T(x)$ which agree with the sample values at the sample locations. The spread of all the T values at x_0 around the average is then determined by the kriging variance, σ_T^2, which is

$$\sigma_T^2(x_0) = \sum_{j=1}^{n} a_{0j}\gamma(x_i - x_0) + \varepsilon \tag{45}$$

Notice that the variance is not a measure of the deviation of the estimate from the true T value at x_0. However, the smaller the variance, the greater the reliability of the estimate $\overline{T}(x_0)$. Conversely, an estimate with a large associated variance must be utilized with caution. In practices, the kriging variance can be used to determine the optimal location for additional field tests. For example, hypothetical locations can be added to the actual data base to calculate the reduction in the kriging variance. Thus, kriging is considered as a valuable tool in quantifying uncertainty in interpolated data and in assessing the value of additional data during any site characterization.

Kriging is different from other interpolation or extrapolation techniques because it considers the spatial structure (variogram) of the variable. It also provides a measure of the probable error associated with estimates of the unknown values. However, for many cases, kriging may have no advantages over polynomial trend surface and may even perform poorly by comparison (Davis, 1973). Unlike some regression models that fit a surface to the data base, kriging preserves the values at points of measurement. Note that the model and objectives behind polynomial or surface fitting differ from those in kriging. For that reason a comparison between them may not be appropriate. In trend surfaces the objective is to fit the mean value while in kriging reconstructing the actual surface is the goal. In this sense kriging includes a kind of conditioning without using the normality assumption. Overall, kriging is a useful tool that has been widely used in groundwater hydrology (e.g., Delhomme, 1979; de Marsily et al., 1984; Neuman, 1984).

In most field problems, hydrologists often have some hydraulic conductivity measurements at some locations but soil water tension, moisture content, soil texture, and geophysical data at other locations. Since all these different information likely have some relationship with each other, it is logical to estimate the variable of our major concern at unsampled locations, using the information of all the variables. For instance, the unknown hydraulic conductivity value at a given point may be estimated by using both measured conductivity and soil texture data at other locations, instead of using conductivity only. This type of parameter estimation technique based on the kriging concept is called cokriging. Similar to kriging, it requires the knowledge of the variogram for each variable but also the cross-variogram between the variables. In general, cokriging requires a much more intensive computational effort but the difficulty in obtaining direct measurements of unsaturated hydraulic conductivities and the convenience in collecting other related data demand the use of such an estimation technique. Therefore, the role of this technique in estimating hydrologic parameters will likely become more important in the future. Detailed discussion of the theory of cokriging is available in de Marsily (1986), Isaaks and Srivastava (1989), and Vauclin et al. (1983).

Cokriging is not new to soil physicists and groundwater hydrologists. Vauclin et al. (1983) applied cokriging to estimate values of gravimetric mater content at 0.3 bar and the available water at a given range of soil-water pressure in a field soil. Yates and Warrick (1987) used moisture content, bare soil surface temperature, and soil texture data sets to estimate moisture content at unrecorded locations using cokriging. Mulla (1988) used cokriging technique based on measured surface temperatures to interpolate values for water content along a transect. Gutjahr and Wilson (1989), Kitanidis and Vomvoris (1983), and Hoeksema and Kitanidis (1984) developed cokriging techniques that used both transmissivity and hydraulic head data to estimate

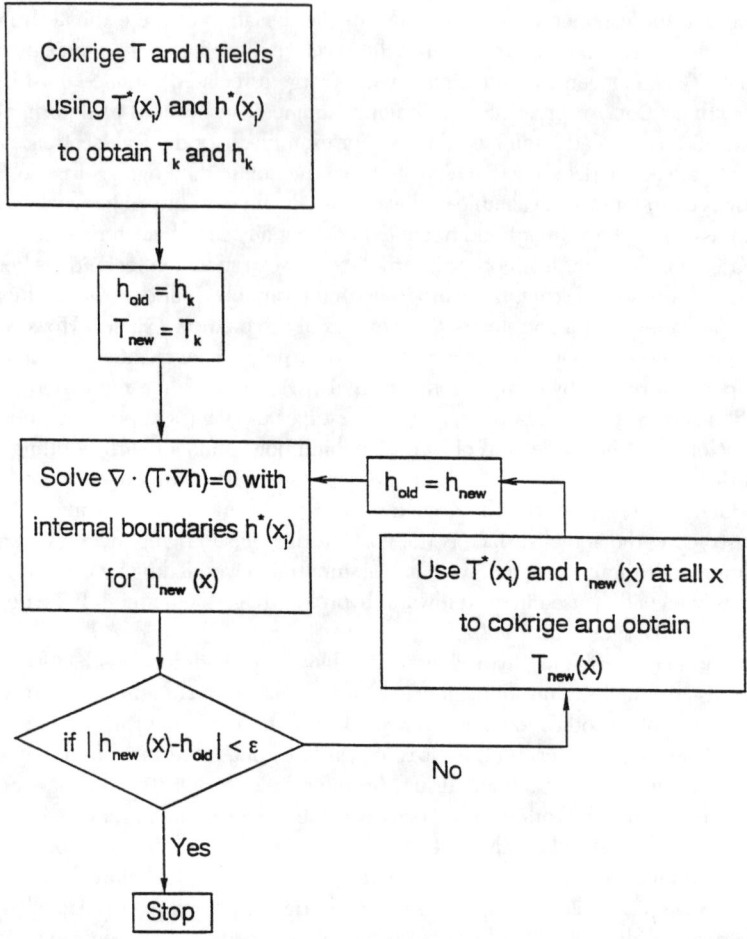

Figure 9. Schematic illustration of an iterative cokriging-like technique for groundwater flow simulation.

transmissivity of a large aquifer. Since the hydraulic head is a response of the aquifer to the transmissivity, using the response and the cokriging technique to estimate transmissivity is sometimes referred as geostatistical inverse approach. The geostatistical inverse approach, however, suffers from the limitations of the linear estimator and the approximate nature of the theoretical cross-variogram between the head and transmissivity. Moreover, the geostatistical inverse approach does not comply with the continuity principle. As a result, Harter and Yeh (1993) and Yeh et al. (1995) showed that such a geostatistical inverse approach may lead to erroneous estimates of groundwater velocity fields.

To overcome this problem, Yeh et al. (1995) developed an iterative cokriging-like approach to estimate transmissivity based on the observations of transmissivities and hydraulic heads in the saturated aquifer. The flow chart in Figure 9 illustrates the principle of the technique. It involves using classical cokriging to estimate hydraulic head and hydraulic conductivity distribution based on some measurements of the head and conductivity. Then, the estimated conductivity field is used in a numerical flow model, MMOC2 (Yeh et al., 1993) to solve for a head distribution that satisfies specified boundary conditions. In this numerical solution, observed head values are specified as internal constant head boundaries. Subsequently, the new head distribution from the model is used in conjunction with the measure conductivity values to estimate a new conductivity field, using classical cokriging. This process is repeated until the difference between the successive head fields is smaller than a specified value. Preliminary results of this approach demonstrated that head measurements may provide useful information for predicting solute movement in the aquifer. A similar concept is under development for water flow and solute transport in the vadose zone since soil-water pressure or water content data are always more abundant and easier to measure than the unsaturated hydraulic conductivity data at most field sites.

Monte Carlo Simulation

The most intuitive approach to dealing with spatial variability in a stochastic sense is the Monte Carlo simulation. Although it is classified as a heterogeneous approach in the sense that the hydraulic property values at every point in the aquifer are specified, it is, in principle, equivalent to the effective parameter approach: they both derive the mean and variance of the pressure head or concentration field. However, Monte Carlo simulation provides numerical results and is more flexible than the stochastic effective parameter approach. It can be used to investigate transient phenomena, nonuniform flows, bounded domains, multiscale heterogeneities, etc. The principle of Monte Carlo simulation is straightforward. First, it assumes that the joint probability distributions of hydraulic properties of any given vadose zone can be inferred from a limited amount of data collected from the given site. Since the probability distributions do not give any information about the parameter value at any point in space, many possible realizations of hydraulic parameter values that conform to the assumed distribution must be created. This creation of random hydraulic property fields is generally done by using some special pseudorandom number generator (such as a fast fourier transform method by Gutjahr, 1989; a matrix inversion method by Smith and Freeze, 1979; and a turning band method by Mantoglou and Wilson, 1982). Each realization of the generated parameter values is, then, input to flow and transport equations which are solved by numerical methods. If there are N realizations of input parameters used, there are N realizations of output from solving the governing equations. Analysis of the output of all the realizations for expected value, variance, covariance, and distribution provides a means to assess the uncertainty in the prediction, resulting from spatial variability.

During the past few decades, Monte Carlo simulation has been used by many researchers to investigate effects of heterogeneity on flow and solute transport in groundwater systems. For example, it was applied by Clifton et al. (1985) to analyze the uncertainty in predicting groundwater travel times and flow paths. Smith and Schwartz (1980, 1981a,b) used Monte Carlo simulation to determine uncertainties in solute transport predictions. Ababou et al. (1988) and Tompson and Gelhar (1990) conducted three-dimensional Monte Carlo simulations of flow and solute transport in heterogeneous geologic formations. For flow and solute transport in the vadose zone, Anderson and Sharpiro (1983), Yeh (1989), and Ünlü et al. (1990) conducted simulation of flow in the one-dimensional heterogeneous unsaturated zone.

Monte Carlo simulation is a powerful tool since it is not restricted to stationary processes with small variances as in the spectral-perturbation approach. It has, however, many other problems, especially for flow and transport in the vadose zone. The governing equation for flow in the vadose zone, the Richards equation, is nonlinear and the degree of nonlinearity increases with the degree of heterogeneity. Analytical solutions may be available for few special cases (see Yeh, 1989) but numerical methods are necessary for realistic cases. Because the equation is nonlinear, numerical methods must rely on iterative schemes and many iterations are required for each solution. As a result, the CPU time required for solving the Richards equation is much greater than that for saturated flow equation. In addition, meaningful statistics from the output during the Monte Carlo simulation can only be obtained if a large number of simulations are employed. The large amount of CPU time required for each simulation and the large number of simulations needed virtually make the Monte Carlo simulation of flow and transport in the vadose zone a computationally expensive adventure. The Monte Carlo simulation of flow and transport in the vadose zone is further hindered by the fact that the numerical iterative scheme does not guarantee the convergence of the solution during each simulation. All of these obstacles make Monte Carlo simulation of flow and solute transport in multidimensional vadose zones an extremely difficult and unattractive approach. Nevertheless, Harter and Yeh (1993) recently demonstrated that the CPU time required for solving the Richards equation can be significantly reduced (a hundred-fold reduction in CPU time) by using a first-order, spectral-perturbation solution as an initial guess for the numerical solution. With such a new numerical technique, Monte Carlo simulation of multidimensional flow and solute transport in the vadose finally becomes a potentially useful tool.

Despite advances in numerical techniques, the major drawback of Monte Carlo simulation stems from the fact that no explicit relationship between the statistics of input parameters and output parameters can be easily derived as in the analytical approach. More importantly, the mean contaminant concentration distribution (ensemble average concentration) derived from Monte Carlo simulation is quite different from the concentration in reality especially for a highly heterogeneous vadose zone. Figure 10 shows the simulated concentration plumes in three hypothetical vadose zones with three variances of unsaturated hydraulic conductivity (0.7, 1.5, and 3.2). Also shown are the mean concentration plumes and the concentration variances derived from Monte Carlo simulations. It is clear from this figure that the agreement between the mean plume and the plume in a single realization deteriorated as the heterogeneity of the vadose zone increases. The usefulness of the Monte Carlo simulation for flow and contaminant transport in highly heterogeneous vadose zone, thus, becomes skeptical.

Conditional Simulation

Conditional simulation is a special kind of Monte Carlo simulation technique but unlike the Monte Carlo simulation, it imposes sample values at sample locations. Therefore, in each realization of the generated random hydrologic parameter field, hydrologic parameter values are kept constant and equal to the measured values at

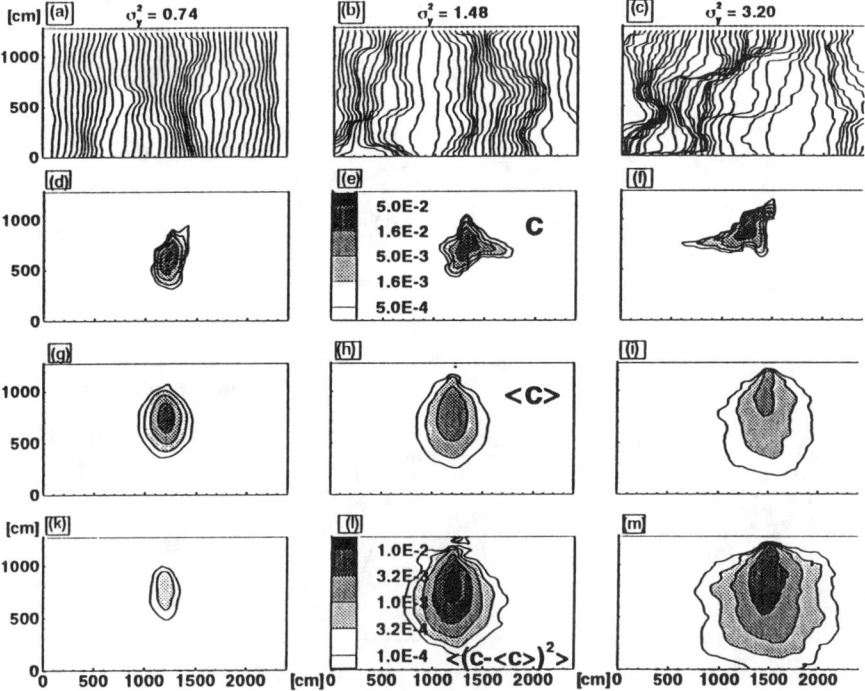

Figure 10. Simulated streamlines (a, b, and c) and plume distributions (e, f, and g) at three vadose zones of different variances of unsaturated hydraulic conductivity ($\sigma y2$ = 0.74, 1.48, and 3.20) and the corresponding mean concentration distributions (g, h, and f) and variance distributions (k, l, and m).

sample locations and the parameter values at other locations are considered random. Thus, there will be no uncertainty in the parameter value at the measurement points, other than measurement errors. By doing so, those possible realizations of the hydraulic parameter value which do not agree with data at sample locations are eliminated. As a result, we expect that the variance of output from the conditional simulation is smaller than that from the Monte Carlo simulation and the conditional simulation can bring us a step closer to the reality.

Techniques for the conditioning are generally based on the conditional expectation and kriging technique discussed earlier. The complete theory of a conditional simulation procedure based on kriging is given by Matheron (1973) and Journel and Huijbregts (1978). A schematic illustration of the conditional simulation concept is shown in Figure 11. Briefly, the procedures of the conditional simulation are: (1) to generate nonconditional simulations, that is, to synthesize different realizations of the random field of hydraulic properties which maintain the actual covariance function that has been inferred from the data, and (2) to condition the simulations obtained in the first step by making the realizations consistent with the measured sample values. The first step is identical to the Monte Carlo simulation. For the second step, one can employ kriging. From the actual sample values, kriging yields a hydraulic conductivity estimate, $\overline{T}(x)$, at the unsampled location which is simply the average of all

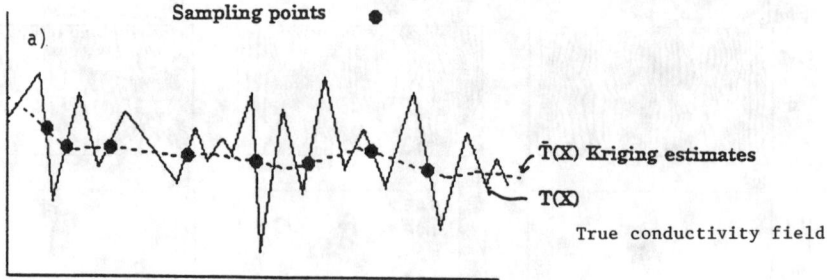
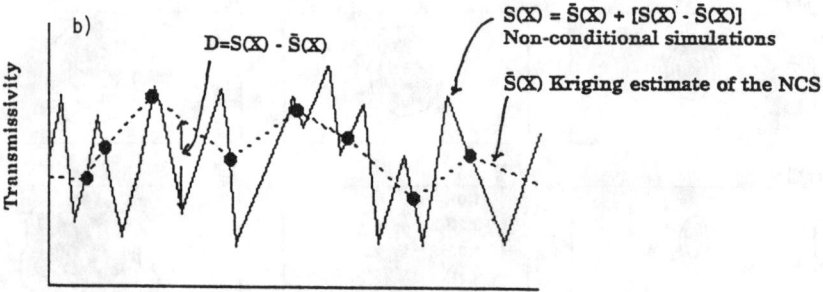
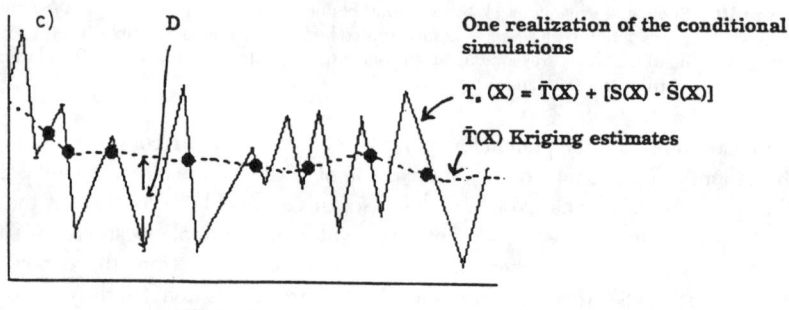

Figure 11. A schematic illustration of conditional simulation.

possible values at point x of the conditioned subset (Figure 11). The true value, $T(x)$, however, equals the estimate plus the error in the estimate, $[T(x) - \overline{T}(x)]$. That is,

$$T(x) = \overline{T}(x) + [T(x) - \overline{T}(x)] \qquad (46)$$

Since the kriging error $[T(x) - \overline{T}(x)]$ is unknown we cannot evaluate this expression exactly. However, we may approximate it by the use of kriging and nonconditional simulation. In other words, in a given nonconditional simulation (one realization), kriging can be performed using the measured values at the actual sample locations as data to derive the kriging estimate, $\overline{S}$ (Figure 11b). The transmissivity value of the given nonconditional simulation $S(x)$ thus can be decomposed as the sum of the kriging estimate $\overline{S}(x)$ and the kriging error, i.e.,

$$S(x) = \overline{S} + [S(x) - \overline{S}(x)] \qquad (47)$$

Since this is derived from the nonconditional simulation, all terms are known. Note that $[S(x) - \overline{S}(x)] = 0$ at the sample locations. By substituting $[S(x) - \overline{S}(x)]$ for $[T(x) - \overline{T}(x)]$ in Equation (46), the $T_s(x)$ for the conditional simulation is thus defined as:

$$T_s(x) = \overline{T}(x) + [S(x) - \overline{S}(x)] \qquad (48)$$

Therefore, $T_s(x)$ is consistent with the measured values at the sample points; $T_s(x)$ and $T(x)$ have the same covariance functions. At unsampled locations, $[S(x) - \overline{S}(x)]$ does not equal zero and is different among realizations (or nonconditional simulation). Thus, conditional simulations create a conditioned subset of the ensemble, which agrees with the measured value at sample locations. The average of many conditional simulations at a given point x is the kriging estimate, and their variance is the kriging variance.

Generally speaking, hydraulic property fields resulting from conditional simulation are (1) smoother than unconditioned fields because it is a conditioned subset of the ensemble, but (2) more variable than kriged fields which essentially represent the conditional expected values (or average).

The conditional simulation incorporates the data value at a sample location and is generally regarded as a more realistic approach than Monte Carlo simulation although it is subject to the same operational difficulties as those in the Monte Carlo simulation. Applications of such conditional simulations to problems in flow and transport in saturated zones are abundant. For example, Delhomme (1979) and Smith and Schwartz (1981a,b) used conditional simulations to investigate the effect of hydraulic conductivity measurements on the reduction of uncertainty in predicting groundwater flow and solute transport. They found that such conditioning does not reduce uncertainty significantly even when measurements are spaced as close as two log hydraulic conductivity correlation lengths.

A more elaborate conditional approach for simulating solute transport was conducted by Wagner and Gorelick (1989). The approach was based on the inverse method developed by Kitanidis and Vomvoris (1983) and Hoeksema and Kitanidis (1985) which involves estimating the average, yet spatially variable, hydraulic conductivity field using both hydraulic conductivity and head measurements. Then, conditional realizations with the same degree of variability as the true hydraulic conductivity fields are generated and used as input to a solute transport model to obtain solute distributions. The advantage of this type of conditional simulation is that the conductivity field is more close to reality due to the additional head measurements.

Graham and McLaughlin (1989) presented an approach to conditional simulations. Instead of computing the mean and variance of the concentration distribution from many simulations with conditioned random parameter fields as input, the mean and the covariance of the concentration distribution are solved directly from approximated moment propagation equations with a numerical method. The Kalman filter is used to update the moments when new measurements of the head, log conductivity, and concentration fields become available. Through two synthetic problems, Graham and McLaughlin (1989) demonstrated that reasonably good estimates of the solute concentration distributions can be obtained by conditioning the ensemble moments on a small number of measurements located in regions of high concentration uncertainty.

Although their methodology is promising, it requires concentration measurements which generally are not available at many proposed landfill sites. Thus, it does not serve as a good predictive tool but it is an attractive method for delineating existing plumes. In the case that the concentration measurement is not available, their approach should yield results similar to those by the classical condition simulation concept using conductivity and head measurements. Furthermore, their moment propagation equations are first-order approximations which imply that the method is also limited to relatively homogeneous aquifers. From the computational efficiency point of view, the method may not be superior to the classical Monte Carlo and conditional simulations for a fully three-dimensional analysis due to cumbersomeness of the Kalman filter. Regardless of these drawbacks, their methodology may provide a valuable tool for developing a sampling strategy to reduce uncertainties in characterizing existing contaminant plumes.

The concept of conditional simulation is appealing but it is difficult to implement this approach for flow and solute transport in the vadose zone, due to nonlinearity of the Richards equation. Subsequently, few applications of the conditional simulation have been used in vadose zone hydrology. Recently, Harter and Yeh (1993) combined a spectral coconditioning simulator for saturated flow (Gutjahr et al., 1992), the spectral solution for unsaturated flow (Yeh et al., 1985), and a numerical model for flow and solute transport in variably saturated porous media, MMOC2 (Yeh et al., 1993), to investigate the effect of measurements of hydraulic conductivity and pressure heads on solute transport in the vadose zone. Figure 12a shows the evolution of a tracer plume at different dimensionless times (4, 8, 16, and 31) in a hypothetical vadose zone with variance of unsaturated hydraulic conductivity equal to 3.2. The effect of 320 saturated conductivity measurements on the simulated mean plume using the conditional approach is illustrated in Figure 12b. The effect of 320 head measurements is shown in Figure 12c. The result in this figure demonstrates that the head measurement can improve the prediction of the plume distribution but the improvement is not as significant as that using the conductivity measurement. The combined measurements of saturated conductivity and pressure head, however, provide the best prediction of the plume distribution at all times as illustrated in Figure 12d.

The results by Harter and Yeh (1993) may be preliminary but demonstrate that conditional simulation appears to be the only way to bring our prediction of flow and solute transport in the vadose zone close to reality and to address uncertainties associated with the prediction.

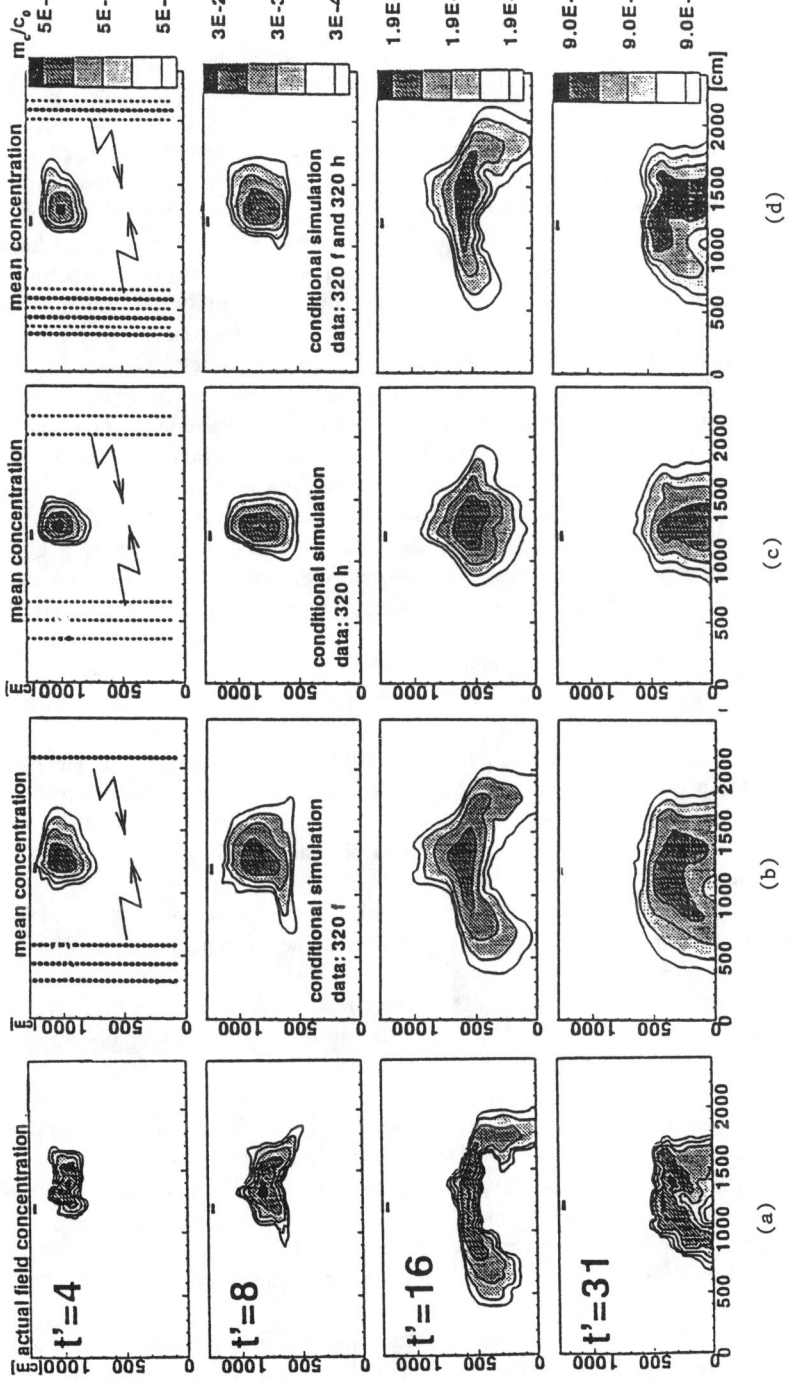

Figure 12. Illustration of effects of conditioning using saturated hydraulic conductivity and soil-water pressure head measurements on the prediction of solute plume distributions.

SUMMARY AND CONCLUSIONS

Predicting solute transport in the vadose zone is a much more difficult task than in fully saturated porous media. This difficulty can be attributed to the fact that flow and transport processes are far more complex in unsaturated porous media than in saturated media. Such complexity further hinders the development of simple and reliable monitoring techniques for flow and solute transport in unsaturated porous media. Without such monitoring techniques, it is virtually impossible to characterize the large-scale vadose zone in detail and to predict flow and solute transport at high resolutions. Although geological information is useful for delineating large structures, means of acquiring more detailed distributions of hydrologic properties are imperative in order to predict highly nonlinear processes in the vadose zone. However, such techniques may not exist in the future unless there are some technological breakthroughs in field testing. Until then, we may have to rely on stochastic approaches to obtain probabilistic results. Many of these approaches are available including geostatistics, the effective parameter approach, Monte Carlo approach, and conditional simulation. Each of these methods has limitations and drawbacks. Nevertheless, these approaches may at least provide us with some estimates of uncertainties in the predictions which may be crucial for regulatory and decision-making purposes. One should bear in mind, however, that the uncertainty estimates afforded by the stochastic models are themselves uncertain (e.g., Smith and Freeze, 1979; de Marsily, 1986).

The approach using effective parameters with the macrodispersion concept may be attractive for relatively simple hydrogeologic systems. Recent tracer experiments in fully saturated aquifers (Sudicky, 1986; Freyberg, 1986; Garabedian, 1987) show that mean travel times and paths of plumes in sandy and glacial outwash aquifers can be adequately predicted by the effective hydraulic conductivity formula developed by Gelhar and Axness (1983). The spatial displacement variances of the observed plumes are also in reasonable agreement with the stochastic results by Dagan (1987) and Naff et al. (1988).

However, few large-scale and well-controlled tracer experiments in the vadose zone have been conducted in the past to rigorously test the effect parameter approach. Results of the field experiments by McCord et al. (1991) seem to support the moisture-dependent anisotropy concept of the effective unsaturated hydraulic conductivity developed by Yeh et al. (1985) but the verification is still very qualitative. Similarly, soil-water pressure data collected by Yeh et al. (1986) also qualitatively support the concept of mean dependence in the variability of the soil-water pressure, derived from the stochastic analysis. An application of the effective unsaturated parameter approach (Polmann et al., 1991) to the experiment at Las Cruces trench site (Wierenga et al., 1989) seems successful but the soil heterogeneity at the site is rather mild. Many tracer experiments conducted in the vadose zone were designed for testing the transfer function model (e.g., Roth et al., 1991; Ellsworth et al., 1991; Butters and Jury, 1989) and data sets collected are not suitable for testing the classical approach.

In spite of the scarcity of field data sets for testing the effective approach, it should be clear that the spatial displacement variances [Equations (18) through (25)] may provide a means to predict the relative size of the plume but the shape of the plume still remains unknown. Furthermore, in the highly heterogeneous vadose zone, a contaminant plume likely evolves into several individual plumes. As a result, the classical moment analysis (spatial variance) clearly becomes inadequate.

Nevertheless, one should look at the bright side of the effective approach and recognize the essence of any stochastic theories: A stochastic predictor will do better than others on the average in many trials under uncertainties. The effective parameter approach may, thus, serve as a practical tool for preliminary analyses (e.g., Gelhar, 1986), especially for cases where little information on hydrologic properties of the vadose zone is available.

After all, we should always ask ourselves: Are we interested in predicting the mean concentration distribution in highly heterogeneous vadose zones in alluvial deposits or fractured rock formations? Can we accept predictions with large degrees of uncertainty, say more than 50%? In other words, under what degree of heterogeneity can the effective parameter approach be used to make predictions that provide us with an acceptable degree of uncertainty? The formula for the concentration variance estimate developed by Vomvoris and Gelhar (1990) may furnish a quick answer to this question. Indeed, this is the major advantage of the stochastic effective parameter approach.

Kriging is a useful mathematical interpolation and extrapolation tool for spatial data. It preserves the data values at sample locations and provides best-linear estimates at the locations where samples are not available. Cokriging appears to be an even more useful tool if additional information, having a direct relationship with the estimated hydrologic property, is readily available. For example, data sets from a detailed geophysical survey (e.g., ground-penetrating radar) may be used in conjunction with sparse hydraulic conductivity measurements to fill in the missing conductivity values at unsampled locations. The iterative cokriging-like approach (Yeh et al., 1995), using both hydraulic head and conductivity measurements still needs to be refined and generalized, preliminary results indicate that the technique appears promising in terms of predicting bulk flow pattern and path. After all, a conditioned mean conductivity field is always a step closer to reality than the effective conductivity, although this type of approach does not address the uncertainty in predicting migration of contaminant plumes.

The Monte Carlo simulation technique may be superior to the stochastic effective parameter approach since it involves less assumptions, but it suffers the same major difficulties as the effective approach. Monte Carlo simulation also produces means and variances of all the possible concentration distributions. Again, we are facing the same question as in the effective parameter approach. Do we need to conduct the Monte Carlo simulation if the aquifer is highly heterogeneous? Unfortunately, one has to conduct many tedious and time-consuming simulations in order to answer this question, if the effective parameter approach is not used.

Conditional simulations, which use available information at sample locations and eliminate many possible realizations which disagree with observations, will enhance our ability to model reality. Although the results of conditional simulations by Delhomme (1979) indicated that using the known transmissivity values at sample locations may not reduce the uncertainties in the prediction of groundwater heads significantly, using both head information and transmissivity values as constraints for the simulation may be useful. In addition, the results of the conditional simulations by Smith and Schwartz (1981) showed that locations of sample data used in the conditional simulation may have important impacts on the reduction of uncertainties. Preliminary results by Harter and Yeh (1993) indicate that coconditioning using information on unsaturated hydraulic conductivity and soil-water pressure improves the prediction of solute plume significantly, considering the moderate effort required to take a large amount of soil-water pressure measurements. Coconditional simulation using unsaturated hydraulic conductivity, soil-water pressure, moisture content, and concentration, if possible, certainly deserves further development.

Regardless of the amount of uncertainty that conditional simulations can reduce, it is rational to include all of the available parameter values in the prediction. It is clear that uncertainty in the simulation will be gradually reduced as more and more data become available. Thus, the conditional simulation seems to be a promising mathematical tool that can bring us a step closer to reality.

Finally, one should recognize that all the stochastic methods are simply tools that we use to overcome our inability to portray the variability of hydrologic properties in the vadose zone. Therefore, it should be clear that no matter how elegant and immaculate a stochastic theory is, the stochastic theory never reduces the uncertainty in our predictions. The only way to tackle the root of the spatial variability problem is to improve our knowledge of the spatial distribution of hydrologic properties in the vadose zone. Thus, to advance our predictive ability, simple, rapid, and accurate monitoring techniques must be developed so that the variability of the vadose zone can be easily characterized. In addition, more efficient and fast numerical algorithms for solving nonlinear equations or the faster and inexpensive computational tools are critical to our ability to predict flow and solute transport in the large-scale vadose zone. Without such technological breakthroughs in these areas, our ability to predict solute movement in the vadose zone will remain at its infancy.

ACKNOWLEDGEMENT

This study is supported by the Subsurface Science Program, Environmental Sciences Division, U.S. Department of Energy, under Grant No. DE-FG02–91-ER61199.

REFERENCES

Ababou, R., McLaughlin, D. B., and Gelhar, L. W., Three-dimensional groundwater flow in random media, Tech. Rep. 318, R. M. Parsons Lab., Dept. of Civil Engineering, Massachusetts Institute of Technology, Cambridge, 1988.

Anderson, M. P., Hydrogeologic facies models to delineate large-scale spatial trends in glacial and glaciofluvial sediments, *Geol. Soc. Am. Bull.*, 101, 501, 1989.

Anderson, S. H. and Cassel, D. K., Statistical and autoregressive analysis of soil physical properties of Portsmouth sandy loam, *Soil Sci. Soc. Am. J.*, 49, 1096, 1986.

Anderson, J. and Shapiro, A. M., Stochastic analysis of one-dimensional steady-state unsaturated flow: A comparison of Monte Carlo and perturbation methods, *Water Resour. Res.*, 19, 121, 1983.

Bakr, A. A., Stochastic Analysis of the Effect of Spatial Variations in Hydraulic Conductivity on Groundwater Flow, Ph.D. dissertation, New Mexico Institute of Mining and Technology, Socorro, NM, 1976.

Bakr, A. A., Gelhar, L. W., Gutjahr, A. L., and MacMillan, J. R., Stochastic analysis of spatial variability in subsurface flows, 1. Comparison of one- and three-dimensional flows, *Water Resour. Res.*, 14, 263, 1978.

Bear, J., *Dynamics of Fluids in Porous Media*, Dover, New York, 1972, p. 76.

Bear, J., *Hydraulics of Groundwater*, McGraw-Hill, New York, 1979, p. 569.

Bennion, D. W. and Griffiths, J. C., A stochastic model for predicting variations in reservoir rock properties, *Trans. AIME*, 237, Part 2, 9–16, 1969.

Black, T. C. and Freyberg, D. L., Stochastic modeling of vertically averaged concentration uncertainty in a perfectly stratified aquifer, *Water Resour. Res.*, 23, 997, 1987.

Bulness, A. C., An application of statistical methods to core analysis data of dolomitic limestone, *Trans. AIME*, 165, 223, 1946.

Butters, G. L., Jury, W. A., and Ernst, F. F., Field scale transport of bromide in an unsaturated soil 1. Experimental methodology and results, *Water Resour. Res.*, 25, 1575, 1989.

Butters, G. L. and Jury, W. A., Field scale transport of bromide in an unsaturated soil, 2. Dispersion modeling, *Water Resour. Res.*, 25, 1583, 1989.

Byers, E. and Stephens, D. B., Statistical and stochastic analysis of hydraulic conductivity and particle size in a fluivial sand, *Soil Sci. Soc. Am. J.*, 47, 1072, 1983.

Clifton, P. M., Sagar, B., and Baca, R. G. Stochastic groundwater traveltime modeling using a Monte Carlo technique, in International Association of Hydrogeologists Memoires: Hydrogeology of Rocks of Low Permeability, Tucson, AZ, 17, No. 1, 319, 1985.

Csanady, G. T., *Turbulent Diffusion in the Environment*, D. Reidel, Norwell, MA, 1973.

Dagan, G., Solute transport in heterogeneous porous formation, *J. Fluid Mech.*, 145, 151, 1984.

Dagan, G., Statistical theory of groundwater flow and transport: pore to laboratory, laboratory to formation and formation to regional scale, *Water Resour. Res.*, 22, 120S, 1986.

Dagan, G., Theory of solute transport by groundwater, *Ann. Rev. Fluid Mech.*, 19, 183, 1987.

Dagan, G., Time-dependent macrodispersion for solute transport in anisotropic heterogeneous aquifers, *Water Resour. Res.*, 24, 1491, 1988.

Dagan, G., Comment on a note on the natural gradient test at the borden site by R. L. Naff, T.-C. Jim Yeh and M. W. Kemblowski, *Water Resour. Res.*, 25, 2521, 1989.

Dagan, G., Transport in heterogeneous porous formations: spatial moments, ergodicity and effective dispersion, *Water Resour. Res.*, 26(6), 1281, 1990.

Davis, J. C., *Statistics and Data Analysis in Geology*, John Wiley & Sons, New York, 1973, p. 550.

Delhomme, J. P., Spatial variability and uncertainty in groundwater flow parameters: a geostatistical approach, *Water Resour. Res.*, 15, 269, 1979.

de Marsily, G., *Quantitative Hydrogeology: Groundwater Hydrogeology for Engineers*, Academic Press, Inc., Orlando, FL, 1986, p. 440.

de Marsily, G., Lauedan, G., Boucher, M., and Fasanino, G., Interpretation of interference tests in a well field using geostatistical techniques to fit the permeability distribution in a reservoir model, in *Geostatistics for Natural Resources Characterization*, Proc. NATO-ASI, Berly et al., Eds. Part 2, Reidel Pub. Co., Dordrecht, The Netherlands, 1984.

Ellsworth, T. R., Jury, W. A., Ernst, F. F., and Shouse, P. J., A three-dimensional field study of solute transport through unsaturated, layered, porous media, 1. Methodology, mass recovery, and mean transport, *Water Resour. Res.*, 27, 951, 1991.

Fisher, H. B., List, E. J., Koh, R. C. Y., Imberger, J., and Brooks, N. H., *Mixing in Island and Coastal Waters*, Academic Press, San Diego, 1979, p. 483.

Freeze, R. A., A stochastic-conceptual analysis of one-dimensional groundwater flow in non-uniform homogeneous media, *Water Resour. Res.*, 9, 725, 1975.

Freyberg, D. L., A natural gradient experiment of solute transport in a sand aquifer 2. Spatial moments and the advection and dispersion of nonreactive tracers, *Water Resour. Res.*, 22, 2031, 1986.

Garabedian, S. P., Large-Scale Dispersive Transport in Aquifers: Field Experiments and Tractive Transport Theory, Ph.D. thesis, Dept. of Civil Engineering, MIT, Cambridge, MA, 1987.

Garabedian, S. P., LeBlanc, D. R., Gelhar, L. W., and Celia, M. A., Large-scale natural gradient tracer test in sand and gravel, Cape Cod, Massachusetts, 2. Analysis of spatial moments for a nonreactive tracer, *Water Resour. Res.*, 27, 911, 1991.

Gardner, W. R., Some steady state solutions of unsaturated moisture flow equations with applications to evaporation from a water table, *Soil Sci.*, 85, 228, 1958.

Gelhar, L. W., Effects of hydraulic conductivity variations on groundwater flows, in Proceedings of the second international IAHR symposium on stochastic hydraulics, *Int. Ass. of Hydraul. Res.*, Lund, Sweden, 1976.

Gelhar, L. W., Stochastic subsurface hydrology from theory to applications, *Water Resour. Res.*, 22, 135S, 1986.

Gelhar, L. W. and Axness, C. L., Three-dimensional stochastic analysis of macrodispersion in aquifers, *Water Resour., Res.*, 19, 161, 1983.

Gelhar, L. W., Gutjahr, A. L., and Naff, R. L., Stochastic analysis of macrodispersion in a stratified aquifer, *Water Resour. Res.*, 15, 1387, 1979.

Graham, W. D. and McLaughlin, D., Stochastic analysis of nonstationary subsurface solute transport, 2. Conditional moments, *Water Resour. Res.*, 25, 2331, 1989.

Greenholtz, D. E., Yeh, T.-C. J., Nash, M. S. B., and Wierenga, P. J., Geostatistical analysis of soil hydrologic properties in a field plot, *J. Contam. Hydr.*, 3, 227, 1988.

Greminger, P. J., Sud, Y. K., and Nielsen, D. R., Spatial variability of field-measured soil-water characteristics, *Soil Sci. Soc. Am. J.*, 49, 1075, 1985.

Gutjahr, A. L., Fast fourier transforms for random field generation, Project Report, 1989, p. 106.

Gutjahr, A. L. and Wilson, J. L., Co-kriging for stochastic models, *Transport in Porous Media*, 4(6), 585, 1989.

Gutjahr, A. L., Bai Q., and Hatch, S., Conditional simulation applied to contaminant flow modeling, Tech. Completion Ret., Dept. of Math., New Mexico Tech, Socorro, New Mexico, 1992.

Gutjahr, A. L., Gelhar, L. W., Bakr, A. A., and MacMillan, J. R., Stochastic analysis of spatial variability in subsurface flows, 2: Evaluation and applications, *Water Resour. Res.*, 14, 953, 1978.

Harter, Th. and Yeh, T.-C. J., An efficient method for simulating steady unsaturated flow in random porous media: using an analytical perturbation solution as initial guess to a numerical model, *Water Resour. Res.*, 29, 4139, 1993.

Herkelrath, W. N., Hamburg, S. P., and Murphy, F., Automatic, real-time monitoring of soil moisture in a remote field area with time-domain reflectometry, *Water Resour. Res.*, 27, 857, 1991.

Hoeksema, R. J. and Kitanidis, P. K., An application of the geostatistical approach to the inverse problem in two-dimensional groundwater modeling, *Water Resour. Res.*, 20, 1003, 1984.

Hoeksema, R. J. and Kitanidis, P. K., Analysis of the spatial structure of properties of selected aquifers, *Water Resour. Res.*, 21, 563, 1985.

Isaaks, E. H. and Srivastava, R. M., *An Introduction to Applied Geostatistics*, Oxford University Press, 1989.

Jensen, J. L., Hinkley, D. V., and Lake, L. W., A statistical study of reservoir permeability: distribution, correlation and averages, Soc. Petr. Eng. Formation Evaluation, 1987, pp. 461–468.

Journel, A. G. and Huijbregts, C. H. J., *Mining Geostatistics*, Academic Press, London, 1978, p. 600.

Jury, W. A., Sposito, G., and White, R. E., A transfer function model of solute transport through soil, 1. Fundamental concepts, *Water Resour. Res.*, 22, 243, 1986.

Jury, W. A., Gardner, W. R., and Gardner, W. H., *Soil Physics*, John Wiley, New York, 1991.

Kemblowski, M. W., Comment on a natural gradient experiment on solute transport in a sandy aquifer: spatial variability of hydraulic conductivity and its role in the dispersion process by E. A. Sudicky, *Water Resour. Res.*, 2, 315, 1988.

Kitanidis, P. K. and Vomvoris, E. G., A geostatistical approach to the inverse problem in groundwater modeling (steady-state) and one-dimensional simulations, *Water Resour. Res.*, 19, 677, 1983.

Law, J., A statistical approach to the interstitial heterogeneity of sand reservoirs, *Trans AIME*, 155, 202, 1944.

Lumley, J. L. and Panofsky, H. A., *The Structure of Atmospheric Turbulence*, John Wiley, New York, 1964, p. 239.

Mantoglou, A. and Gelhar, L. W., Large-scale models of transient unsaturated flow and transport systems, *Water Resour. Res.*, 23, 37, 1985.

Mantoglou, A. and Gelhar, L. W., Stochastic modeling of large-scale transient unsaturated flow systems, *Water Resour. Res.*, 23, 37, 1987a.

Mantoglou, A. and Gelhar, L. W., Capillary tension head variance, mean soil moisture content, and effective specific moisture capacity of transient unsaturate flow in stratified soils, *Water Resour. Res.*, 23, 47, 1987b.

Mantoglou, A. and Gelhar, L. W., Effective hydraulic conductivities of transient unsaturated flow in stratified soils, *Water Resour. Res.*, 23, 57, 1987c.

Mantoglou, A. and Gelhar, L. W., Large scale models of transient unsaturated flow and transport, Dept. of Civil Engineering, R. M. Parsons Laboratory Technical Report 299, Massachusetts Institute of Technology, Cambridge, MA, 1985.

Mantoglou, A. and Wilson, J. L., The turning bands method for simulation of random fields using line generation by a spectral method, *Water Resour. Res.*, 18, 1379, 1982.

Mas-Pla, J., Yeh, T.-C. J., Williams, T. M., and McCarthy, J. F., Field tracer tests on the mobility of natural organic matter and chloride in a sandy aquifer: slug tests analyses and aquifer heterogeneity, submitted to *Water Resour. Res.*, Sept. 1994.

Matheron, G. The intrinsic random functions and their applications, *Advan. Appl. Probab.*, 5, 438, 1973.

McCord, J. T., Stephens, D. B., and Wilson, J. L., The importance of hysteresis and state-dependent anisotropy in modeling variably saturated flow, *Water Resour. Res.*, 27(7), 1991.

Mizell, S. A., Gutjahr, A. L., and Gelhar, L. W., Stochastic analysis of spatial variability in two-dimensional steady groundwater flow assuming stationary and non-stationary heads, *Water Resour. Res.*, 19, 1853, 1982.

Mualem, Y., A new model for predicting the hydraulic conductivity of unsaturated porous media, *Water Resour. Res.*, 12, 513, 1976.

Mulla, D. J., Estimating spatial patterns in water content, matric suction, and hydraulic conductivity, *Soil Sci. Soc. Am. J.*, 52, 1547, 1988.

Naff, R. L., Yeh, T.-C. J., and Kemblowski, M. W., A note on the recent natural gradient tracer test at the Borden site, *Water Resour. Res.*, 24, 2099, 1988.

Naff, R. L., Yeh, T.-C. J., and Kemblowski, M. W., Reply to the comment on a note on the recent natural gradient tracer test at the Borden site by G. Dagan, *Water Resour. Res.*, 25, 2523, 1989.

Neuman, S. P., Role of geostatistics in subsurface hydrology, in Verly, G., David, M., Journel, A. G., and Marechal, A., Eds., *Geostatistics for Natural Resources Characterization*, Part 2, 787–816, D. Roede, Publ. Co., Boston, MA, 1984, p. 1072.

Neuman, S. P., Stochastic continuum representation of fractured rock permeability as an alternative to the REV and fracture network concepts, Proceedings of the 28th U.S. Symposium on Rock Mechanics, Tucson, AZ, 1987, pp. 533–559.

Neuman, S. P. and Zhang, Y-K., A quasi-linear theory of non-fickian and fickian subsurface dispersion, 1. Theoretical analysis with application to isotropic media, *Water Resour. Res.*, 26, 887, 1990.

Polmann, D. J., McLaughlin, D., Luis, S., Gelhar, L. W., and Ababou, R., Stochastic modeling of large-scale flow in heterogeneous unsaturated soils, *Water Resour. Res.*, 27, 1447, 1991.

Priestley, M. B., *Spectral Analysis and Time Series*, Academic Press, San Diego, 1981, p. 890.

Rajaram, H. and Gelhar, L. W., Plume scale-dependent dispersion in heterogeneous aquifers, 1. Lagrangian analysis in a stratified aquifer, *Water Resour. Res.*, 29(9), 3249, 1993a.

Rajaram, H. and Gelhar, L. W., {lume scale-dependent dispersion in heterogeneous aquifers, 2. Eulerian analysis and three-dimensional aquifers, *Water Resour. Res.*, 29(9), 3261, 1993b.

Roth, K., Jury, W. A., Flühler H., and Attinger, W., Transport of chloride through an unsaturated field soil, *Water Resour. Res.*, 27, 2533, 1991.

Russo, D., Stochastic modeling of macrodispersion for solute transport in a heterogeneous unsaturated porous formation, *Water Resour. Res.*, 29, 383, 1993.

Saddiq, M. H., Wierenga, P. J., Hendricks, M. H., and Hussain, M. Y., Spatial variability of soil water tension in an irrigated soil, *Soil Sci.* 140, 126, 1985.

Smith, L. and Freeze, R. A., Stochastic analysis of steady state groundwater flow in a bounded domain. 2. Two-dimensional simulations, *Water Resour. Res.*, 15, 1543, 1979.

Smith, L. and Schwartz, F. W., Mass transport, 1. A stochastic analysis of macroscopic dispersion, *Water Resour. Res.*, 16, 303, 1980.

Smith, L. and Schwartz, F. W., Mass transport, 2. Analysis of uncertainty in prediction, *Water Resour. Res.*, 17, 351, 1981a.

Smith, L. and Schwartz, F. W., Mass transport, 3. Role of hydraulic conductivity in prediction, *Water Resour. Res.*, 17, 1463, 1981b.

Sudicky, E. A., A natural gradient experiment on solute transport in a sand aquifer: spatial variability of hydraulic conductivity and its role in the dispersion process, *Water Resour. Res.*, 22, 2069, 1986.

Taylor, G. I., *Diffusion by Continuous Movements*, Proc. London. Math. Soc. Ser., Vol. A20, 1922, pp. 196–211.

Tompson, A. F. W. and Gelhar, L. W., Numerical simulation of solute transport in three-dimensional, randomly heterogeneous porous media, *Water Resour. Res.*, 26, 2541, 1990.

Ünlü, J., Nielsen, D. R., and Biggar, J. W., Stochastic analysis of unsaturated flow: one-dimensional Monte Carlo simulations and comparison with spectral perturbation analysis and field observations, *Water Resour. Res.*, 26, 2207, 1990.

van Genuchten, M. Th., A closed-form equation for predicting the hydraulic conductivity of unsaturated soils, *Soil Sci. Soc. Am. J.*, 44, 892, 1980.

van Genuchten, M. and Nielson, D. R., On describing and predicting the hydraulic properties of unsaturated soils, *Ann. Geophys.*, 3, 615, 1985.

Vauclin, M., Vieira, D. R., Vachaud, G., and Nielsen, D. R., The use of cokriging with limited field soil observations, *Soil Sci. Soc. Am. J.*, 47, 175, 1983.

Vomvoris, E. G. and Gelhar, L. W., Stochastic analysis of the concentration variability in a three-dimensional heterogeneous aquifer, *Water Resour. Res.*, 26, 2591, 1990.

Wagner, B. J. and Gorelick, S. M. Reliable aquifer remediation in the presence of spatially variable hydraulic conductivity: from data to design, *Water Resour. Res.*, 25, 2211, 1989.

Warren, J. E., Skiba, F. F., and Price, H. S., An evaluation of the significance of permeability measurements, *J. Pet. Technol.*, 13, 739, 1961.

White, I., Comment on a natural gradient experiment on solute transport in a sandy aquifer: spatial variability of hydraulic conductivity and its role in the dispersion process, by E. A. Sudicky, *Water Resour. Res.*, 892, 1988.

White, I. and Sully, M. J., On the variability and use of the hydraulic conductivity alpha parameter in stochastic treatments of unsaturated flow, *Water Resour. Res.*, 28, 209, 1992.

Wierenga, P. J., Toorman, A. F., Hudson, D. B., Vinson, J., Nash, M., and Hills, R. G., Soil Physical Properties at the Las Cruces Trench Site, NUREG/CR-5441, 1989.

Yates, S. R. and Warrick, A. W., Estimating soil water content using cokriging, *Soil Sci. Soc. Am. J.*, 51, 23, 1987.

Yeh, T.-C. J., Comment on modeling of scale-dependent dispersion in hydrogelogic systems, by J. F. Pickens and G. E. Grisak, *Water Resour. Res.*, 23, 522, 1987.

Yeh, T.-C. J., A review of the scale problem and applications of stochastic methods to determine groundwater travel time and path, Tech. Rep. HWR 89–010, Dept. of Hydr. and Water Resour., Univ. of Arizona, Tucson, AZ, 1989, p. 199.

Yeh, T.-C. J., Gelhar L. W., and Gutjahr, A. L., Stochastic analysis of unsaturated flow in heterogeneous soils, 1. Statistically isotropic media, *Water Resour. Res.*, 21, 447, 1985a.

Yeh, T.-C. J., Gelhar, L. W., and Gutjahr, A. L., Stochastic analysis of unsaturated flow in heterogeneous soils, 2. Statistically anisotropic media with variable alpha, *Water Resour. Res.*, 21, 457, 1985b.

Yeh, T.-C. J., Gelhar, L. W., and Gutjahr, A. L., Stochastic analysis of unsaturated flow in heterogeneous soils, 3. Observations and applications, *Water Resour. Res.*, 21, 465, 1985c.

Yeh, T.-C. J., Gelhar, L. W., and Wierenga, P. J., Observations of spatial variability of soil-water pressure in a field soil, *Soil Sci.*, 142, 7, 1986.

Yeh, T.-C. J., Mas-Pla, Williams, T. M., and McCarthy, J. F., Field tracer tests on the mobility of natural organic matter and chloride in a sandy aquifer: observation and simulation of three-dimensional chloride plumes, submitted to *Water Resour. Res.*, Sept. 1994.

Yeh, T.-C. J., Srivastava, R., Guzman, A., and Harter, T., A numerical model for water flow and chemical transport in variably saturated porous media, *Ground Water*, 31, 634, 1993.

Yeh, T.-C. J., Gutjahr, A. L., and Jin, M. H., An iterative cokriging-like technique for ground water flow modeling, *Ground Water*, 33(1), Jan.-Feb. 1995.

Zhang, Y-K. and Neuman, S. P. A quasi-linear theory of non-fickian and fickian subsurface dispersion, 2: An application to anisotropic media and the Borden site, *Water Resour. Res.*, 26, 903, 1990.

CHAPTER 13

Modeling Multiphase Contaminant Flow in Groundwater Aquifers

M. Yavuz Corapcioglu, Kiran K. R. Kambham, and Rajasekhar Lingam

INTRODUCTION

Multiphase contaminants such as petroleum products and chlorinated hydrocarbons enter soils and groundwater through leaks at chemical waste disposal sites and accidents. In many countries around the world, groundwater contamination by NAPLs such as gasoline, fuel oil, trichloroethylene (TCE) and tetrachloroethylene (PCE) has become a serious environmental problem due to adverse health effects.

Management of soil and groundwater contamination by multiphase contaminants requires information on the response of the managed system to the implementation of the management decisions. The tool for providing the management with the required input is the model. In recent years, modeling has been a widely accepted and powerful tool to solve quantitative groundwater contamination problems. The most important step in modeling of the subsurface contamination by multiphase pollutants is the construction of a conceptual model. The complexities of the subsurface and detailed level to which the simulation is required determine the extent to which simplifying assumptions can be made. With these considerations in mind, transport and fate of multiphase contaminants in the subsurface can be modeled in five categories: (1) multiphase multicomponent models, (2) multiphase models with interphase partitioning between phases, (3) multiphase models with capillary effects, (4) sharp interface models, and (5) the single cell model.

In each of these categories, we make certain assumptions to develop the conceptual model. Next the conceptual model is translated into a mathematical model expressed in terms of a set of governing differential equations. It is the objective of this paper to present the first and the fourth modeling approaches in detail, discuss governing equations including balance equations for phases and components, flux equations, and constitutive relations defining the behavior of the contaminant and properties of the subsurface. Special emphasis will be given to recent developments. Following the development of theoretical background, a solution to a contamination problem will be discussed. The preferable method of solution is the analytical one. Especially, analytical expressions obtained by solving governing equations of sharp interface approach are quite attractive in routine exposure assessment studies. Furthermore, they can be utilized without a great deal of data in comparison to the types

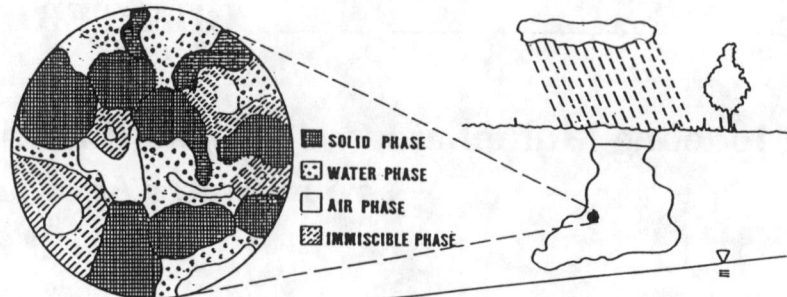

Figure 1. Conceptualization of a soil contaminated by an immiscible fluid. (From Corapcioglu, M. Y. and Baehr, A. L., *Water Resour. Res.*, 23(1), 191, 1987.)

of models which require extensive databases. However, in most cases, this is not possible due to the irregular shape of the boundaries, the heterogeneity of the subsurface, and the uncertainty of various model parameters. Therefore, numerical techniques are being used for field predictions.

MULTIPHASE MULTICOMPONENT MODELS

Mass balance equations for transport processes in reservoirs and in porous media have been developed at various degrees of complexity by a number of researchers. Corapcioglu and Baehr (1987), Abriola and Pinder (1985a,b), Kazemi et al. (1978), Ngheim et al. (1973), Crookston et al. (1979), Coats (1980), Young and Stephenson (1983), and Youngren (1980), to name a few, have developed compositional simulators to a high degree of sophistication, as is necessary to model the multiphase, multicomponent flow. For a review of the subject, the reader is referred to Corapcioglu and Panday (1991a). A compositional simulator describes mass balances for components present in the reservoir fluids rather than the phases and, therefore, the complexity of the system is higher. In the definition of governing equations, our starting point is the three-dimensional macroscopic conservation of mass equations for each component in each phase, i.e., water, gas, oil, and soil solids (Figure 1).

The conservation of mass equations for any component (i) present in the water phase can be expressed as

$$\frac{\partial}{\partial t}(C_w^i \rho_w n S_w + \nabla \cdot (C_w^i \rho_w J_w - D_w^i \nabla \rho_w n S_w C_w^i)$$
$$= (R_{ow}^i - R_{wo}^i) + (R_{gw}^i - R_{wg}^i) + (R_{sw}^i - R_{ws}^i) + R_{source_w}^i - R_{sink_w}^i \quad (1)$$

where C_w^i is the mass fraction of component i ($i = 1, \ldots, NC$, where NC is the total number of components present in the system) in the water phase (w). The superscript denotes a component, while the subscript refers to a phase. ρ_w is the density of the water phase, S_w the degree of water saturation, n the porosity of the medium, J_w the volumetric flux of the water phase, and D_w^i is the hydrodynamic dispersion coefficient of component (i) in the water phase. Each component (i) in the water phase is transported by the convective flux of the water phase and by the diffusive and dispersive flux of the component (i) within the water phase. We assume that the solid grains (s) of the porous medium are nonreactive to other phases (i.e., $C_w^s = 0$) and, hence, in Equation (1) a component term for solid grains is neglected. This implies that grains do not dissolve in either water or oil, or vaporize in the gas phase. However, in the presence of substances like hydrofluoric acids, sandstone particles can react (Hekim et al., 1982). We must also note that in dealing with reservoirs like salt caverns, the reservoir solids might dissolve in water. All the terms on the right-hand side of Equation (1) are either sources (the positive terms) or sinks (the negative terms) of mass in the water phase. R_{ow}^i is the rate of transfer of mass of (i) from oil phase to water phase and R_{wo}^i is the mass transfer rate of component (i) from water to oil phase. Similarly, R_{gw}^i and R_{wg}^i denote mass transfer rates from gas to water, and water to gas phases, respectively. R_{sw}^i and R_{ws}^i are the desorption and adsorption terms which denote mass transfer rates of component (i) between the water phase and the soil solids. The last two terms are source and sink terms due to fluid injection or withdrawal or mass loss due to a degradation process (e.g., biodegradation). We must note that the summation of the equations for each component in the water phase would give the conservation of mass equation for the water phase.

The conservation of mass equation for component (i) ($i = 1, \ldots, NC$) in the gas phase is

$$\frac{\partial}{\partial t}(C_g^i \rho_g n S_g) + \nabla \cdot (C_g^i \rho_g J_g - D_g^i \nabla \rho_g n S_g C_g^i)$$
$$= (R_{og}^i - R_{go}^i) + (R_{wg}^i - R_{gw}^i) + (R_{sg}^i - R_{gs}^i) + R_{source_g}^i - R_{sink_g}^i \tag{2}$$

where C_g^i is the mass fraction of component i in the gas phase, ρ_g is the density of the gas phase, S_g the gas saturation, J_g the volumetric flow of the gas phase, and D_g^i is the hydrodynamic dispersion coefficient of component (i) in the gas phase (g). R_{og}^i, R_{go}^i, R_{sg}^i, and R_{gs}^i are the mass transfer rates of (i) from oil to gas, from gas to oil, from soil solids to gas, and from gas onto soil solids, respectively. $R_{source_g}^i$ and $R_{sink_g}^i$ are source-sink terms for the gas phase, due to injection or withdrawal. Note that $C_g^s = 0$.

The conservation of mass equation for component (i) existing in oil phase is written as

$$\frac{\partial}{\partial t}(C_o^i \rho_o n S_o) + \nabla \cdot (C_o^i \rho_o J_o - D_o^i \nabla \rho_o n S_o C_o^i)$$
$$= (R_{wo}^i - R_{ow}^i) + (R_{go}^i - R_{og}^i) + (R_{so}^i - R_{os}^i) + R_{source_o}^i - R_{sink_o}^i \tag{3}$$

where C_o^i is the mass fraction of mass component (i) in the oil phase, ρ_o is the oil phase density, S_o the oil saturation, J_o the volumetric flux of the oil phase, and R_{os}^i and R_{so}^i are the adsorption and desorption rates from the oil phase onto soil solids and vice-versa, respectively. D_o^i is the hydrodynamic dispersion coefficient of component (i) in the oil phase (o). $R_{source_o}^i$ and $R_{sink_o}^i$ are the source and sink terms for the oil phase. Note that $C_o^s = 0$. Finally, for a rigid porous medium, the mass balance equation of a component (i) in the solid phase (soil solids) is expressed as

$$\frac{\partial}{\partial t}(C_s^i \rho_s \{1-n\}) = (R_{ws}^i - R_{sw}^i) + (R_{gs}^i - R_{sg}^i) + (R_{os}^i - R_{so}^i) \tag{4}$$

where C_s^i is the mass of component (i) per unit mass of soil solid particles (i.e., mass fraction), n is the porosity, and ρ_s is the density of the soil solids. In the absence of any adsorption and reaction $C_s^s = 1$. In writing Equation (4), we neglected the diffusion and dispersion of components within the solid phase. This may not be true in cases like activated carbon adsorption. In addition, adsorbed components might diffuse within the secondary pores of carbon particles. Equations (1) to (4) have been discussed in greater detail in relation to gasoline components in soil water and air by Corapcioglu and Baehr (1987).

The mass balance of Equations (1) to (3) for the fluid phases are constrained by the total volume of void spaces existing within the matrix. Hence

$$S_w + S_g + S_o = 1 \tag{5}$$

Further, fully compositional models have mass fraction constraints which state that the total mass per unit volume of a phase is equal to the sum of the masses of all components comprising that phase per unit volume of the phase.

The volumetric flux of a fluid phase (f) in a rigid porous medium is obtained by Darcy's law as

$$J_f = -\frac{\rho_f g}{\mu_f} k_r k_o \nabla \left(\frac{p_f}{\rho_f g} - z \right) f = o, w, g \tag{6}$$

where g is the gravitational acceleration, μ_f the viscosity of the fluid, k_{r_f} the relative permeability to fluid f, k_o the absolute permeability, p_f the pressure in fluid (f), and z the vertical coordinate positive downwards (i.e., depth). In the literature, based on the wetability concepts, it is generally assumed that the relative permeability to water is a function of water saturation only. Similarly, the gas relative permeability is a function of gas saturation only, and the oil relative permeability is a function of both gas and water saturations due to Equation (5). Interactions that affect the relative permeability curves due to the other phases present are hence assumed to be small.

The pressures in each of the fluid phases present are related to each other by the capillary pressure which is the pressure difference across the interface of any two fluids and is taken to be positive when the pressure in a wetting phase is subtracted from the pressure in a nonwetting phase. Thus, we have the definition of capillary pressure between the air and water phase as

$$p_{c_{aw}} = p_a - p_w \tag{7}$$

capillary pressure between the oil and water phases

$$p_{c_{ow}} = p_o - p_w \tag{8}$$

and between the air and oil phases

$$p_{c_{ao}} = p_a - p_o = p_{c_{aw}} - p_{c_{ow}} \tag{9}$$

Note that Equation (9) is not independent of Equations (7) and (8). $p_{c\alpha\beta}$ is the capillary pressure across two phases, α and β, and can be represented as a function of any one of the phase saturations and the porosity. In many cases, it is assumed that p_c is fairly independent of porosity or the state of consolidation. This function, however, exhibits hysteresis during drainage and imbibition.

The sum of Equations (1) to (4) for each component $i = 1, \ldots, NC$ over all the phases yields

$$\frac{\partial}{\partial t}(C_i^w \rho_w n S_w + C_g^i \rho_g n S_g + C_o^i \rho_o n S_o + C_s^i \rho_s\{1 - n\}) + \nabla \cdot (C_w^i J_w \rho_w + C_g^i J_g \rho_g + C_o^i J_o \rho_o - D_g^i \nabla \rho_g n S_g C_g^i - D_o^i \nabla \rho_o n S_o C_o^i - D_w^i \nabla \rho_w n S_w C_w^i) = R_{source}^i - R_{sink}^i \tag{10}$$

where R_{source}^i and R_{sink}^i are the total source and sink terms for component (i) over all phases, and the relationships between the concentrations can be determined using equilibrium thermodynamics. Equation (10) gives one equation for each component (i) distributed among all phases. The distribution between the phases is calculated efficiently, by assuming the ideal case laws to hold. For treatment of nonideal cases occurring in compositional reservoirs, the reader is referred to Corapcioglu and Panday (1991a). Henry's law partitions masses between the liquid phases. Raoult's law depicts the equilibrium between the gas and aqueous phases, and an adsorption isotherm may be used to partition the mass onto the soil solids. The mass partition coefficients are functions of pressure, temperature, and composition of each of the phases as discussed in the next section.

Application of the Multicomponent Approach to a Gasoline Spill Problem

Baehr and Corapcioglu (1987) have solved Equation (10) to simulate a gasoline spill problem in an unsaturated soil (Figure 2). Gasoline, except for minute amounts of compounds containing sulfur, oxygen, or nitrogen, is a mixture of hydrocarbons. There are several hundred different hydrocarbon components in various proportions in any single commercial gasoline. Certain components of gasoline are shown to be hazardous. Benzene, for example, has been determined to be a human carcinogen. The partitioning of components among phases is achieved by assuming that the equilibrium approximation provides a meaningful working assumption. Jennings and Kirkner (1983) presented a criterion for evaluating the appropriateness of using the equilibrium assumption for solute transport, utilizing the Damköhler number which is the ratio of chemical reaction rates to the bulk flow rate. They defined a critical Damköhler number beyond which local equilibrium is valid.

Corapcioglu and Baehr (1987) employed the Raoult's law to quantify the ideal reference state for equilibrium between the oil and air phases. Raoult's law states that the ideal vapor pressure of a component solution is proportional to its volatility, as measured by the vapor pressure over the pure component, and its relative abundance in the oil phase as measured by the mole fraction of the component in the oil phase.

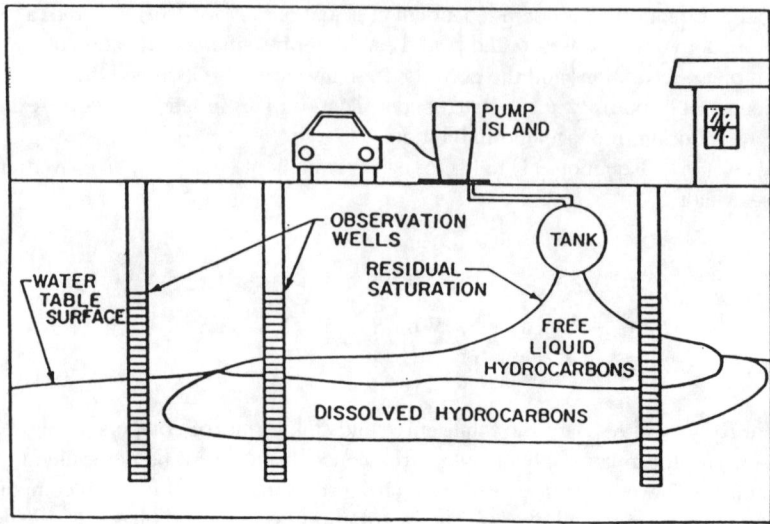

Figure 2. A schematic diagram of gasoline contamination of groundwater.

By assuming that air phase behaves as an ideal gas, Corapcioglu and Baehr obtained a relationship between gas and oil phase concentrations

$$C_g^i \rho_g = H_{go}^i \gamma_i \left(\frac{1}{\omega_i \sum_{j=1}^{N} \frac{1}{\omega_j} C_o^i} \right) C_o^i \tag{11}$$

where ω_i is the molecular weight of the ith component, H_{go}^i is the equilibrium partition coefficient of the ith component between the gas and oil phases, and γ_i is the activity coefficient of the ith component adjusting for nonideality.

Corapcioglu and Baehr (1987) applied Henry's law to express the equilibrium between the gas and water phases for the contaminant components. Henry's law is obeyed very well for sparingly soluble, nonelectrolyte constituents assuming an ideal gas phase. Henry's law states that the partial pressure of the ith component above the water phase is proportional to the concentration of the ith component in the water phase. This yields the relationship between the equilibrium concentrations of the oil and water phases,

$$C_w^i \rho_w = H_{wo}^i \gamma_i \left(\frac{1}{\omega_i \sum_{j=1}^{N} \frac{1}{\omega_j} C_o^i} \right) C_o^i \tag{12}$$

where H_{wo}^i is the partition coefficient of the ith component between the water and oil phases.

Equations (11) and (12) suggest the following relationship between the equilibrium concentrations of the oil and adsorbed phases:

$$C_s^i = \left[H_{so}^i \gamma_i \left(\frac{1}{\omega_i \sum_{j=1}^{N} \frac{1}{\omega_j} C_o^i} \right) \right] C_o^i \qquad (13)$$

where H_{si}^k is the partition coefficient of the ith component between solid and oil phases. Note that Equation (13) holds if, and only if, $C_s^i = (H_{so}^i/H_{wo}^i)C_w^i\rho_w$. H_{so}^i depends on the nature of the solid surfaces and unlike other partitioning coefficients, depends on the type of the porous media. At this point, it should be noted that fluid phase densities are functions of reservoir temperatures, fluid compositions as well as fluid pressures. Corapcioglu and Baehr (1987) suggest

$$\rho_f = \rho_{fo} \exp\left[\alpha_f(T - T_o) + \beta_f(p_f - p_{fo}) + \sum_i \bar{v}^i(M_i^f - M_{io}^f) \right] \qquad (14)$$

where

$$\bar{v}^i = \frac{1}{\rho_f}\left[\frac{\partial \rho_f}{\partial M_i^f}\right]_{T,p,M_i^h \neq M_i^f} \qquad (15)$$

M_f^i is the mass of component i in phase f, and $\bar{v}^i$ is the partial molar density, as defined above. In the above two expressions, subscript o denotes a reference level (e.g., the surface condition). Liquid viscosities can be assumed to be functions of temperature only, i.e., $\mu_f = \mu_f(T)$.

Baehr and Corapcioglu (1987) solved Equation (10) for an immobilized gasoline column in unsaturated soil at residual saturations (i.e., $J_o = 0$; $S_o = S_{o_{res}}$). The gasoline was assumed to consist of eight hydrocarbon components, approximating the gasoline composition reported by Jamison et al. (1976). Benzene, toluene, 1-hexane, cyclohexane, and n-hexane were selected as individual components to analyze the fate of components. Aromatics such as benzene and toluene have relatively low air-water partition coefficients and thus will tend to favor the water phase, while alkanes such as n-hexane have higher air-water partition coefficients and favor the air phase. Cyclohexane (naphthene) and 1-hexene (an alkene) represent hydrocarbons of intermediate partition coefficient. Three composite component aromatics (besides benzene and toluene), alkanes (besides n-hexane), and "heavy ends" (any molecule with more than eight carbon atoms) complete the characterization of the gasoline. Values chosen for the thermodynamic properties of composite constituents were obtained by averaging the properties of individual hydrocarbons in each class. The transport of each component is coupled to the other components via the mole fraction only. Corapcioglu and Baehr (1987) also took into consideration the biodegradation of gasoline components in the water phase. This was achieved by relating the biodegradation rate (R_{sink}^i) to the oxygen concentration in the gas and water phases by a first-order kinetic equation. Baehr and Corapcioglu (1987) first obtained conservative predictions by neglecting the diffusion mechanism in the soil-gas phase.

Figure 3 illustrates the partitioning of the total hydrocarbon mass, obtained by summing over the eight hydrocarbons, into various fates (remaining in soil, leached down into the aquifer or biodegraded) as a function of time. With the high estimate

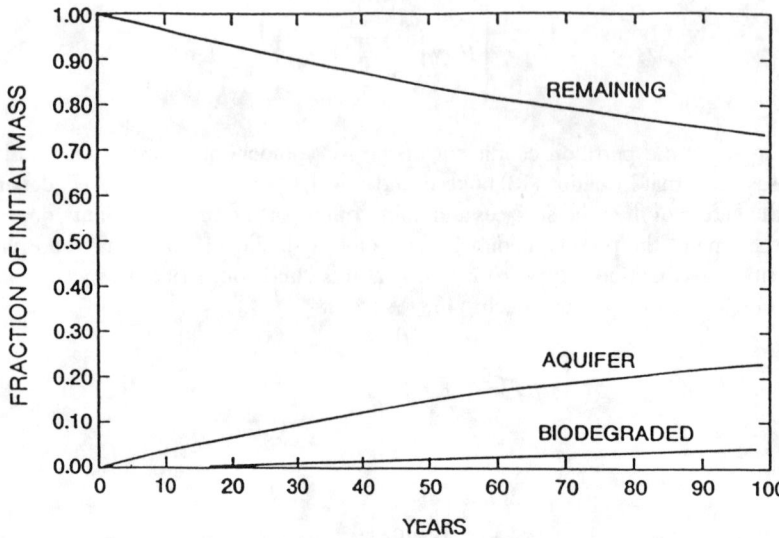

Figure 3. Total hydrocarbon distribution with no diffusive transport. (From Baehr, A. L. and Corapcioglu, M. Y., *Water Resour. Res.*, 23(1), 201, 1987.)

of average recharge, $J_w = 105.6$ cm/year, representative of precipitation rates of the northeastern United States, one finds that at 100 years, 23% of the total initial hydrocarbon mass has entered the aquifer. Aerobic biodegradation at 100 years cannot account for more than 4% of the initial hydrocarbon mass trapped in the soil under these conditions. Figure 4 presents the rates at which the constituent mass enters the groundwater aquifer. Fluxes corresponding to cyclohexane, hexene, hexane, and heavy ends do not appear, since they remain less than 0.2 mg/cm^{-2}/year^{-1}. Thus, non-aromatic components are essentially rendered immobile when movement is due to solute transport only. Figure 4 illustrates that approximately 82% of the total hydrocarbon flux into the aquifer is due to the three aromatic components, benzene, toluene, and other aromatics, which comprise 27% of the initial hydrocarbon mass trapped in the soil. The flux of total hydrocarbons experiences an undulating decline with time as more soluble components are leached out of the soil column. The points of inflection on the total curve correspond to the times when benzene and toluene are exhausted from the column.

Baehr and Corapcioglu (1987) also predicted the fate of gasoline components by allowing the diffusive transport in the air phase. Predictions obtained assuming non-zero values for soil diffusion coefficients allow for hydrocarbon mass to leave the soil column at the ground surface, entering the atmosphere. Also, oxygen recharge rates were increased as a diffusive component supplements the recharge due to dissolved oxygen in infiltrating water. This increase in available oxygen implies higher aerobic biodegradation rates. Thus, allowing for diffusive transport reduces the total mass available for leaching down to groundwater. Figure 5 illustrates the partitioning of hydrocarbon mass into various fates as a function of time. Referring to the total distribution of the initial hydrocarbon mass left in the soil column after 70 years of which

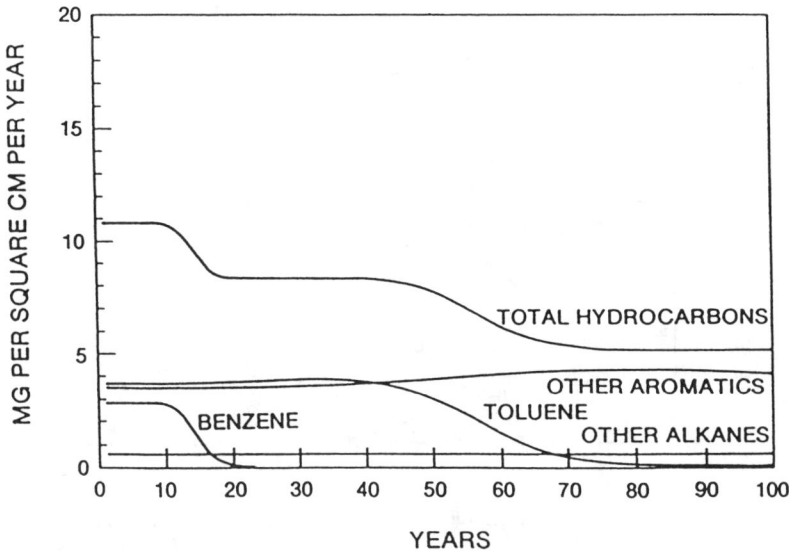

Figure 4. Contaminant flux into aquifer with no diffusive transport.

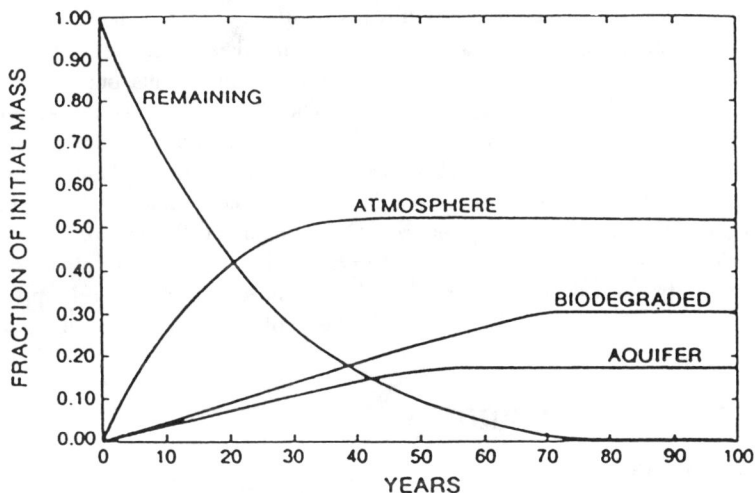

Figure 5. Total hydrocarbon distribution with a diffusive transport.

17% entered the aquifer, 52% entered the atmosphere, and 31% was biodegraded. Figure 6 presents the rates at which component mass enters the aquifer with $J_w =$ 105.6 cm/year. The three aromatic components comprise at least 99% of the total hydrocarbon mass entering the aquifer. Nonaromatic components (except for heavy ends), for the most part, ultimately enter the atmosphere.

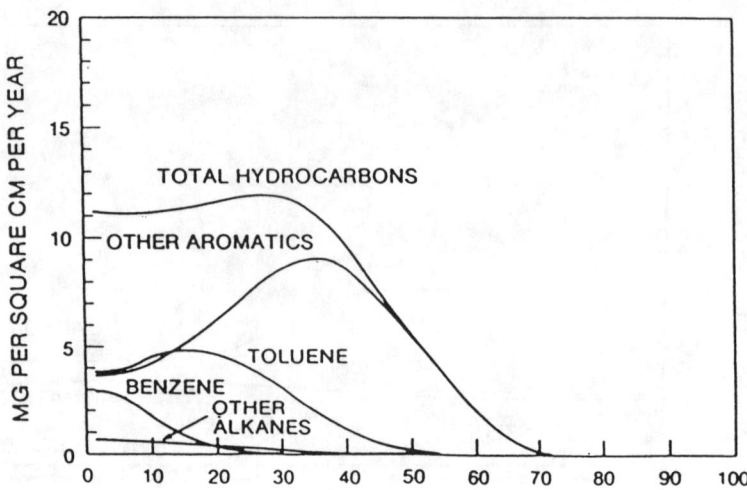

Figure 6. Contaminant fluxes into aquifer with a diffusive transport.

Other Compositional Approaches

As discussed in the early part of the paper, Corapcioglu and Baehr's (1987) model represented the transport of a NAPL phase with eight components. Abriola and Pinder (1985a,b) also had a model with an assumption of local equilibrium. The NAPL phase was composed of, at most, two distinct components, one of which may be volatile and slightly water soluble and the other is both nonvolatile and insoluble in water. Falta et al. (1992) presented a multicomponent model of steam injection for the removal of NAPL from the subsurface. Falta et al.'s model contained three mass components, three mass balance equations, and an energy balance equation. Corapcioglu and Panday (1991b) generalized Corapcioglu and Baehr's (1987) model for a nonisothermal system and applied it to predict the migration of petroleum components in frozen soils. In another paper, Panday and Corapcioglu (1991) applied the fundamentals of their formulation to study the solute rejection in freezing soils.

SHARP INTERFACE MODELS

Abriola and Reeves (1990) noted that although sharp interface models have some limitations, such as negligence of capillary forces in comparison to pressure and gravity forces, they "may prove useful as screening or site assessment tools because of their relative simplicity". They may be employed for source identification in oil-contaminated groundwater. Especially, analytical expressions obtained by solving governing equations are quite attractive in routine exposure assessment studies. Therefore it is the objective of this section to present an analytical solution to calculate the variation of free product thickness and volume of free product in a two-pump recovery operation by using a sharp interface model.

Recovery of oil accumulated on water table can be accomplished by creating a cone of depression with a pumping recovery well. Free product is induced to flow into the recovery well by the influence of water table gradient established by pumping. Once the well is in place and the cone of depression is formed, a scavenger oil recovery unit is placed in the well on top of the oil layer and the oil is pumped to a recovery tank.

Although sharp interface models have been investigated by a number of researchers, analytical solutions have been attempted in very few of them. Lateral spreading of free product on water table has been studied by van Dam (1967), Mull (1971, 1978), Dracos (1978), Greulich and Kaergaard (1984), Greulich (1985), Holzer (1976), Levy et al (1990), El-Kadi (1992, 1994), Schiegg (1977), and Hochmuth and Sunada (1985). van Dam (1967) obtained an expression to estimate the areal extent of an oil lens at equilibrium by assuming a constant oil saturation within the lens equal to half residual saturation and a stationary water table. Mull (1971, 1978) presented an analysis by assuming the oil lens to cease spreading when the lens thickness at the spill site matches the capillary rise. Greulich (1985) [also presented in Greulich and Kaergaard (1984)] represented the oil lens as a combination of a cylinder at the center and a thin circular disc outside. Holzer (1976) applied Hantush's (1968) theory by making an analogy between the decay of an oil lens on a horizontal water table and the movement of a freshwater lens in an unconfined saline aquifer. Dracos (1978) has proposed an approach to estimate the spread of an oil slick in soils and presented expressions to calculate the maximum spread of the oil-polluted area. Levy et al. (1990) used Hantush's methodology to estimate the rate and volume of oil leakage from a tank. Hochmuth and Sunada (1985) developed a two-dimensional numerical finite-element model to simulate the movement of a hydrocarbon and groundwater by solving the Boussinesq equation stated for each phase. Schiegg (1977) [also presented in Schiegg and Schwille (1991)] presents a numerical lateral spreading solution by using a semianalytical vertical infiltration expression as an initial condition. Schiegg assumes that spreading starts after the infiltration of oil ceases.

Formulation of the Sharp Interface Approach

The first step in the mathematical formulation of any transport phenomena involves the balance equations. The lateral spreading of oil is simulated by considering the migration of oil and water phases simultaneously. Since both phases are mobile, the starting point is the mass balance equations of oil and water phases in a porous medium

$$\nabla \cdot \rho_o \, q_o + \frac{\partial (\rho_o S_o n)}{\partial t} = 0 \tag{16}$$

$$\nabla \cdot \rho_w \, q_w + \frac{\partial (\rho_w S_w n)}{\partial t} = 0 \tag{17}$$

where ρ_o and ρ_w are the densities of oil and water phases, respectively, S_o and S_w are the phase saturations of oil and water phases, respectively, t is the time, and n is the porosity. Note that in a two-phase system $S_o + S_w = 1$. q_o and q_w are specific discharges of oil and water phases, respectively.

Governing equations to estimate oil thickness L, and water elevation η, as a function of space and time, can be obtained by averaging Equations (16) and (17) along the vertical in their respective regions, by applying appropriate boundary conditions on top and bottom surfaces. The averaging procedure is carried out by applying the Leibnitz' rules. Then the governing equation for oil phase can be obtained by integrating the mass balance equation

$$\int_{\eta(r,t)}^{h(r,t)} [\nabla \cdot \rho_o\, q_o + \frac{\partial (\rho_o S_o n)}{\partial t}]\, dz = 0 \tag{18}$$

We also need boundary conditions at the top and bottom of the oil lens. Boundary conditions are defined by the particular physical conditions imposed during a particular process. In our problem, although the lower boundary at the groundwater/bedrock interface is stationary, oil/groundwater and unsaturated soil/oil interfaces are moving boundaries. In general, boundary conditions state that at any point on the boundary, the mass flux of a phase normal to the boundary surface must be equal on both sides of the boundary if there is no sink or source of mass on the surface. Let us denote the unsaturated soil by subscript un, oil lens by o and groundwater by w. Then we can write at the unsaturated soil/oil interface, oil surface h,

$$[\rho_o q_o|_o - n\rho_o S_o U] \cdot \nabla F_2 = [\rho_o q_o|_{un} - n\rho_o S_{o_{un}} U] \cdot \nabla F_2 \tag{19}$$

where U is the velocity of the moving boundary, and $F_2 = 0$ is the equation of the oil surface. In our case, $F_2 = z - h(r,t)$ where h is the elevation of oil surface measured from a reference level, z is the vertical coordinate, and r is the radial coordinate. Similarly, at the oil/groundwater interface (water table), η

$$[\rho_o q_o|_o - n\rho_o S'_{oo} U] \cdot \nabla F_1 = [\rho_o q_o|_w - n\rho_o S'_{ow} U] \cdot \nabla F_1 \tag{20}$$

where $F_1 = 0$ is the equation of the water table. For this problem $F_1 = z - \eta(r,t)$, where η is the elevation of the water table.

If we assume that the water table is impervious to oil flux, then, Equation (19) yields

$$\rho_o q_o|_h \cdot \nabla F_2 = -\rho_o R_o\, \delta(r-a) + \rho_o[S'_{oo} - S'_{oun}] U \cdot \nabla F_2 \tag{21}$$

where $-R_o$ is the distributed rate of oil leak at $r=a$, $\delta(r-a)$ is Dirac Delta function located at $r=a$, and S'_{oo} and S'_{oun} are the degrees of oil saturation in the oil lens and in the unsaturated soil at corresponding sites of the oil surface, respectively. In an oil recovery operation with declining oil surface $S_{oo} = 1 - S_{wres}$, $S_{oun} = S_{ores}$ where S_{wres} is the degree of residual water saturation in an air/water system and S_{ores} is the degree of residual oil saturation in an air/water system. At the water surface

$$\rho_o q_o|_\eta \cdot \nabla F_1 = \rho_o n [S'_{oo} - S'_{ow}]_\eta U \cdot \nabla F_1 \tag{22}$$

where S'_{oo} and S'_{ow} are the degrees of oil saturation in the oil lens and in groundwater respectively at corresponding sides of the water table. Since in a recovery operation the oil surface and water table will be lowered $S'_{oo} = 1 - S'_{wres}$, $S'_{ow} = 0$ where S'_{wres} is the degree of residual water saturation in an oil/water system.

By substituting the boundary conditions into Equation (18) and applying the averaging rules we obtain a linearized equation for L and η.

$$\nabla^2 L + \nabla^2 \eta = \pm \frac{R_o}{K_o L_o} \delta(x-\xi)\delta(y-\zeta) + \frac{n[S_{oo} - S_{oun}]}{K_o L_o} \frac{\partial L}{\partial t} - \frac{n[S_{oun} - S_{ow}]}{K_o L_o} \frac{\partial \eta}{\partial t} \tag{23}$$

Similarly, water-phase equation is derived by using a similar procedure followed in the derivation of oil-phase equation. The mass balance equation for water is integrated between impervious bed rock and water table and appropriate boundary conditions are applied. Integration of water balance equation means

$$\int_{b(x,y)}^{\eta(x,y,t)} [\nabla \cdot \rho_w q_w + \frac{\partial (\rho_w S_w n)}{\partial t}] \, dz = 0 \tag{24}$$

Note that the bottom surface, F_3, is stationary. Substitution of approximate boundary conditions and averaging rules, yields an equation in L and η as

$$\nabla^2 \eta + \nabla L^2 \left(\frac{\rho_o}{\rho_w}\right) = \pm \frac{Q_w}{K_w \eta_o} \delta(x-\xi')\delta(x-\zeta') + \frac{n[S_{ww} - S_{wo}]}{K_w \eta_o} \frac{\partial \eta}{\partial t} \tag{25}$$

In order to obtain an expression for free product thickness, the oil-phase equation is solved uncoupled. Uncoupling of the oil-phase equation is done by solving the unconfined groundwater flow equation independently for η and then substituting it into the oil-phase equation. The governing oil-phase Equation (23), can be rewritten in radial coordinates as

$$\frac{\partial^2 L(r,t)}{\partial r^2} + \frac{1}{r} \frac{\partial L(r,t)}{\partial r} = \frac{n(S_{oo} - S_{oun})}{K_o L_o} \frac{\partial L}{\partial t} - \left[\frac{\partial \eta^2}{\partial r^2} + \frac{1}{r} \frac{\partial \eta}{\partial r} + \frac{n(S_{oun} - S_{ow})}{K_o L_o} \frac{\partial \eta}{\partial t}\right] \tag{26}$$

The unconfined groundwater flow equation, known as Boussinesq's equation, can be expressed as

$$\frac{\partial^2 \eta}{\partial r^2} + \frac{1}{r} \frac{\partial \eta}{\partial r} = \frac{S_y}{K_w \eta_o} \frac{\partial \eta}{\partial t} \tag{27}$$

where S_y is the specific yield of the phreatic aquifer. The solution of Equation (27) for an aquifer of infinite extent is given by the Theis equation as

$$\eta_o - \eta(r,t) = \frac{Q_w}{4\pi K_w \eta_o} W(u) \tag{28}$$

where

$$u = \frac{r^2 S_y}{4K_w \eta_o^t} \tag{29}$$

$W(u)$ is an exponential integral known as the Well function in groundwater hydrology. Substitution of Equations (27) and (28) in Equation (26) yields

$$\frac{\partial^2 L(r,t)}{\partial r^2} + \frac{1}{r}\frac{\partial L(r,t)}{\partial r} - \frac{n(S_{oo} - S_{oun})}{K_o L_o}\frac{\partial L}{\partial t}$$

$$= \frac{Q_w}{4\pi K_w \eta_o}\left[\frac{S_y}{K_w \eta_o} + \frac{n(S_{oun} - S_{ow})}{K_o L_o}\right]\frac{e^{-\frac{r^2 S_y}{4K_w \eta_o^t}}}{t} \tag{30}$$

A Case Study: Recovery of an Established Oil Mound on the Water Table

In this section, we seek a solution to the recovery problem of an established oil mound on an initially horizontal water table (see Figure 7). The initial condition can be expressed as

$$L(r,0) = \frac{L_o}{\ln\left(\frac{r_o}{r_w}\right)} \ln\left(\frac{r_o}{r}\right) \tag{31}$$

where r_w and r_o are the radii of the recovery well and the oil lens, respectively. In a two-pump recovery system as seen in Figure 8, one pump creates a cone of depression by pumping groundwater and it draws groundwater, and the free product floating on it, into the recovery well. A second pump at the water table skims off the free product. The condition at the recovery well is given by

$$\lim_{r \to 0} r \frac{\partial L(r,t)}{\partial r} = \frac{1}{2\pi}\left(\frac{Q_o}{K_o L_o} - \frac{Q_w}{K_w \eta_o}\right) \tag{32}$$

where Q_o and Q_w are oil and water pumpage rates, respectively. At the outer edge of the oil lens a no-flux boundary condition is defined as

$$\frac{\partial L}{\partial r}\bigg|_{r=r_o} = \frac{Q_w}{2\pi k_w n_o r_o} e^{-\frac{ar_o^2}{4t}} \tag{33}$$

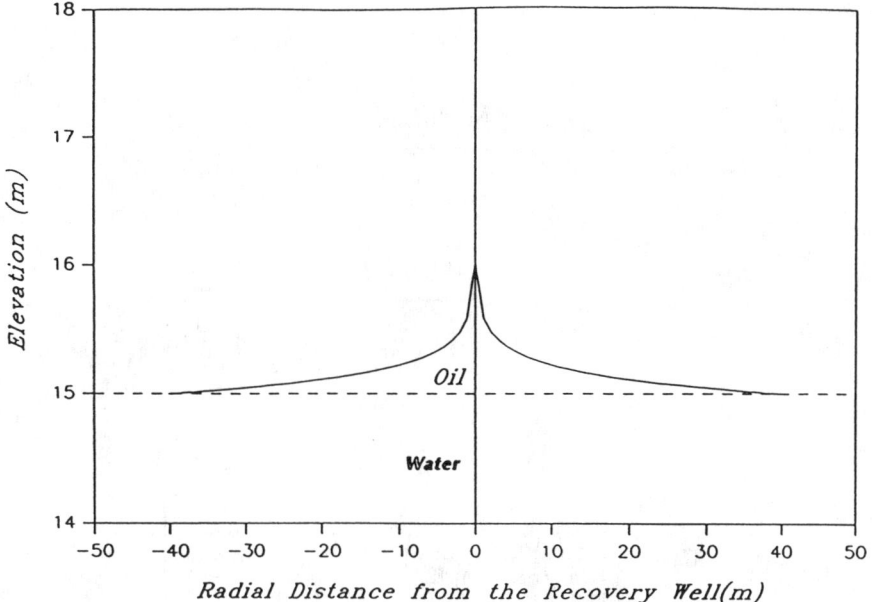

Figure 7. Initial profile of an oil mound on an initially horizontal water table.

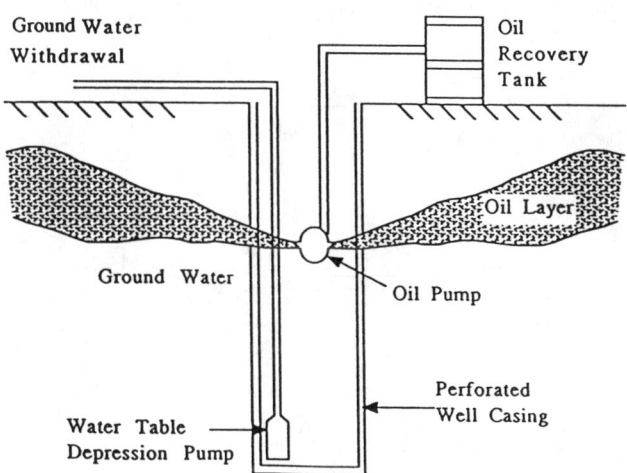

Figure 8. A two-pump recovery system for free product recovery.

where a is given in Equation (39). Solution of Equation (30) using the initial and boundary conditions is obtained by Laplace transformation technique (Corapcioglu et al., 1994). It can be expressed as

$$L(r,t) = C_1^\circ I_0\left(r\sqrt{\frac{A}{2t}}\right) + \frac{C_2^\circ}{2} W(u_1) + \frac{2B}{(a-A)} \frac{1}{2} W(u_2) + \frac{L_o}{\ln\left(\frac{r_o}{r_w}\right)} \ln\left(\frac{r_o}{r}\right) \quad (34)$$

where

$$C_1^o = \left(-\frac{Q_w}{2\pi K_w \eta_o} + \frac{2B}{(a-A)}\right)\sqrt{\frac{a}{A}} \frac{K_1\left(r_o\sqrt{\frac{a}{2t}}\right)}{I_1\left(r_o\sqrt{\frac{A}{2t}}\right)}$$

$$+ C_2 \frac{K_1\left(r_o\sqrt{\frac{A}{2t}}\right)}{I_1\left(r_o\sqrt{\frac{A}{2t}}\right)} + \frac{L_o}{r_o \ln\left(\frac{r_o}{r_w}\right)\sqrt{\frac{A}{2t}} I_1\left(r_o\sqrt{\frac{A}{2t}}\right)} \quad (35)$$

and

$$C_2^o = -\left[\frac{1}{2\pi}\left(\frac{Q_o}{L_o K_o} - \frac{Q_w}{\eta_o K_w}\right) + \frac{L_o}{\ln\left(\frac{r_o}{r_w}\right)} + \frac{2B}{(a-A)}\right] \quad (36)$$

where A, B, a, u_1 and u_2 are defined as

$$A = \frac{n(S_{oo} - S_{oun})}{K_o L_o} \quad (37)$$

$$B = \frac{Q_w}{4\pi K_w \eta_o}\left[\frac{S_y}{K_w \eta_o} + \frac{n(S_{oun} - S_{ow})}{K_o L_o}\right] \quad (38)$$

$$a = \frac{S_y}{K_w \eta_o}; \; u_1 = \frac{r^2 A}{4t}; \; u_2 = \frac{r^2 a}{4t} \quad (39)$$

The variation of free product thickness with time is illustrated in Figure 9. As seen in this figure, an initially 1 m thick oil lens at the recovery well is reduced to 0.17 m after a month of oil pumping at a rate of 4.25 m³/day. Figure 10 illustrates a graphical procedure to estimate the recovery efficiencies based on the readings at the monitoring wells. For example, a 20 cm free product thickness in an observation well 10 m from the recovery well indicates 40% removal efficiency.

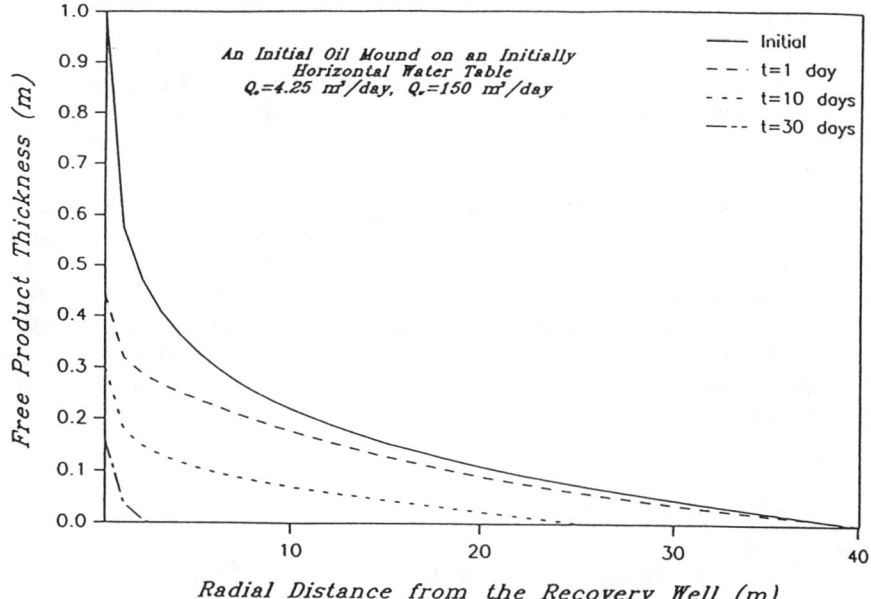

Figure 9. Variation of free product thickness.

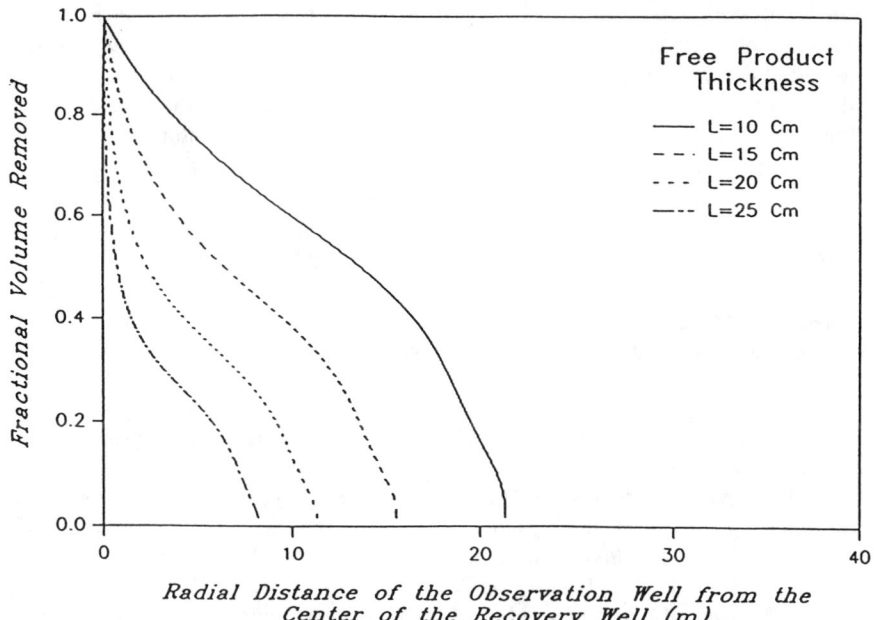

Figure 10. Estimation of recovery efficiency based on readings at the monitoring wells.

CONCLUSIONS

In this paper we presented a summary of predictive models of multiphase fluid flow in subsurface environments. The mathematical models reviewed were categorized in various groups. Each group of models requires a different set of input parameters describing the subsurface and fluid(s). The level of complexity of the information is directly related to the level of sophistication of the model. The choice of modeling approach depends on the level of accuracy required and the availability of data. Usually the availability of information limits the choice of the model. However, since models provide relatively inexpensive and useful tools to design and test effective remediation schemes, they are critical in our clean-up efforts. Understanding and prediction of hydrocarbon migration in subsurface environments will advance effective recovery and clean-up techniques of these contaminants, as well as help to protect existing water resources.

ACKNOWLEDGEMENT

The research presented in this paper has been partially supported by grants from National Science Foundation CEE-8401438, American Chemical Society/The Petroleum Research Fund PRF #15890-AC5, and U.S. Department of Interior G-897/02. The research reported in Section 4 has been funded in part with federal funds as part of the program of the Gulf Coast Hazardous Substance Research Center, which is supported under cooperative agreement R 815197 with the United States Environmental Protection Agency (U.S. EPA), and in part with funds from the State of Texas as part of the program of the Texas Hazardous Waste Research Center. The contents do not necessarily reflect the views and policies of the U.S. EPA or the State or Texas, nor does the mention of trade names or commercial products constitute endorsement or recommendation for use.

REFERENCES

Abriola, L. M. and Pinder, G. F., A multiphase approach to the modeling of porous media contamination by organic compounds, 1, Equation development, *Water Resour. Res.*, 21(1), 11, 1985a.

Abriola, L. M. and Pinder, G. F., A multiphase approach to the modeling of porous media contamination by organic compounds, 2, Numerical simulation, *Water Resour. Res.*, 21(1), 19, 1985b.

Abriola, L. M. and Reeves, H. W., Slightly miscible organic chemical migration in porous media: Present and future directions in modeling, *Proc. Env. Res. Conf. Groundwater Quality and Waste Disposal*, Murarka, I. P. and Cordle, S. S., Eds., EPRI, Palo Alto, CA, 1990, 15.1–15.24.

Baehr, A. L. and Corapcioglu, M. Y., A compositional multiphase model for groundwater contamination by petroleum products, 2, Numerical solution, *Water Resour. Res.*, 23(1), 201, 1987.

Coats, K. H. In situ combustion model, *Soc. Petrol. Eng. J.*, 533, 1980.

Corapcioglu, M. Y. and Baehr, A. L. A compositional multiphase model for groundwater contamination by petroleum products, 1. Theoretical considerations, *Water Resour. Res.*, 23(1), 191, 1987.

Corapcioglu, M. Y. and Panday, S., Compositional Multiphase Flow Models, in Corapcioglu, M. Y., Ed., *Advances in Porous Media*, Vol. 1, Elsevier, Amsterdam, 1991a, 1–60.

Corapcioglu, M. Y. and Panday, S., Soil and groundwater contamination by petroleum products in frozen soils, *Proc. Third International Symp. on Cold Regions Heat Transfer*, June 11–14, University of Alaska, 1991b, 303–310.

Corapcioglu, M. Y., Tuncay, K., Lingam, R., and Kambham, K. K. R., Analytical expressions to estimate the free product recovery in oil-contaminated aquifers, *Water Resour. Res.*, 30(12), 1994.

Crookston, R. B., Culham, W. E., and Chen, W. H. A numerical simulation model for thermal recovery processes, *Soc. Petrol. Eng. J.*, 37, 1979.

Dracos, T., Theoretical considerations and practical implications in the filtration of hydrocarbons in aquifers, in *Proc. International Symp. on Ground Water Pollution by Oil Hydrocarbons*, Stavebn'i Geologie Praha, Prague, Czechoslovakia, 1978, 127–137.

El-Kadi, A. I., Applicability of sharp-interface models for NAPL transport. 1. Infiltration, *Ground Water*, 30(6), 849, 1992.

El-Kadi, A. I., Applicability of sharp-interface models for NAPL transport. 2. Spreading of a LNAPL, *Ground Water*, 32(5), 784, 1994.

Falta, R. W., Pruess, K., Javandel, I., and Witherspoon, P. A., Numerical modeling of steam injection for the removal of nonaqueous phase liquids from the subsurface, Numerical formulation, *Water Resour. Res.*, 28, 433, 1992.

Greulich, R. H., Groundwater contamination from oil: behavior, control and treatment, *Water Supply*, 3, 233, 1985.

Greulich, R. and Kaergaard, H., The movement of a continuously growing body of oil on the groundwater table, *Nordic Hyd.*, 15, 265, 1984.

Hantush, M. S., Unsteady movement of fresh water in thick unconfined saline aquifers, *Bull. Internat. Assoc. Sci. Hydrology*, 13, 40, 1968.

Hekim, Y., Fogler, H. S., and McCune, C. C., The radial movement of permeability fronts and multiple reaction zones in porous media, *Soc. Petrol. Eng. J.*, 99, 1982.

Hochmuth, D. P. and Sunada, D. K., Ground water model of two-phase immiscible flow in coarse material, *Ground Water*, 23, 617, 1985.

Holzer, T. L., Application of groundwater flow theory to a subsurface oil spill, *Ground Water*, 14, 138, 1976.

Jamison, V. W., Raymond, R. L., and Hudson, J. O., Biodegradation of high octane gasoline, in *Proceedings, Third International Biodegradation Symposium*, Sharpley, J. M. and Kaplan, A. M., Eds., Applied Science Publishers, Englewood, NJ, 1976, 187–196.

Jennings, A. A. and Kirkner, D. J., Criteria for selecting equilibrium or kinetic sorption descriptions in groundwater quality models, in *Frontiers in Hydraulic Engineering*, Shen, H. T., Ed., American Society of Civil Engineers, 1983, 42–47.

Kazemi, H., Vestal, C. R., and Shank, G. D., An efficient multicomponent numerical simulator, *Soc. Petrol. Eng. J.*, 355, 1978.

Levy, B. S., Riordan, P. J., and Schreiber, R. P., Estimation of leak rates from underground storage tanks, *Ground Water*, 28, 378, 1990.

Mull, R., Migration of oil products in the subsoil with regard to groundwater pollution by oil, *Advances in Water Pollution Research,* Jenkins, S. H., Ed., Pergamon Press, Oxford, 1971, 2Ha-7a, 1–8.

Mull, R. Calculations and experimental investigations of the migration of oil products on natural soils, *Proc. Int. Symp. on Groundwater Pollution by Oil Hydrocarbons,* Prague, June 5–9, Int. Assoc. Hydrogeol., 1978, 167–181.

Nghiem, L. X., Fong, D. K., and Azis, K., Compositional modeling with an equation of state, *Soc. Petrol. Eng. J.,* 687, 1973.

Panday, S. and Corapcioglu, M. Y., Solute rejection in freezing soils, *Water Resour. Res.,* 27, 99, 1991.

Schiegg, H. O., Methode zur abschatzung der ausbreitung von erdol derivaten in mit wasser and luft erfullten boden, *Mitteilung der Versuchanstalt fur Wasserbau, Hydrologie and Glaziologie an der ETH Zurich,* 1977, 22.

Schiegg, H. O. and Schwille, F., Hydrocarbons in porous media, in *Transport Processes in Porous Media,* Bear, J. and Corapcioglu, M. Y., Eds., Kluwer, 1991, 69–202.

van Dam, J., The migration of hydrocarbons in a water-bearing stratum, *Joint Problems of the Oil and Water Industries,* Hepple, P., Ed., The Inst. of Petroleum, London, 55–96, 1967.

Young, L. C. and Stephenson, R. E., A generalized compositional approach for reservoir simulation, *Soc. Petrol. Eng. J.,* 727, 1983.

Youngren, G. K., Development and application of an in situ combustion reservoir simulator, *Soc. Petrol. Eng. J.,* 39, 1980.

SECTION 4

On Island Modeling

CHAPTER 14

Groundwater Modeling in Hawaii: A Historical Perspective

L. Stephen Lau and John F. Mink

INTRODUCTION

Water supply has been a dominating feature in the evolution of the Hawaii economy since the first settlement of the islands more than a millennium ago. The early Hawaiians relied on water from streams, springs, and shallow excavations to irrigate crops, taro in particular. Water use was regulated by the Hawaiian chiefs and their land managers according to strict rules. The source and delivery systems were so well designed that they served the expanding economy for a century after the opening of the archipelago to the western world by Captain James Cook in 1778.

The prediscovery Hawaii economy was highly organized and successful. When Cook arrived the population of the islands was about 300,000, but a century later it had declined to about 60,000 as a result of devastation caused by western diseases and the breakdown of a unifying culture. The demand for water increased to meet the needs of new agricultural initiatives, especially sugarcane, exceeding the supply available from the intricate Hawaiian distribution network. Collection works and transmission systems were constructed by western entrepreneurs. By 1910 virtually every major surface water resource had been seized for plantation agriculture.

Still, not enough water was available to satisfy the thirst of the arid lands planted in sugarcane. The greatest impetus to the advancing agricultural economy came with the discovery of artesian groundwater in the Ewa Plain in southwestern Oahu in 1879. Widespread drilling followed, proving the existence of vast groundwater resources. The experience was quickly repeated on all of the major islands. The arid leeward plains of each island, blessed with a bounty of sunshine, became the premier agricultural lands through irrigation with groundwater. Honolulu, served by an unreliable surface water supply for all of the 19th century, prospered and grew after voluminous artesian groundwater resources were discovered beneath the coastal plain. The city became the uncontested urban and commercial center of the island chain.

Groundwater became the first choice for municipal drinking water, and as the public appreciated its purity and reliability it also worried about its sustainability. A severe drought in 1926 magnified the concern, leading to the creation of the City and County of Honolulu Board of Water Supply. Basic investigations by the U.S. Geological Survey, which had started a few years earlier, were expanded.

The adequacy of the groundwater resources of southern Oahu was severely tested in World War II when a large military population was grafted on the existing civilian and agricultural economy. Then in 1959 Hawaii became a state, setting off a wave of economic activity which concurrently expanded demand for water. By the 1970s the combined agricultural, military, and civil economy water demands in southern Oahu were rapidly approaching sustainability of the aquifers. Elsewhere in the islands economic expansion also strained water source and distribution systems. The adequacy of the groundwater supply persisted as an unusually widespread concern among the public. Finally, after nearly a decade of legislative attempts, a State Water Code regulating all water development became law in 1987.

GEOLOGICAL AND HYDROLOGICAL ENVIRONMENT

Hydrological Character of Hawaiian Rocks

All of the Hawaiian islands originated as the effusion of lavas from fissures overlying a "hot spot" in the Pacific oceanic plate. The loci of eruptions had a limited extent and consisted characteristically of a central caldera from which rift zones radiated. The accumulation of thousands of thin lava layers formed broad, elongate volcanic shields gently sloping at 3 to 10° away from the eruption zones.

The volcanos of all of the islands followed a similar evolution. The erupting rocks were basaltic in composition and consisted of highly fluid flows that lithified into discontinuous, inhomogeneous layers averaging less than 3.05 m (10 ft) thick. These primitive basalts account for more than 95% of the total rock mass of the islands and compose an even larger share of the subsurface geology in the zone of saturation. Not until after the primary epoch of island building, during which the essential shape and size of the volcanic shields were established, were the primitive basalts succeeded by other eruptives differentiating from the basic magma.

In the early, most voluminous eruptive stage, each volcano rose as a monolithologic pile of lava units that accumulated too rapidly for significant interflow erosion to occur. In many of the volcanos, an intermediate stage of eruption, transitional from the early stage, followed, during which some of the lavas were chemically and mineralogically differentiated from the basic primitive basalt to form andesitic and trachytic rocks. For some volcanos, a mild erosional hiatus preceded the intermediate eruptive phase, but for others the transition was unbroken. The early and intermediate stages were succeeded by a long period of quiescence, during which profound erosion of the volcanic pile took place. Deep valleys, narrow ridges, and a highly dissected terrain were formed, and vast quantities of terrestrial sediments were laid down in the lower reaches of the valleys and along the coasts. Subsidence accompanied the initial erosion, and subsequent changes in relative sea level created a complicated succession of terrestrial and marine sediments at lower elevations.

A final phase of eruptive activity occurred at a few of the older volcanos. On Hawaii it was extensive, but elsewhere, such as in southeast Oahu, activity was moderate in scale and extent. The resulting rocks range from ordinary basalt to basic nephelinitic and associated basalts. This posterosional stage of volcanism, which is highly significant in the eastern half of Kauai, is relatively minor in overall hydrological importance on Oahu in contrast to the earlier eruptive periods and the interval of profound erosion. Concurrently with the main period of volcanism and for a long time afterward, the emplaced lavas in calderas and along parts of the rift zones were affected by hydrothermal alteration in places so intense that the mineralogy and chemistry of the rocks were profoundly altered. The resulting metamorphic rocks are restricted to former zones of volcanic eruptive activity, in particular the near vicinity of the caldera where they control hydrological behavior.

The geological evolution of the Hawaiian islands has resulted in a lithology which, although dominated by volcanic rocks, includes sedimentary and metamorphic sequences. All of the principal aquifers are composed primarily of the primitive volcanic series and secondarily of the intermediate series. Other smaller and less important aquifers are composed of sedimentary rocks, but sedimentary sequences also serve as aquitards. The metamorphic rocks have such poor hydrological characteristics that they are not considered aquifers within the framework of Hawaii hydrology.

The volcanic rocks are either extrusive or intrusive. The bulk of the volume consists of the extrusive variety that effused from fissures as molten lava onto the surface before solidifying into flow units ranging from highly fragmented piles of debris to dense, massive layers. A much smaller amount was blown out of vents as pyroclastics. Intrusive rocks solidified below the surface into more homogeneous units than the eruptive series. Lava flows occur in all three of the principal structural features of volcanic shields (i.e., calderas, rift zones, and flanks), while intrusive rocks are commonly restricted to calderas and rift zones—the spatially most limited terrains. Pyroclastic material is most noticeable along the eruptive zones, but it is also found within the flank sequences, often as a result of wind dispersal of ejecta from erupting sites.

Extrusive Rocks

The chemical composition of basic basaltic magma gives rise to fluid lava flows that travel easily down slopes for considerable distances before congealing. These primitive lavas form lithologic sequences in which identifiable units are normally less than 3.05 m (10 ft) thick on the volcano flanks. Lavas of the intermediate series (hawaiite, mugearite, and other andesitic—trachytic forms) are more viscous and solidify into thicker, more massive units.

The final structure of a lava unit is largely determined by the grade down which the fluid magma flowed. In the Hawaiian islands practically all of the lavas of hydrologic interest are subaerially extruded; few submarine lavas are exposed or have been encountered in borings. The flanks of the volcanic shields have slopes usually greater than 3° and less than 10°, a grade permitting rapid flow and subsequent cooling into heterogeneous piles of pahoehoe and aa-clinker, the principal extrusive rock forms. In calderas, pit craters in rift zones, and topographic depressions from which effusions could not escape, layers of dense and massive lavas have accumulated. However, along the rift zones the lavas are mainly of the flank type.

Pahoehoe refers to flows having smooth, hummocky, glassy surfaces surrounding highly vesicular interiors. Although highly porous, pahoehoe has little intergranular-type permeability but, en masse, may be highly permeable because of structural features. The permeability arises as a result of units that are unconformably matched, leaving openings between the units; the presence of lava tubes, originally conduits from which lava drained before cooling; and cooling joints and miscellaneous fractures. Lava tubes are the largest single permeability element, having diameters most frequently on the order of 0.3 m (1 ft) but occasionally as large as 6.1 m (20 ft). They are not, however, as common as the other permeability elements. Pahoehoe is more characteristic of primitive basalts than of basalt differentiates.

In contrast to the smoothness of pahoehoe, the aa-clinker association consists of a dense, massive, discontinuous central phase, the aa core, bounded by spiny, fragmented lava breccia called "clinker". As in pahoehoe, virtually all permeability in aa-clinker results from structural features created in the course of emplacement and cooling of the fluid lavas. The openings in clinker beds are probably the most effective of the common permeability elements in extrusive rocks. In the massive aa phase, cooling yields jointing, also an important permeability element. The vertical component of permeability is enhanced by frequent bridging of clinker across flows. Andesitic and trachytic lavas normally congeal into the aa-clinker rock variety.

Pyroclastic rocks account for only a small fraction, no more than 5% and probably less than 1%, of the total extruded mass. On an areal scale the dominant pyroclastic forms are ash and tuff. Constituents, particularly glass, alter at normal temperatures into complex hydrates called palagonite. Palagonitized ash and tuff display extremely low permeability. Cinder and larger ejecta retain a higher fraction of their original permeability.

Only in a large-scale sense can statistically describable aquifer parameters be assigned to volcanic rocks. Aquifer hydraulics of extrusive rocks are controlled by regional values of hydraulic conductivity and effective porosity, but local features may control the behavior of individual wells and other engineering constructions. Also, a tremendous difference exists between essentially unweathered flows and overlying in situ weathered zones. All of the great aquifers of Hawaii are composed of extrusive rocks, principally of primitive basalts and olivine basalts.

Laboratory determinations of hydraulic conductivity and porosity of extrusive rocks are not very meaningful because only a tiny piece of the heterogeneous rock mass can be tested at one time. Consequently, few laboratory measurements have been made. Wentworth (1951) found a range of porosity of 5.2 to 51.4% in drill-hole cores of the Koolau basalt on Oahu. Gravity surveys in shafts and tunnels on Oahu, Maui, and Hawaii indicate that the regional porosity of unweathered primitive lavas is about 20% (Huber and Adams, 1971). This overall porosity includes unconnected vesicles that do not contribute to effective porosity, which normally is less than 10%.

Regional hydraulic properties of extrusive rocks are commonly deduced from analysis of pump tests or from estimates of total groundwater flow. Pump tests give transmissivity from which hydraulic conductivity is inferred. However, it is impossible to stipulate categorically the depth of flow in which wells and galleries are only partially penetrating a Ghyben-Herzberg lens that is not confined below. Usually depth of flow in an unconfined lens is taken as coincident with the theoretical static thickness of the fresh water, calculated as 41 times the freshwater head in the ideal Ghyben-Herzberg case.

Wentworth (1938) conducted the first regional aquifer test in the Hawaiian islands and computed the hydraulic conductivity of the Koolau basalt as ranging from 554 to 1072 m/day (1818 to 3516 ft/day). Many similar tests have been conducted since then, and a summary of the most informative ones is given in Williams and Soroos (1973). Summarizing their work, a total of 51 pumping tests in wells on Oahu and Maui showed an average transmissivity of 1.02 E+5 m²/day (1.1 E+6 ft²/day), equivalent to a hydraulic conductivity of 366 m/day (1200 ft/day) if the theoretical thickness of the lens is taken as depth of flow. Mink (1980) analyzed the behavior of the Pearl Harbor aquifer in southern Oahu in response to the instantaneous cessation of all major pumping brought about by a labor strike on the sugar plantations. He calculated hydraulic conductivity as 457 m/day (1500 ft/day) and specific yield as 5%.

Although not definable as a precise, unvarying number for any of the aquifers, the regional hydraulic conductivity of the major aquifers consisting of unweathered primitive basalts and olivine basalts that were laid down as flank flows dipping between 3 and 10° lies in the range of 305 to 1524 m/day (1000 to 5000 ft/day), with most probable values centering around 457 m/day (1500 ft/day). The intermediate extrusive rocks could be expected to have a lower overall hydraulic conductivity because of the greater thickness of the massive parts of the flows. The posterosional lavas congealed frequently on gentle slopes, resulting in dense and massive rocks of comparatively low permeability.

Manifestly, the regional hydraulic conductivity of extrusive rocks constituting the major aquifers is extremely high. The layered sequencing of flow units suggests that the horizontal component of conductivity should be greater than the vertical component, and in modeling, this relationship is normally assumed. It is debatable as to what degree such an assumption is justified. The uncertainty allows for manipulation in calibrating models. Extrusive rocks that have been weathered, especially those underlying a blanket of older alluvium and marine sediments, exhibit hydraulic characteristics grossly inferior to those of fresher rock. The differences are so great that weathered zones are treated as distinct from fresh rock aquifers. The effects of weathering are most pronounced in valleys and in coastal plains where a column of saturated sediments overlies the parent basalt. The weathered mantle (saprolite) in situ on parent basalt is usually more than 9.2 m (30 ft) thick. The original mineralogy is altered by hydration, and the permeability elements are clogged by clays and colloids. Wentworth's (1938) laboratory determinations of hydraulic conductivity for four samples of weathered Koolau basalt showed a range of 0.025 to 0.039 m/day (0.083 to 0.128 ft/day). A pump test in saprolite in central Oahu (Mink, 1982) indicated hydraulic conductivity of 0.152 to 0.305 m/day (0.5 to 1.0 ft/day). Laboratory measurement of borehole cores of saprolite in another part of central Oahu (Lau et al., 1987) indicated hydraulic conductivity of 0.13 to 3.65 m/day (0.42 to 12.0 ft/day). Such weathered sections are effective aquitards and, indeed, are probably the most effective confining member of the sedimentary sequence forming the caprock of the coastal plain.

Intrusive Rocks

Intrusive rocks of Hawaii volcanos consist almost entirely of magma congealed in the rock structures through which the magma was moving but did not effuse at the surface. There are no large plutonic masses in the islands. Individual intrusive units are small-scale features, but because they are concentrated along rift zones and in the caldera region, they play a fundamental role in the hydrology of each island.

Normally, the extrusive zones of a Hawaii volcano are comprised of a caldera and several narrow rift zones radiating from it. Calderas are usually less than 11.26 km (7 mi) in diameter and contain an intrusive assemblage of dikes, stocks, and sills mixed with collapse breccia and pyroclastics. Rift zones are normally less than 4.83 km (3 mi) wide but extend for tens of kilometers along linear trends. Their dominant intrusive rocks are dikes. Sills, occurring mostly as the horizontal expression of dikes, are frequent but limited in dimensions. Stocks and other small intrusive bodies are rare and hydrologically insignificant.

Dikes of the rift zone are the most widespread intrusive rocks. Volumetrically they account for only a tiny portion of the volcanic masses, but their hydrological significance is immensely greater in degree. The quasi-vertical dikes are poorly permeable barriers to the flow of groundwater in contrast to the extraordinary flow characteristics of the layered rocks. In the main part of the rift zone of the Koolau Range on Oahu the average dike is 1.5 to 1.8 m (5 to 6 ft) thick. Multiple dikes with virtually no extrusive rock between them are common. Between most dikes lie compartments of normal-layered lavas that often form small aquifers.

Metamorphic Rocks

The metamorphic rocks in Hawaii are associated with caldera processes and with hydrothermal activity deep in the rift zones. Surface exposures are limited to calderas; in rift zones these rocks do not commonly occur in regions of groundwater interest.

Basalts altered by volcanic gases and hydrothermal solutions become metamorphic rocks. Typically, olivine converts to serpentine and talc, and pyroxene to chlorite (Macdonald and Abbott, 1970). Precipitation from circulating solutions yields quartz, opal, calcite, and zeolites that fill vesicles and clog permeability elements, greatly reducing the porosity and hydraulic conductivity of the rock mass. Metamorphic rocks are too poorly permeable to behave as aquifers.

Sedimentary Rocks

For the island of Oahu, Wentworth (1951) divided the sedimentary rocks into older, intermediate, and recent alluvial and marine sediments. This classification is most applicable to the older islands of Kauai, Oahu, Molokai, and Maui, and less so to Lanai and Hawaii. Nevertheless it is convenient for descriptive purposes.

The older alluvium consists of detritus created during the initial period of profound erosion. It filled the deeply carved valleys and spread out as deltaic platforms along coastal margins. The older alluvium is composed of variable-sized particles of basalt, from silt to boulders, cemented in a clay matrix. The individual particles, which have been altered by weathering and compaction into a moderately hard coherent mass, retain their original shape.

The older alluvium lies immediately above weathered basalt bedrock in deep valleys and beneath coastal plains. In the valleys the alluvium, referred to as "valley fill", is thickest below an elevation of about 30.5 m (100 ft). Beneath the coastal plains and in lower reaches of valleys it is hundreds of meters thick. The older alluvium is a major component of "caprock" rimming portions of the older islands.

Wentworth (1938) measured parameters of the older alluvium in the laboratory. Eight samples gave a range in porosity of 46.4 to 62.4% and a range in hydraulic conductivity of 0.058 to 0.113 m/day (0.19 to 0.37 ft/day), approximately ten thousand times less than the hydraulic conductivity of fresh basalt. The most effective confining layering in the caprock evidently consists of weathered basalt overlain by older alluvium.

The older marine sediments interfinger with and overlie the older alluvium. The basal part of the marine section consists predominantly of estuarine and lagoonal type mud, silt, and sand. Fossil coral reefs and associated detritus typically appear in the upper portion of the section. The older marine sediments grade without noticeable unconformity into the intermediate marine sediments, which reach to an elevation of about 30.5 m (100 ft) above sea level. Coral reef deposits are more common in the intermediate sequence.

The clays, muds, and silts of the marine sediments may be as poorly permeable as the older alluvium, but fossil coral reefs are as permeable as unweathered basalt. These reefs form small, occasionally intercalated but frequently isolated aquifers in the sedimentary column. The whole caprock is saturated with salty to fresh water, but freshwater flow is most dynamic in the limestone sections.

The intermediate alluvium embraces the detritus that formed subsequent to the major erosional stage but prior to contemporary activity. It includes colluvium, talluvium, and, in general, the slope wash mantle overlying bedrock in the mountain areas. Neither as voluminous nor as stable as older alluvium, it serves as aquifers on only a small scale and in unusual circumstances.

The recent alluvium is even less important hydrologically than the intermediate alluvium. It consists of the clays, sands, gravel, and boulders deposited since the last high sea stand at about 7.6 m (25 ft) approximately 125,000 years ago. Except in unique topographic situations, its accumulation is less than 3.1 m (10 ft) thick. Similarly, the recent marine sediments comprise the detritus and coral reefs laid down on and along the coasts since the last high sea stand. They are unimportant in groundwater hydrology.

Aquifers occur within the sedimentary column, yet sediments as a whole are hydrologically important as aquitards confining volcanic rock aquifers. Only relatively thick fossil coral in the uppermost part of the sequence constitute aquifers capable of yielding fresh to brackish water.

Groundwater Occurrence

In the Hawaiian islands groundwater occurs either in "basal" aquifers or in "high-level" aquifers. This fundamental discrimination separates groundwater that is continuous with underlying seawater (a basal lens) from groundwater that is not hydraulically connected to seawater or groundwater whose interface with seawater is so deep as to be unimportant in ordering the course of groundwater development.

Basal groundwater is the most common groundwater occurrence. Normally, lavas forming the flanks of volcanos beyond the intrusives of the rift zones contain basal water. Many of the basal lenses are very large, and where coastal plains exist these lenses attain depths of more than 305 m (1000 ft) in response to the obstruction to outflow caused by the sedimentary caprock column. Heads in basal lenses range from

0.9 m (3 ft) where the outflow front is open, such as along 161 km (100 mi) of the west coast of the island of Hawaii, to as much as 12.8 m (42 ft), such as at the original head in Honolulu. The caprock confines the basaltic aquifer from above, but lenses are unconfined from below.

High-level groundwater, although not as really extensive as basal groundwater, plays a very important role in the water supply of Hawaii. The most familiar high-level aquifers are those composed of extrusive lavas trapped between intrusive dikes. The size of individual compartmented aquifers depends on where in the rift zone they occur. In the marginal dike zone, where dikes constitute less than 5% of the rock mass, the aquifers are large, whereas in the dike complex, defined as the zone containing more than 10% dike rock, the individual aquifers are small. The dikes are magnitudes less permeable than the lavas, but hydraulic connection among an array of dike aquifers allows large areas of a rift zone to be treated as single aquifers in which the controlling permeability is that of the quasi-vertical dikes.

Another common type of high-level groundwater occurs in permeable strata resting on poorly permeable layers such as ash, tuff, erosional unconformities, and a really extensive massive lava flows.

In the sediments of the caprock groundwater is normally basal, but parabasal conditions occasionally occur. Parabasal refers to groundwater hydraulically continuous with a basal lens but which itself is not underlain by salt water.

Aquifer Identification and Classification

To standardize references concerning the occurrence of groundwater, a framework within which groundwater regions on each island are identified and classified has been established (Mink and Lau, 1990). The classification is based on a hierarchy of descriptors beginning with location by island and sector, to which belongs a set of aquifer systems, within which are a variety of aquifer types. An aquifer sector is a large region containing a variety of hydrogeological environments that are generally continuous with each other. An aquifer system is an area within a sector in which hydrogeological conditions may vary but groundwater is hydraulically continuous. An aquifer type is a portion of a system displaying similar hydrogeological features. Finally, an aquifer unit can be added to the classification. The unit is an identifiable aquifer.

In the absence of identified aquifer units, the aquifer type is the category for which aquifer conditions are described. Groundwater is either basal or high level and may be confined or unconfined. Aquifers are classified as flank (horizontally extensive), dike (dike compartments), perched (resting on an impermeable layer), and sedimentary (nonvolcanic lithology). The hierarchical classification leads to an aquifer code which succinctly describes essentials of location and type of groundwater occurrence.

HISTORY OF GROUNDWATER INVESTIGATIONS AND MODELS IN HAWAII

History records events that happened, why they happened, and relates them to the present. Events occurred because either opportunities or crises presented themselves. In Hawaii, the following three outstanding motivations have taken place since 1879.

1. Water Supply

    The need for irrigation water for large-scale sugarcane culture provided the necessary economic incentive for prospecting and exploiting large, sustainable water sources since the discovery of the artesian groundwater source in Ewa on Oahu in 1879. The groundwater resources proved to be a bonanza not only for irrigation but for drinking water because of their superior water quality. These two water demands and the fear of seawater encroachment have kept the pressure on groundwater investigations (Fujimura and Chang, 1981).

2. Protection from Contamination

    In 1983, organic chemicals were discovered in a few drinking water wells in the Pearl Harbor aquifer, the largest water source on Oahu, after one century of intense groundwater exploitation (Lau and Mink, 1987). The crisis created a new dimension in water quality concerns and has prompted intensive investigations on chemical leaching and groundwater contamination—both remedial and preventive measures. This crisis should be viewed in the context of the excellent record of groundwater protection that was initiated in the 1920s and has been made ever more stringent since.

3. Sustenance of Water Environment

    Environmental concerns caught up with groundwater management in Hawaii in the late 1970s inasmuch as groundwater interacts with surface waters, both freshwater and coastal water. They are reflected in such events as (1) taro culture that depends on groundwater discharges (*Reppun v. Board of Water Supply*, 1979), (2) minimum stream flow that is regulated by the state, and (3) anchialine ponds, which are exposures of the groundwater table, and the pond ecosystems along the Kona coast on the island of Hawaii. In general, they all fall under wetland protection and have posed additional criteria for groundwater extraction and management.

Investigation of real-world situations invariably involves fact finding—acquisition of site-specific data. Analysis of the data attempts to explain why and how. Often, models are either used as a tool incorporating data or created as a result of the absence of data. The temptation to predict with models must weigh carefully with the uncertainty resulting from insufficient data and hence the risk. Subsequent operations or investigations produce more facts. The added facts coupled with applicable scientific theories either validate or falsify the model(s); in the latter event, new or better models are created.

Hawaii investigations in groundwater flow and transport have involved the whole range of models: conceptual, analytical, physical, numerical, and even stochastic. Limited by length, this paper attempts to be representative rather than comprehensive. It focuses on basal lenses—flow dynamics and organic contamination; high-level water is included in the potpourri.

Early Recognitions of Groundwater Occurrence

In the course of searching for sources of water supply, major types of groundwater occurrence in Hawaii were recognized early through a succession of discoveries. The discovery of artesian water in Hawaii was resulted from the Pioneer Well drilled for J. Campbell in Ewa on Oahu in 1879 (Lau, 1981). At a depth of 67.1 m (220 ft), a flowing artesian condition was encountered; the water level under a no-flow condition rose to 9.8 m (32 ft) above sea level. The drilled site was located about 4.6 m (15 ft) above sea level.

It was not until 1909 that the phenomenon of buoyancy of freshwater on underlying saltwater was recognized by C. B. Andrews in Hawaii. He worked out the mathematical expression of the ratio without knowledge of the work done in Europe by W. Badon-Ghyben (1888) and A. Herzberg (1901). The first recognition of the transition zone was probably credited to H. S. Palmer in 1927; his field data presented a gradual rather than an abrupt increase of salinity with depth. The nomenclature of basal water was introduced by O. E. Meinzer in 1930.

The two main types of high-level water were also discovered early. In 1912, J. B. Cox already recognized that volcanic dikes had something to do with the high-level water in Honokohau on Maui (Cox, 1981). In 1912, W. O. Clark determined that the high-level springs in Kau on Hawaii were fed by groundwater perched on low permeability ash layers (Cox, 1981).

Early Flow Models

Treating the Hawaii groundwater system by a distributive dynamic approach has been the model of many investigations since the 1930s. This approach requires the use of a fluid dynamics law governing the flow of fluids in a porous medium. M. Muskat in 1937 and M. K. Hubbert in 1940 already firmly demonstrated the utility of Darcy's law for groundwater flow. Is Darcy's law valid for Hawaii basalt?

In Hawaii in 1933, S. T. Hoyt conducted a "mechanical testing program" involving tests that consisted of pumping a well at different constant rates and measuring the accompanying drawdowns (Mink and Lau, 1980). He then solved the following equation for K and n, $H = KQ^n$, in which H is drawdown, Q is pumpage, and K and n are constants. Hoyt's program could have deduced a unity value for n, thus pointing to an equation like Darcy's law. Unfortunately, the head was measured in the vicinity of the well face where the flow is nearly turbulent rather than laminar. Hoyt obtained the value of n between 1 and 2.

In 1946, Wentworth reanalyzed and supplemented Hoyt's data with the Halawa Station data and was able to show that, except for the first few meters around large wells pumped at substantial rates, laminar flow prevailed. The value of n was found to be unity. Wentworth's results verified Darcy's law for Hawaii basalt aquifers and firmly established the first Hawaii groundwater model.

Flow Dynamics In Thick Lenses

Conceptual Models

Aquifers and groundwater, for the most part, are invisible; they offer special challenges to the human imagination. Several conceptual models in Hawaii inspired subsequent quantitative studies.

1. Bottom Storage of Thick Lenses

    Between 1942 and 1951, Wentworth (1942, unpublished data) originated the bottom storage concept to rationalize aquifer flow behavior in the Honolulu thick lenses. He hypothesized a substantial "lag" time for the interface to respond fully to a sustained step-change in the freshwater head, given initial dynamic equilibrium of the thick lens. However, he did not formulate mathematical equations describing the hydraulics of the system itself. Instead, he opted to use and narrate a "reservoir-pipe" analogy of the groundwater flow system. A thick lens offers great flexibility for groundwater extraction and management; this was perhaps the motivation for gaining understanding of the flow system. The "lag" enigma continued to attract studies (Essaid, 1986; Ogata and Lau, 1990), but it remains not totally resolved today.

2. Transition Zone

    Another conceptual model was the rinsing hypothesis advanced by Wentworth in 1948 to explain the creation and expansion of the transition zone of the interface between freshwater and saltwater. His explanation using discretized mixing layers was an inspiration in the derivation of modern dispersion theory.

3. Flowline Patterns

    Explicit expression of the distributive flow pattern present in the basal lens did not appear until the early 1960s. Conceptual patterns of flow lines in a freshwater lens appear in the summary of preliminary findings in groundwater studies of southern Oahu. Head measurements by the Board of Water Supply and other agencies, when reduced to equipotential lines, can portray flow patterns on a horizontal plane. This was done apparently for the first time in Hawaii for the eastern system of the Pearl Harbor aquifer and was reproduced in the calibration of a physical sandbox model.

Pioneering Investigations

A few pioneering investigations introduced new methods and models to blaze the trail for later efforts. Probing of the freshwater–saltwater interface by geophysics—electrical resistivity method—was successfully performed by Swartz on several Hawaiian islands in 1937. The sequel appeared in the mid-1960s in several geophysical studies on groundwater occurrence conducted by the University of Hawaii Water Resources Research Center (UH-WRRC).

In 1936, a gradual transition between freshwater above and saltwater below was demonstrated under the guidance of H. T. Stearns using water samples drawn at various depths from a test hole drilled on Maui (Cox, 1981). In 1960 the first instrumented measurement was performed for a thick lens near the inner shore of Pearl Harbor on Oahu (Visher and Mink, 1960). The latter provided not only the physical evidence of a thick transition zone in a major basal spring area but also the field data required to calibrate subsequent modeling of the thick lens (Lau, 1962; Souza and Voss, 1987).

Geochemical mixing phenomena near the top of the lens were observed in the return percolation water from sugarcane irrigation by Mink in 1962; he later modeled the observation as mixing cells. Many years later, organic contamination of basal lens was modeled with a more elaborate mixing cell model (Orr and Lau, 1987).

Physical Models

Physical modeling was introduced in a timely way to collaborate with field observations. Flow distributions of freshwater and seawater in and beneath a lens were visualized in a series of sandbox experiments utilizing dye-produced flow patterns by Lau in 1960. In 1962, he developed a sandbox simulation model of Kalauao Springs and the surrounding basal lens in southern Oahu for the purpose of determining the optimum allowable extraction of groundwater. The model, which was calibrated for flow and transport, provided some insight to the dynamic behavior of a thick lens.

Analytical Models

Analytical models with closed-form solutions show a clear functional relationship between variables and the significance of parameters; they are simple to use and easy to understand. Their use can be justified for planning and management purpose if they are validated by field data.

A robust analytical flow model was developed by Mink in 1980 for a vertical section of a thick lens with a sharp interface. Its pseudo-transient solution is obtained for a substantial period during which the draft is kept constant. The model was successfully tested with data from southern Oahu and at present is used by the state of Hawaii for estimating aquifer sustainable yields.

A simplified transport model with an analytical solution was developed by C. C. K. Liu and Q. Huang in 1992 for the thick lens in the "Beretania" aquifer, now classified as the Nuuanu aquifer system in Honolulu to predict the position of isochlors in the lens, especially the 2% isochlor that approximates the recommended upper limit of chlorides for potability (Huang, 1992). The basic assumptions are predominance of horizontal flow and transverse dispersion. The present version showed some promise when compared with Huang's SUTRA simulation results and field data of the Nuuanu aquifer.

Numerical Models

The first numerical model was devised for the thick lens in Honolulu by GE-Tempo in 1968 (Meyers et al., 1974). Many have followed since, including state-of-the-art models such as SUTRA used for several aquifers in Hawaii. The Tempo model simulates advection in a thick basal lens in Area 1 of Honolulu, now classified as the Palolo aquifer system. The long-term (40 year) record of head was successfully simulated with fitted parameter values by treating the transition zone as a single sharp interface. The fitting process required the use of a lower permeability value than normally accepted, yet it still encountered difficulty for the period of high demand during the World War II years.

The inclusion of dispersion with advection in modeling was soon followed by S. W. Wheatcraft in 1979 for a quite different reason—wastewater injection into and beneath the lens. The model features two solutes—chlorides and a conservative contaminant—in a vertical section; validation of the numerical results was performed with sandbox experiments. This effort not only confirmed buoyancy as a hydraulic expectation but visually showed simultaneous but different dispersion phenomena of two conservative solutes. The first quasi-three-dimensional modeling was conducted for the entire basal lens in the Pearl Harbor aquifer as an understanding exercise (Liu et al., 1983). The long-term records of head and draft were simulated for a sharp interface.

The next advancement was the inclusion of saltwater flow, yet retention of a sharp interface in a quasi-three-dimensional model for the Waialae aquifer system in Honolulu (Essaid, 1986). The long-term head record was claimed to be better approximated by simulation with two fluid flows than with just one as done by Eyre in 1985. The effort included a theoretical discourse on the transient rise of the interface.

The SUTRA model is a numerical two-dimensional model coupling flow and transport phenomena of a varying density fluid. It was used to simulate a vertical section of the thick lens in the Pearl Harbor aquifer, allowing a mixing zone to occur at the interface as a result of fluid-density difference (Voss and Souza, 1987; Souza and Voss, 1987). The simulation of long-term head and dispersion in a deep well required extensive fitting with existing data. Some simulation is made consistent with a priori values such as hydraulic conductivity and transverse dispersivity. Other values (e.g., longitudinal dispersivity, anisotropy of hydraulic conductivity, caprock hydraulic conductivity, specific yield, aquifer rock compressibility, and hydraulic impedance) were identified as being necessary to simulate existing data. The model velocity distribution appears to support Dupuit approximation. The SUTRA model did not address temperature phenomena observed in a thick lens. Nevertheless, the model successfully simulates the evolution of a transition zone.

Organic Contamination of Basal Lenses

In 1983, organic chemical contamination in the Pearl Harbor aquifer created a crisis and engendered massive remedial actions of the potable groundwater sources in which they were detected in trace amounts (Lau and Mink, 1987). The harbinger was the discovery of 1,2-dibromo-3-chloropropane (DBCP) and ethylene dibromide (EDB) in the Del Monte Corporation Kunia well in 1977 (Mink, 1982). The level of contamination, resulting from a large spill and numerous minor incidents of dripping over the years, was very high. The discovery of these two chemicals and another organic chemical, 1,2,3-trichloropropane (TCP), all of which were applied as nematocides in extensive pineapple fields since the 1940s, culminated in the crisis of July 1983. Although the discovery of these organics can have an agricultural impact, the possibility of nonagricultural impact cannot be excluded. Nine municipal water wells were closed in the all-important Pearl Harbor aquifer for a total loss of 0.57 m^3/s (13 mgd) of yield. The suspected health risks are carcinogenic and other toxic effects, but MCLs—levels requiring regulatory actions—were nonexistent at the time and have remained controversial for many years. In 1986 trichloroethylene (TCE) from unknown sources was discovered in drinking water wells in the Wahiawa aquifer system at Schofield on Oahu. In 1990 this site was included in the National Priorities List (superfund). The present state MCLs for some of these chemicals are more stringent than the U.S. EPA's; the resulting impact is that all natural attenuative processes should be assessed, including those of lesser importance. Petroleum hydrocarbon compounds have not been the major cause of contamination of potable groundwater sources. However, their discovery in soils and nonpotable aquifers has resulted in significant remediation.

An investigation conducted by the University of Hawaii in 1985 revealed that an extensive region [163 km^2 (63 mi^2)] of basal water in the Pearl Harbor, Central, and North Aquifer Systems on Oahu contains organics at detectable levels beneath and downgradient of pineapple fields (Lau et al., 1987). The nature of nonpoint source pollution that was characterized severely limited the range of remediation measures to point-of-use treatment and natural attenuation. The extensiveness of the contamination persisted with some downgradient migration toward uncontaminated potable wells, as evident in the 1993 monitoring investigation (Lau et al., 1993a). These symptoms were not unexpected; they were generally predicted in 1987 by two groundwater transport models for the Mililani area—multiple mixing cells and method of characteristics.

The 1985 investigation also revealed that the surface soils and deep saprolite [up to 45.7 m (150 ft)] act as exceedingly effective natural barriers against transport of these organics by leaching. For example, in the case of DBCP, the level is reduced from 10 mg/kg as applied to surface soil to 0.1 µg/l as percolate beneath the root zone and reduced by only one more order of magnitude after transport in the basal lens. Further, the 1985 investigation indicates that the DBCP residue remaining in the surface soil does not pose a significant source of further groundwater contamination—partially because of the high sorption and especially due to the extremely low desorption. These findings in the unsaturated zone guided the strategies—focus on the near surface layers or solum—for investigations and modeling in subsequent years (Green et al., 1988). The specific purposes have been developing (1) a leaching index for ranking various pesticides for a given site and (2) a dynamic model for predicting pesticide leaching concentration in soils.

When coupled with a geographic information system, Rao's attenuation factor was tested and considered satisfactory as a mobility index for ranking and selecting pesticides for the Pearl Harbor basin (Rao et al., 1985). For example, atrazine appears very unlikely to leach whereas diuron appears likely. However, the quality of the U.S. Soil Survey database was considered inadequate as first-order uncertainty analysis showed significant uncertainty in the mobility estimates. Better characterization of the parameter values for retardation and transformation is required (Loague et al., 1989c, 1990).

A transport dynamic model, PRZM, was tested satisfactorily for use in Hawaii after modification and extension by skilled persons but not for general use. Again, the tested soil survey database was considered inadequate in terms of soil hydraulic and retardation properties (Loague et al., 1989a,b).

The many inherent difficulties encountered in using the traditional physically based simulation of solute transport in upper soils have prompted examination of system theory as an alternative modeling approach (Liu et al., 1991). In this approach, the relations between the chemical input and the subsequent leaching are represented by system response functions that do not require the full knowledge of the transport system. Initial testing with the pesticide fenamiphos in Hawaii soils produced encouraging results.

The investigations and models of the unsaturated zone thus far conclude that the transport phenomena are inadequately described by existing models. This may imply the existence and hence the importance of preferential flow, which has not yet been included. University of Hawaii investigations have produced some physical evidences that would permit preferential flow.

It should be noted that groundwater investigations and modeling in Hawaii have been hampered by the great depth of volcanic rocks that must be penetrated before reaching the basal lens. This problem has rendered monitoring in potable groundwater sources difficult, expensive, and scarce. If it were not for this problem, monitoring of the huge, organics-contaminated groundwater bodies would present a golden opportunity for understanding lens behavior and managing the groundwater resources.

Aided by the development of a precise yet practical helium detector, injected helium was successfully field tested in a basal lens in a basaltic aquifer in Waipahu on Oahu (Gupta et al., 1990). Although not designed for modeling, field test results yielded dispersivity values of basalts that are within the realm of acceptability. With further definement, injected helium can be a useful tool for transport studies in the saturated zone.

A large-scale investigation at Waiawa, Oahu, cited later as a planning model, essentially represents a culmination of knowledge to deal with potential nonpoint source pollution in the Hawaii environment from either agricultural or urban areas where the use of chemicals is unavoidable. A resulting strategy for prevention of nonpoint source pollution is the selection of chemicals with respect to their interaction with environmental factors that characterize their retardation and transformation. With this information, other strategies can be easily designed, including soil management (conservation and amendment), reduction of water percolation, and scheduling of irrigation and chemical application. These strategies must be coupled with wellhead protection area delineation—another investigation and model already completed in Hawaii—in order to deal with the realities of social and economic development (Lau et al., 1993b).

As a caveat, some organic chemicals discovered in Hawaii groundwater appear to be background. While it is premature to comment on anything else about background organics other than their detection, their presence seriously interfered with the activated carbon treatment of the contaminated groundwater in the Pearl Harbor aquifer (Dugan et al., 1992).

Also, a 20-year investigative research program on the reuse of wastewater indicates that when properly applied on land, effluents from publicly owned wastewater treatment facilities are not likely sources of groundwater contamination or environment problems in the Hawaii environment (Lau et al., 1975, 1980, 1989).

Potpourri

Potpourri simply suggests a mixture of several groundwater issues—each of considerable importance—that collectively are unique to Hawaii.

High-Level Water

In the interest of water supply, investigations of high-level water have been many but models have been few. Modeling extraction of the dike-impounded water from a high-level tunnel is derived from the desire to predict (1) free-flow discharge responding to varying hydrological inputs and (2) controlled discharge under a bulkheaded condition. The early models and field data for Waihee and Kahana tunnels on Oahu were reanalyzed as laminar and turbulent flow models, for which analytical solutions were given (Takasaki and Mink, 1985).

The Schofield high-level water body was modeled to predict a seemingly mundane question—the amount of lowering of the groundwater table resulting from a specified additional draft (Dale and Takasaki, 1976). This water body is unique and important, serving as a major subsurface recharge to the all-important Pearl Harbor aquifer on Oahu. It differs from basal water but its hydrogeological nature is not adequately understood. The model was necessarily conceptual with many assumptions; it is lump-parametered and calibrated with a 40-year record of head.

Virus Monitoring

The microbiological quality of Hawaii's potable groundwater sources is recognized as excellent, as evidenced by the regulatory records that show virtually no bacterial indicators in these sources and no epidemics attributable to drinking from these sources. These records have allayed concerns by some over the practice of cesspools that was outlawed as a private wastewater facility in 1991.

On-site virus assessment that is essential to wastewater reuse was conducted by UH-WRRC; it shows that the human enteric viruses identified in secondary chlorinated sewage effluent applied as irrigation water are effectively retained and inactivated by the 1.52-m (5-ft) deep soil in a percolate lysimeter (Lau et al., 1975, 1980). Environmental factors that enhance virus inactivation, including desiccation, temperature, and UV light, have been identified in field experiments.

Special monitoring of potable groundwater sources for human enteric viruses produced negative results for three different sites in connection with municipal or domestic wastewater—two instances of disposal (Fujioka and Lau, 1982, 1983) and one of reuse.

The same UH-WRRC field research program on wastewater reuse for irrigation and recharge conducted over a span of 20 years has firmly established that, under established guidelines, vegetated soil mantle of 1.52 m (5 ft) thickness can retain and inactivate all bacterial indicators as well as remove virtually all other contaminants, including nitrogen, phosphorus, BOD, and suspended solids, to below MCL levels (Lau et al., 1980).

Planning Models

Models have been used recently for assessing effects of proposed development on groundwater recharge and quality on Oahu. A situation was created by anticipated urban use of chemicals in a proposed large-scale residential land development tributary to the Waiawa infiltration gallery on Oahu, a large [0.66 m^3/s (15 mgd)] sole source for drinking water supply for the U.S. Navy situated in a basal lens some 122 m (400 ft) beneath the proposed urban site (Oki et al., 1991). A calibrated coupled model—PRZM for the unsaturated zone and MOC for the basal lens—indicates potential transport of trace amount of chemicals to the facility but not at a level exceeding the applicable MCL, provided best management practice of chemicals, land, and water is performed in the land development.

Another situation was recently created for the planners and public officials on the proposal to raise the permissible level of urban development in central Oahu (Ridgley and Giambelluca, 1990). Opposition was presented on the grounds that it threatens agricultural land and the sustainability of the basal lens. To evaluate the impact, multiobjective programming models were tested, coupled with groundwater recharge and water balance models. Although validation will be difficult, the model offers insights to the geohydrological effects of a range of proposed land use patterns.

Stochastic Subsurface Hydrology

Treating groundwater systems as stochastic in Hawaii has been quite rare. As a forerunner, Wentworth (1951) performed multiple correlation of head, rainfall, and draft for the Honolulu aquifers. The effort was abandoned because the model failed to validate reliable estimates of natural water flux in the aquifers. It was not until 1982 that geostatistical concepts were applied to design a sampling scheme to characterize a space-dependent hydraulic property of the surface soils in the Pearl Harbor basin (Bresler and Green, 1982).

Contemporaneity of Hawaii research work is the study of scale-dependent macrodispersion in unsaturated heterogeneous porous media with numerical experiments in synthetic soil columns (Liu et al., 1991). This study helps support the contention that the unsaturated zone in the Hawaii environment lends itself to all modes of modeling and stands a good chance for fast scientific advancement in contrast to on-site research of the saturated zone that for the most part is in deep hard rock not easily subject to validation.

CURRENT AND FUTURE ROLE OF MODELS IN HAWAII

The historical events associated respectively with the Pioneer Well in 1879 and the Del Monte Kunia Well in 1977 have engendered many subsurface investigations and models in Hawaii. In the cases of water supply, investigations have discovered and established the basic knowledge of Hawaii hydrogeology and the occurrence of Hawaii groundwater systems as highlighted in this paper. As it stands today, this body of knowledge is fundamental and irrefutable as the base of all investigations and modeling.

Nature's complexities in Hawaii are an exciting challenge to normal scientific advancement of groundwater hydrology. The first constraint is the great depth of rock which must be penetrated to reach the saturated zone, rendering monitoring difficult, expensive, and scarce. The second is the great heterogeneity of the numerous thin layers of hard rocks constituting the aquifer—heterogeneous by layer and within layer. While studies involving a regional scale can produce reliable results by lumping, local-scale studies, especially those involving transport, encounter substantial uncertainties.

Fortunately, operational experience of major waterworks for groundwater extractions has provided a strong database to enhance geohydrological investigations. The data not only reflect how the groundwater systems actually responded to various hydrological stresses but also offer guidelines for subsequent exploitation and understanding of the systems.

Models have rarely, if ever, been the sole determining factor in decision making partly because of the intent of modeling not for such purpose and partly because of the difficulties encountered in model calibration and validation. Models are commonly used in arguments supporting or opposing alternative means of groundwater exploitation and in evaluating proposed land developments. For the purpose of general planning of water resources a validated analytical model is used at present to help determine aquifer sustainable yield.

As testimony of the combined value of investigations and operational experience, the groundwater resources in Hawaii continue to be robust and, on Oahu, are capable of yielding a sustainable water supply of about 19.7 m^3/s or 450 mgd. Having proved its worth, this strategy for acquiring the needed knowledge should continue. However, additional basic knowledge about Hawaii groundwater systems that are needed may be efficiently augmented by carefully planned investigative research projects. They need to be promoted and conducted at sites that are economically feasible and that can yield transferable information.

A caveat about unreasonable expectation of models, and especially the numerical models now in vogue, is obvious yet appropriate. Paraphrasing J. Bear, no model is perfect; no model suits all purposes (Bear, 1979). Numerical models have yet to be engaged in establishing groundwater development rules anywhere in Hawaii, but they have been successfully employed in describing groundwater systems for which a long-term data record exists. Any push to use numerical models to stipulate governing rules is obviously premature for groundwater sources having no significant record of behavior with which to validate a model, or a clear grasp of aquifer parameters, or even a definition of aquifer boundaries.

In the case of contaminant transport in the unsaturated zone (surface soils and deep saprolites), investigations and modeling have helped to comprehend that vegetated soils in the tropical environment are a remarkable deterrent to contaminant transport—such as bacteria, viruses, nutrients, organics, and common quality parameters in wastewater effluent. Modeling certain processes—such as organic retardation and leaching—has aided prevention of pollution. The current investigative directions and strategies are being proven appropriate and worthy to be further employed in tandem with numerical modeling.

ACKNOWLEDGMENTS

The authors thank the Water Resources Research Center publications office staff for their assistance in the preparation of this manuscript. This is contributed paper CP–94–03 of the Water Resources Research Center, University of Hawaii at Manoa, Hawaii.

REFERENCES

Andrews, C. B., The structure of the southeastern portion of the island of Oahu, Master's thesis, Rose Polytechnic Institute, 1909, 18.

Badon-Ghyben, W., Nota in verband met de voorgenomen putboring nabij Amsterdam (Notes on the probable results of well drilling near Amsterdam), *Tijdschr. Kon. Inst. Ing.,* The Hague, 1888/9, 8–22.

Bear, J., *Hydraulics of Groundwater,* McGraw-Hill, New York, 1979, 437–439.

Bresler, E. and Green, R. E., Assessment of integral soil-water properties for hydrologic characterization of Hawaii soils, Tech. Rep. No. 148, Water Resources Research Center, University of Hawaii at Manoa, Honolulu, 1982, 42.

Cox, D. C., A century of water in Hawai'i, in *Groundwater in Hawai'i: A Century of Progress,* Fujimura, F. N. and Chang, W. B. C., Eds., University Press of Hawaii, Honolulu, 1981, 51–79.

Dale, R. H. and Takasaki, K. J., Probable effects of increasing pumpage from the Schofield ground-water body, Island of Oahu, Hawaii, Water-Resources Investigations 76-47, U.S. Geological Survey, 1976, 45.

Dugan, G. L., Fujioka, R. S., Lau, L. S., Takei, G. H., Gee, H. K., and McParland, T. L., Granular activated carbon treatment of Mililani well water: Phase II study to extend effective life of GAC, 6-month progress report, Spec. Rep. 09.24:92, Water Resources Research Center, University of Hawaii at Manoa, Honolulu, 1992, 39.

Essaid, H. I., A comparison of the coupled fresh water-salt water flow and the Ghyben-Herzberg interface approaches to modeling of transient behavior in coastal aquifer systems, *J. Hydrol.,* 86, 169, 1986.

Eyre, P. R., Simulation of ground-water flow in southeastern Oahu, Hawaii, *Ground Water,* 23(2), 325, 1985.

Fujimura, F. N. and Chang, W. B. C., Eds., *Groundwater in Hawai'i: A Century of Progress,* University Press of Hawaii, Honolulu, 1981, 260.

Fujioka, R. S. and Lau, L. S., Evaluation of possible movement of viruses in Makakilo sewage through the soil strata and into surface waters in the vicinity of Makakilo Well No. 1, Spec. Rep. 4.5:82, Water Resources Research Center, University of Hawaii at Manoa, Honolulu, 1982, 25.

Fujioka, R. S. and Lau, L. S., Haina well water analysis for presence of human enteric viruses and other water quality parameters, island of Hawaii, Spec. Rep. 3.30:83, Water Resources Research Center, University of Hawaii at Manoa, Honolulu, 1983, 9.

Green, R. E., Loague, K. M., and Yost, R. S., Assessment of pesticide leaching using soil survey and taxonomy, International Interactive Workshop on Soil Resources: Their inventory, analysis, and interpretation for use in the 1990's, March 22–24, Minnesota Extension Service, University of Minnesota, St. Paul, 1988, 204.

Gupta, S. K., Lau, L. S., Moravcik, P. S., and El-Kadi, A., Injected helium: A new hydrological tracer, Spec. Rep. 06.01.90, Water Resources Research Center, University of Hawaii at Manoa, Honolulu, 1990, 94.

Herzberg, A., Die wasserversorgung einiger Nordseebaden (The water supply on parts of the North Sea coast in Germany), *Z. Gasbeleucht. Wasserversorg.,* 44, 815, 824, 1901.

Huang, Q., Salt water intrusion in response to the basal water development and estimation of sustainable yield of Beretania aquifer, Oahu, Hawaii, Master's thesis, University of Hawaii at Manoa, Honolulu, 1992, 85.

Hubbert, M. K., The theory of ground water motion, *J. Geol.*, 48, 785, 1940.

Huber, R. P. and Adams, W. H., Density logs from underground gravity surveys in Hawaii, Tech. Rep. No. 45, Water Resources Research Center, University of Hawaii at Manoa, Honolulu, 1971, 39.

Lau, L. S., Laboratory investigation of sea-water into ground-water aquifer, Board of Water Supply, City and County of Honolulu, 1960, 91.

Lau, L. S., Water development of Kalauao basal springs: Hydraulic model studies, Board of Water Supply, City and County of Honolulu, 1962, 102.

Lau, L. S., Development and critique of geohydrological concepts, in *Groundwater in Hawai'i: A Century of Progress*, Fujimura, F. N. and Chang, W. B. C., Eds., University Press of Hawaii, Honolulu, 1981, 81–100.

Lau, L. S., Buxton, D. S., Fujioka, R. S., Gee, H. K., Giambelluca, T. W., Green, R. E., Loague, K. M., Miller, M. E., Mink, J. F., Murabayashi, E. T., Oki, D. S., Orr, S., and Peterson, F. L., Organic chemical contamination of Oahu groundwater, Tech. Rep. No. 181, Water Resources Research Center, University of Hawaii at Manoa, Honolulu, 1987, 153.

Lau, L. S., Dugan, G. L., Ekern, P. C., Young, R. H. F., and Burbank, N. C., Jr., Recycling of sewage effluent by irrigation: A field study of Oahu—Final progress report for August 1971 to June 1975, Tech. Rep. No. 94, Water Resources Research Center, University of Hawaii at Manoa, Honolulu, 1975, 151.

Lau, L. S., Ekern, P. C., Loh, P. C. S., Young, R. H. F., Dugan, G. L., Fujioka, R. S., and How, K. T. S., Recycling of sewage effluent by sugarcane irrigation: A dilution study, October 1976 to October 1979, Phase II-A, Tech. Rep. No. 130, Water Resources Research Center, University of Hawaii at Manoa, Honolulu, 1980, 107.

Lau, L. S., Giambelluca, T. W., Oki, D. S., Takei, G., and Gee, H. K., Assessment of the migration of organics contaminated groundwater in southern, central, and northern Oahu, Hawaii, Project completion report (synopsis), Water Resources Research Center, University of Hawaii at Manoa, Honolulu, 1993a, 8–13.

Lau, L. S., Hardy, W. R., Gee, H. K., Moravcik, P. S., and Dugan, G. L., Recharge with Honouliuli wastewater irrigation: Ewa plain, southern Oahu, Hawaii, Tech. Rep. No. 182, Water Resources Research Center, University of Hawaii at Manoa, Honolulu, 1989, 121.

Lau, L. S. and Mink, J. F., Organic contamination of groundwater: A learning experience, *J. Am. Water Works Assoc.*, 79(8), 37, 1987.

Lau, L. S., Oki, D. S., and Mink, J. F., Wellhead protection strategy for small aquifers: A study of Hawaii, Paper presented at the 1993 Annual Conference, American Water Works Association, San Antonio, Texas, June 6–10, 1993b, 25.

Liu, C. C. K., Feng, J. S., and Chen, W., System modeling approach for solute transport through upper soils, *J. Water Sci. Tech.*, 24(6), 67, 1991.

Liu, C. C. K., Lau, L. S., and Mink, J. F., Groundwater model for a thick freshwater lens, *Ground Water*, 21(3), 293, 1983.

Liu, C. C. K., Loague, K., and Feng, J. S., Fluid flow and solute transport processes in unsaturated heterogeneous soils: Preliminary numerical experiments, *J. Contam. Hydrol.*, 7, 261, 1991.

Loague, K. M., Green, R. E., Liu, C. C. K., and Liang, T. C., Simulation of organic chemical movement in Hawaii soils with PRZM: 1. Preliminary results for ethylene dibromide, *Pac. Sci.*, 43(1), 67, 1989a.

Loague, K. M., Giambelluca, T. W., Green, R. E., Liu, C. C. K., Liang, T. C., and Oki, D. S., Simulation of organic chemical movement in Hawaii soils with PRZM: 2. Predicting deep penetration of DBCP, EDB, and TCP, *Pac. Sci.*, 43(4), 362, 1989b.

Loague, K. M., Yost, R. S., Green, R. E., and Liang, T. C., Uncertainty in a pesticide leaching assessment for Hawaii, *J. Contam. Hydrol.*, 4, 130, 1989c.

Loague, K. M., Green, R. E., Giambelluca, T. W., Liang, T. C., and Yost, R. S., Impact of uncertainty in soil, climate, and chemical information in a pesticide leaching assessment, *J. Contam. Hydrol.*, 5, 171, 1990.

Macdonald, G. A. and Abbott, A. T., *Volcanoes in the sea*, University Press of Hawaii, Honolulu, 1970.

Meinzer, O. E., Ground water in the Hawaiian Islands, Water Supply Paper No. 616, 1930, 10.

Meyers, C. K., Kleinecke, D. C., Todd, D. K., and Ewing, L. E., Mathematical modeling of fresh water aquifers having salt water bottoms, Tech. Rep. GE74 TEM-43, Center for Advanced Studies, TEMPO, General Electric Co., report to Office of Water Resources Research, U.S. Dept. of the Interior, 1974.

Mink, J. F., Excessive irrigation and the soils and groundwater of Oahu, Hawaii, *Science*, 135(3504), 672, 1962.

Mink, J. F., State of the groundwater resources of southern Oahu, Board of Water Supply, City and County of Honolulu, 1980, 83.

Mink, J. F., DBCP and EDB in soil and water, Kunia, Oahu, Hawaii, Report prepared for Del Monte Corporation, Honolulu, 1982, 32.

Mink, J. F. and Lau, L. S., Hawaiian groundwater geology and hydrology, and early mathematical models, Tech. Memo. Rep. No. 62, Water Resources Research Center, University of Hawaii at Manoa, Honolulu, 1980, 75.

Mink, J. F. and Lau, L. S., Aquifer identification and classification for Oahu: Groundwater protection strategy for Hawaii, Tech. Rep. No. 179, Water Resources Research Center, University of Hawaii at Manoa, Honolulu, 1990, 28.

Muskat, M., *The Flow in Homogeneous Fluids through Porous Media*, McGraw-Hill, New York, 1937, 763.

Ogata, A. and Lau, L. S., Vertical movement of saltwater-freshwater interface in a thick groundwater system, Tech. Memo. Rep. No. 82, Water Resources Research Center, University of Hawaii at Manoa, Honolulu, 1990, 9.

Oki, D. S., Lau, L. S., and Green, R. E., Impact of proposed urbanization on groundwater quality: A case study in Hawaii, in *Proc. 1991 Annual Conference of American Water Works Association*, Philadelphia, 1991, 587–599.

Orr, S. and Lau, L. S., Trace organic (DBCP) transport simulation of Pearl Harbor aquifer, Oahu, Hawaii: Multiple mixing-cell model, phase I, Tech. Rep. No. 174, Water Resources Research Center, University of Hawaii at Manoa, Honolulu, 1987, 60.

Palmer, H. S., The geology of the Honolulu artesian system, Supplement to the report of the Honolulu Sewer and Water Commission, 1927, 68.

Rao, P. S. C., Hornsby, A. G., and Jessup, R. E., Indices for ranking the potential for pesticide contamination of groundwater, *Soil Crop Sci. Soc. Florida Proc.*, 44, 1, 1985.

Reppun v. Board of Water Supply, Civ. No. 50121 (Hawaii, 1st Circuit, October 11, 1979), 1979.

Ridgley, M. A. and Giambelluca, T. W., Urbanization, land-use planning, and groundwater management in central Oahu, Hawaii, Spec. Rep. 10.05:90, Water Resources Research Center, University of Hawaii at Manoa, Honolulu, 1990, 30.

Souza, W. R. and Voss, C. I., Analysis of an anisotropic coastal aquifer system using variable-density flow and solute transport simulation, *J. Hydrol.*, 92, 17, 1987.

Swartz, J. H., Resistivity studies of some salt-water boundaries in the Hawaiian islands, *Trans. Am. Geophys. Union*, 18, 387, 1937.

Takasaki, K. J. and Mink, J. F., Evaluation of major dike impounded ground-water reservoirs, island of Oahu, Water-Supply Paper 2217, U.S. Geological Survey, 1985, 77.

Visher, F. N. and Mink, J. F., Summary of preliminary findings in ground-water studies of southern Oahu, Hawaii, Circ. 435, U.S. Geological Survey, 1960, 116.

Voss, C. I. and Souza, W. R., Variable density flow and solute transport simulation of regional aquifers containing a narrow freshwater-saltwater transition zone, *Water Resour. Res.*, 23(10), 1851, 1987.

Wentworth, C. K., Geology and groundwater resources of the Palolo-Waialae district, Report to Board of Water Supply, City and County of Honolulu, 1938, 274.

Wentworth, C. K., Storage consequences of Ghyben-Herzberg theory, *Trans. Am. Geophys. Union*, 23, 683, 1942.

Wentworth, C. K., Laminar flow in the Honolulu aquifer, *Trans. Am. Geophys. Union*, 27, 540, 1946.

Wentworth, C. K., Growth of Ghyben-Herzberg transition zone under a rinsing hypothesis, *Trans. Am. Geophys. Union*, 29(1), 97, 1948.

Wentworth, C. K., Geology and ground-water resources of the Honolulu-Pearl Harbor area, Oahu, Hawaii, Board of Water Supply, City and County of Honolulu, 1951, 111.

Wheatcraft, S. W., Numerical modeling of liquid waste injection into island and coastal ground-water environments, Ph.D. Dissertation, University of Hawaii at Manoa, Honolulu, 1979.

Williams, J. A. and Soroos, R. L., Evaluation of methods of pumping test and analyses for application to Hawaiian aquifers, Tech. Rep. No. 70, Water Resources Research Center, University of Hawaii at Manoa, Honolulu, 1973, 159.

CHAPTER 15

Modeling Atoll Groundwater Systems

Frank L. Peterson and Stephen B. Gingerich

INTRODUCTION

Atolls commonly consist of a ring of small carbonate islets surrounding a shallow seawater lagoon. Under favorable geologic and recharge conditions small quantities of fresh groundwater may be contained within the larger islets. Usually fresh groundwater occurs as a thin, lens-shaped body separated from the underlying seawater by a layer of mixed water called the transition zone (Figure 1). As described by Underwood et al. (1992) and Griggs and Peterson (1993), the size and shape of a fresh groundwater lens within an atoll island are controlled primarily by the geologic framework and hydrodynamic processes resulting when a two-phase fluid system (freshwater and saltwater) is subjected to a combination of recharge, discharge, and tidally driven stresses.

Historically, atoll studies have consisted primarily of field investigations, and as a result of many years of data collection and analysis—including hydrogeologic mapping, borehole drilling, well testing, and groundwater tidal response analysis—the geologic framework of atoll islands has become fairly well understood. Conversely, our understanding of atoll hydrodynamic processes and how they are coupled with the geologic framework has developed much more slowly. It has only been with the application of high-speed computer modeling that we have begun to understand the details of hydrodynamic processes within atoll groundwater systems.

One model used by several investigators to simulate variable-density groundwater systems within atolls is SUTRA (Voss, 1984). In this paper we describe the use of the SUTRA model to evaluate several aspects of fresh groundwater lens dynamics and development on atoll islands. To do this two different atoll models were simulated. The first uses a generic atoll island with hydrogeologic parameters that are a composite of several typical atoll islands in the Pacific Basin to evaluate controls on the size and dynamics of the freshwater lens. The second models development and sustainable yield under various recharge conditions for Roi-Namur Island in Kwajalein Atoll, Republic of the Marshall Islands.

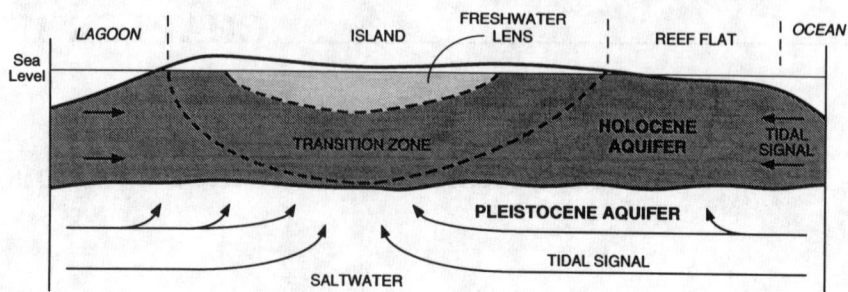

Figure 1. Groundwater lens on two-layered atoll island.

HYDROGEOLOGY OF ATOLL ISLANDS

Atoll groundwater systems have been studied for many decades. Although the geologic complexity and variability of atolls have long been recognized, until recently most investigators treated groundwater systems on atoll islands as single-layer homogeneous aquifers subjected to tidal stresses propagated laterally through the aquifer. In recent years a more realistic model of atoll hydrogeologic systems has evolved from investigations of a number of Pacific Ocean atolls (Buddemeier and Holladay, 1977; Hunt and Peterson, 1980; Wheatcraft and Buddemeier, 1981; Ayers and Vacher, 1986; Vacher, 1988; Anthony et al., 1989; Oberdorfer et al., 1990; Gingerich, 1992; Underwood et al., 1992; Griggs and Peterson, 1993). It is now believed that the hydrogeology of most atoll islands can best be modeled as a multi-aquifer system of up to 30 m of moderately permeable Holocene age sediments overlying much more permeable Pleistocene age deposits which extend hundreds of meters to the volcanic basement. It is recognized that the detailed hydrogeology of individual islands may vary considerably, and often the Holocene deposits are composed of several layers of varying permeability. In this conceptual model the tidal signal is thought to rapidly move laterally through the highly permeable Pleistocene aquifer and then propagate more slowly upward through the less permeable Holocene aquifer (Figure 1). This model generally is supported by groundwater tidal responses in the Holocene aquifers. Data summarized by Peterson (in press) illustrate that except in wells very close to the shoreline where the tidal signal can move most quickly through the Holocene aquifer, groundwater tidal efficiencies (ratio of amplitude of groundwater tidal fluctuation to coastal tidal fluctuation) generally increase with depth as the more permeable Pleistocene aquifer is approached. Groundwater tidal lags (time that groundwater tidal signal lags behind ocean tidal signal) also often decrease with increasing depth.

ATOLL MODELING

Early investigations of atoll and insular groundwater systems (Collins, 1976; Hantush, 1968; Rumer and Shiau, 1968; Vacher, 1974, 1988; Van der Veer, 1977; Wheatcraft and Buddemeier, 1981) used analytical solutions that assume a sharp interface between the freshwater and the saltwater. Henry (1964) was the first to use a semianalytical solution to define the position of the transition zone. In fact, this work has come to be known as Henry's problem, and its solution is often used to verify the accuracy of numerical models.

Recently, investigators have used numerical models to simulate atoll and insular groundwater systems. Numerical models allow the use of complex aquifer and boundary conditions, transient solutions, and density-dependent fluid flow and solute transport, all of which are especially important for atoll systems. However, as described by Underwood et al. (1992), few numerical modeling studies have dealt realistically with atoll groundwater systems. Most studies (Anderson, 1976; Ayers and Vacher, 1983; Chidley and Lloyd, 1977; Falkland, 1983; Fetter, 1972; Herman and Wheatcraft, 1984; Lam, 1974) assume a sharp freshwater–saltwater interface and do not take into account density-dependent flow. Sharp interface models work reasonably well for most large islands and coastal groundwater systems that have a thick freshwater lens overlying a relatively thin transition zone. However, atoll groundwater systems commonly contain a thin freshwater lens overlying a much thicker transition zone, which cannot be realistically represented by a sharp interface model.

During the past several years the authors, as well as several other investigators, have begun to use numerical models that much more closely represent dynamic groundwater conditions within atoll systems, including density-dependent flow, fluctuating tidal boundaries, and multiple aquifers with anisotropic permeabilities and dispersivities. Table 1 provides a summary of previous numerical modeling studies of atoll and small island aquifers.

GENERIC MODELING OF GROUNDWATER LENS DYNAMICS

Numerical modeling of a generic atoll groundwater system was conducted by the senior author and M. Underwood (Underwood et al., 1992; Underwood, 1990) to study factors controlling fresh groundwater lens dynamics. The mathematical model SUTRA (Voss, 1984) was used to simulate variable-density saturated flow and solute transport in a vertical cross-section. It was selected because it can simulate a variable-density cross-section, is numerically robust and accurate, and has proved to be the most applicable documented model for simulating a thin freshwater lens underlain by a thick transition zone. The model is based on a hybridization of finite element and integrated finite difference methods, which preserve finite element geometric flexibility while taking advantage of finite difference efficiency. The original SUTRA code was modified to simulate the storage of water for a water table condition, and a fluctuating ocean tidal boundary was added (see appendix in Underwood, 1990).

Since the purpose of the study was to investigate lens dynamics for atolls in general, a generic rather than site-specific atoll model was used. The model simulated consisted of a dual-aquifer system in which a 15-m thick Holocene aquifer overlaid a Pleistocene aquifer having permeabilities one order of magnitude greater than those of the upper Holocene aquifer. Values used for aquifer and hydraulic parameters and responses were typical of those from Pacific Ocean atolls (as listed in tables 2 and 3 in Underwood, 1990).

Table 1. Summary of Modeling Studies of Island and Atoll Aquifers

Author	Model Type	Model Dimension and Orientation	Calibration Data	Sharp Interface	Dupuit Addumption	Transient Flow	Density-Dependent Flow	Tidal Boundary Condition	Anisotropic Permeabilities
Fetter (1972)	F.D.	Two-dimensional areal	Salinity	x	x				
Lam (1974)	F.D.	Two-dimensional radial	Tides		x			x	
Anderson (1976)	F.D.	One-dimensional vertical	Heads	x	x	x			
Chidley and Lloyd (1977)	F.D.	Two-dimensional areal	Salinity	x	x	x			
Falkland (1983)	F.D.	Two-dimensional areal	Salinity	x	x	x			
Ayers and Vacher (1983)	F.D.	Two-dimensional areal	Heads	x	x	x			
Herman and Wheatcraft (1984)	F.E.	Two-dimensional vertical	Tides			x		x	
Hogan (1988), Oberdorfer et al. (1990)	F.E.	Two-dimensional vertical	Tides			x	x	x	
Griggs (1989), Griggs and Peterson (1989)	F.E.	Two-dimensional vertical	Salinity			x	x		
Underwood (1990), Underwood et al. (1992)	F.E.	Two-dimensional vertical	Tides/salinity			x	x	x	x
Gingerich (1992)	F.E.	Two-dimensional vertical	Tides/salinity			x	x	x	x

Note: Crosses denote use in model; F.D. = finite difference, F.E. = finite elements.

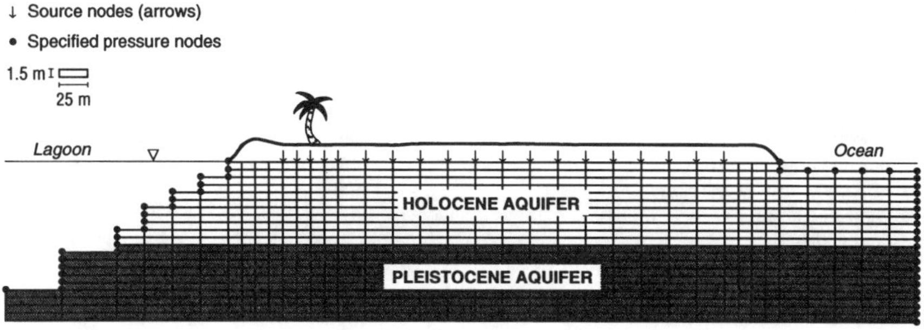

Figure 2. Two-dimensional mesh construction for area beneath atoll island. Note: Not drawn to scale. (From Underwood, M. R., Ph.D. Dissertation, Dept. of Geology Geophysics, Univ. of Hawaii at Manoa, Honolulu, 1990.)

Mesh Design and Boundary Conditions

The entire model mesh extends laterally from the lagoon center to the ocean reef face and vertically from sea level to the volcanic basement. Figure 2 shows the portion of the mesh beneath and immediately adjacent to the island, which contains the fresh groundwater lens. Four different meshes were used in this study, one for each island width of 250, 500, 750, and 1000 m.

Specified pressure values equivalent to hydrostatic seawater and solute mass concentration equivalent to seawater were assigned to all nodes bordering the ocean and lagoon (Figure 2). Furthermore, the ocean and lagoon boundaries were modeled as a fluctuating semidiurnal tide with a period of 12 h and an amplitude up to 0.75 m. Sources having a specified concentration equivalent to rainwater were assigned to nodes at the water table where recharge enters the groundwater system (Figure 2). No-flow boundaries were assigned to the bottom of the mesh and to the three sea-level nodes immediately inland of either shore in order to reduce numerical instabilities (Underwood et al., 1992).

Input Data and Calibration

Data input and model calibration are described in detail by Underwood et al. (1992). The following is a brief summary of their calibration procedures. Initial data input involved distributing aquifer characteristics across the mesh and assigning initial pressure and concentration values to each mesh node. Porosity and aquifer compressibility were assigned single values for the entire domain. Hydraulic conductivities and dispersivities were assigned separately to the upper and lower aquifers. Initial parameter values were chosen to be within the range of those reported for Pacific Ocean atolls. The initial salinity distribution was derived from a simulation of freshwater recharge into a saltwater aquifer run to steady state.

The model calibration procedure, as devised by Underwood (1990), involved selecting a set of aquifer parameters and boundary condition values that met three general criteria: (1) values should be within the range of reported literature values (see Table 2 in Underwood et al., 1992), (2) simulated groundwater salinity profiles should

Table 2. Parameter Values for Calibrated Generic Atoll Groundwater System

Parameter	Value	Unit
Porosity	0.25	—
Aquifer compressibility	10^{-9}	(m s^2)/kg
Holocene aquifer horizontal K	50	m/day
Holocene aquifer vertical K	10	m/day
Pleistocene aquifer horizontal K	500	m/day
Pleistocene aquifer vertical K	100	m/day
Island width	250–1000	m
Tidal range	0.5–1.5	m
Recharge	0.5–2.0	m/year
Holocene aquifer thickness	15.0	m
Horizontal longitudinal dispersivity	6.0–12.0	m
Vertical longitudinal dispersivity	0.01–0.05	m
Transverse dispersivity	0.01	m

(After Underwood, 1990.)

be similar to those reported in the literature, and (3) simulated groundwater tidal responses should be within the range of reported literature values (see Table 3 in Underwood et al., 1992). The final parameter values for the calibrated generic atoll island model are given in Table 2.

Model Results

Underwood (1990) described the entire set of more than a hundred numerical simulations conducted to investigate the hydrogeology and hydrodynamics of atoll groundwater systems. Here we describe only the results of simulations that investigate controls on the size of the freshwater lens and the thickness of the transition zone.

The freshwater lens, which defines the extent of potable groundwater beneath an atoll island, is extremely important in groundwater development and management. In this study the freshwater lens is defined as containing water with chloride concentrations of less than 500 mg/L (2.5% seawater) and the transition zone as containing brackish water with dissolved solids composition ranging from 2.5 to 95% seawater.

Sensitivity analysis demonstrated that the thickness of the freshwater lens is controlled by the following aquifer parameters and boundary conditions: (1) upper aquifer horizontal permeability, (2) upper aquifer vertical longitudinal dispersivity, (3) tidal range, (4) groundwater recharge, and (5) island width. As shown in Figure 3, the thickness of the freshwater lens (2.5% seawater) varies inversely with the horizontal permeability of the upper (Holocene) aquifer (KHH). The freshwater lens also is sensitive to dispersive processes and hence is strongly controlled by the upper aquifer vertical longitudinal dispersivity (α_{Lvert}) and the tidal range. Figure 4 shows the relationship between α_{Lvert} and the freshwater lens thickness and the transition zone thickness. As can be seen, an increase in α_{Lvert} causes the freshwater lens to shrink and the transition zone to thicken. Similarly, Figure 5 shows that an increase in tidal range causes the freshwater lens thickness to decrease.

Computer simulations also demonstrate that groundwater recharge is a critical factor controlling the thickness of atoll island freshwater lenses. Since most atoll

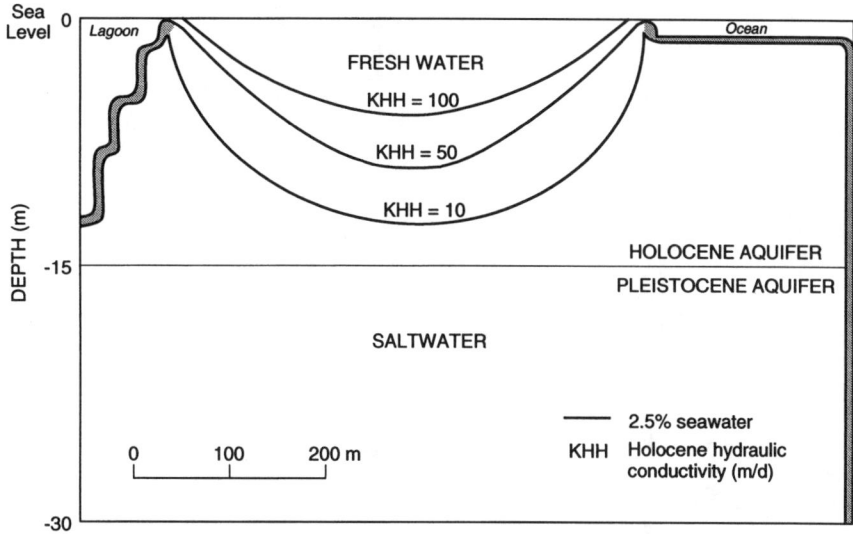

Figure 3. Simulated steady-state position of 2.5% seawater isopleth for different values of Holocene hydraulic conductivity. Recharge was 1 m/year, and all other parameters were held constant.

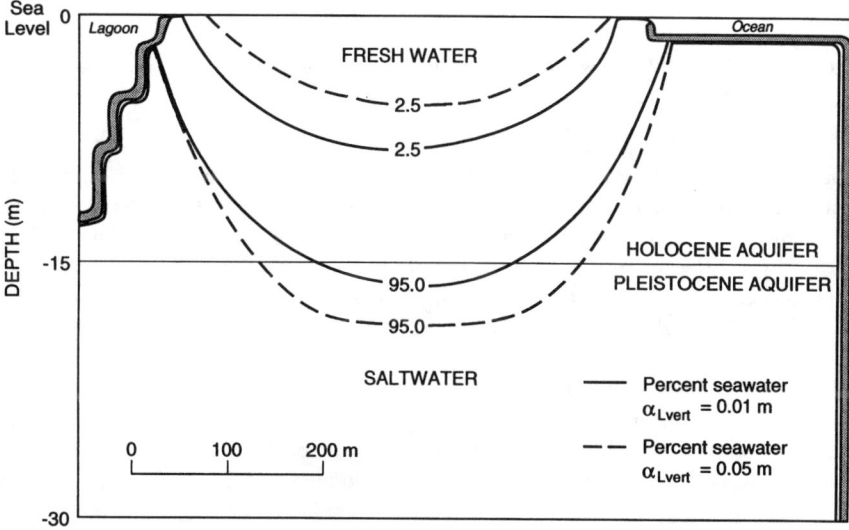

Figure 4. Simulated steady-state salinity profiles having vertical longitudinal dispersivity values of 0.01 and 0.05 m. Used tidal model with tidal range of 1 m and recharge of 1 m/year; all other parameters held constant. (Modified from Underwood, 1990.)

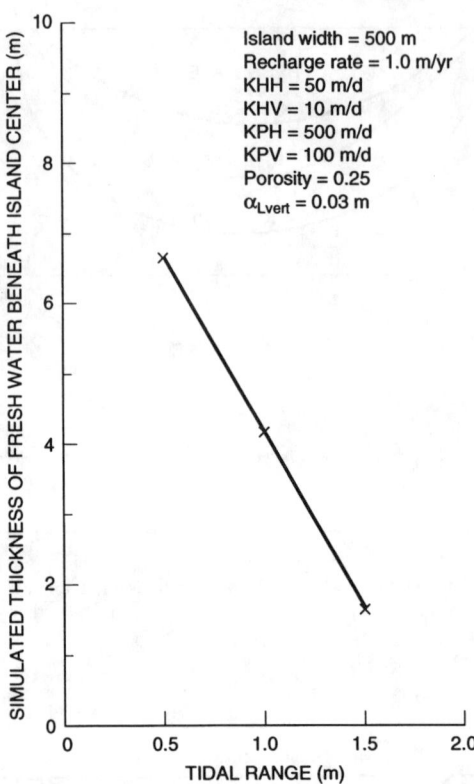

Figure 5. Relationship between tidal range and simulated thickness of freshwater. (Modified from Underwood, 1990.)

islands are elongate, groundwater recharge is a function of both island width and the annual recharge rate. Figure 6 shows the relationship between simulated freshwater thickness and island width for several different recharge rates. Although not all atoll island groundwater systems behave exactly as shown in Figure 6, as described by Underwood et al. (1992), this relationship may serve as a useful reconnaissance tool to evaluate freshwater development potential when more detailed data are unavailable.

The thickness of the transition zone in an atoll island groundwater system is controlled by freshwater and saltwater mixing processes. These in turn are driven by tidal fluctuations in a dispersive aquifer medium. Because tidal fluctuations cause predominately vertical groundwater movement, the dominant dispersive control in tidal mixing is α_{Lvert}. This is shown in Figure 4 where an increase in α_{Lvert} causes the transition zone to expand. Likewise, an increase in tidal range causes the thickness of the transition zone to increase as shown in Figure 7.

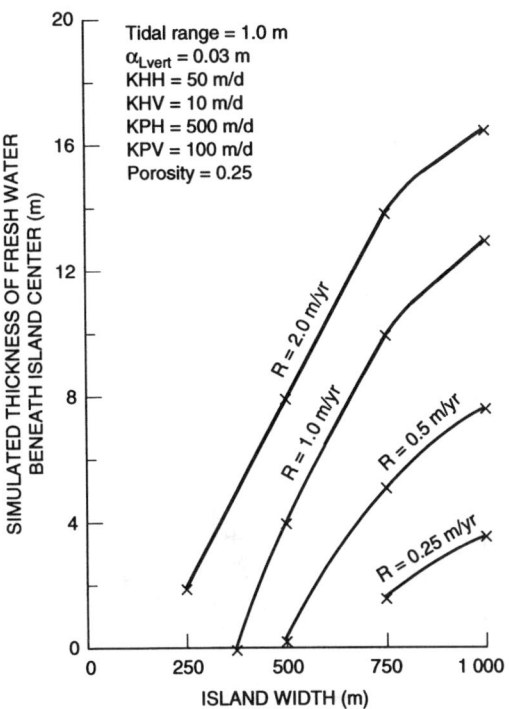

Figure 6. Relationship between island width and simulated thickness of freshwater at island center for four steady recharge rates (R). (Modified from Underwood, 1990.)

DEVELOPMENT AND SUSTAINABLE YIELD MODELING FOR ROI-NAMUR ISLAND

Roi-Namur, located at the northeast tip of Kwajalein Atoll (lat. 9°23′ N, long. 167°28′ E), consists of two roughly circular islets (Roi and Namur) connected by a partially dredge-filled isthmus and covers an area of 1.9 km² (Figure 8). About 20% of Roi is paved, including a 1370-m long asphalt runway and two adjacent concrete-lined rainwater catchment basins. Rainwater catchment currently provides about 70% (29 000 m³/year) of the total freshwater demand, with the remainder (8700 m³/year) being supplied by groundwater pumpage. The groundwater that is used in the water supply system comes from a 1000-m long horizontal skimming well system located adjacent and parallel to the runway on Roi. The salinity of the pumped groundwater has remained well below the U.S. Environmental Protection Agency (USEPA) drinking water standard of 250 mg/L chloride, and hence the ultimate sustainable yield for the groundwater system is unknown. Thus, numerical modeling of the Roi-Namur groundwater system was conducted by Gingerich (1992) to study groundwater development and sustainable yield under varying discharge and recharge scenarios.

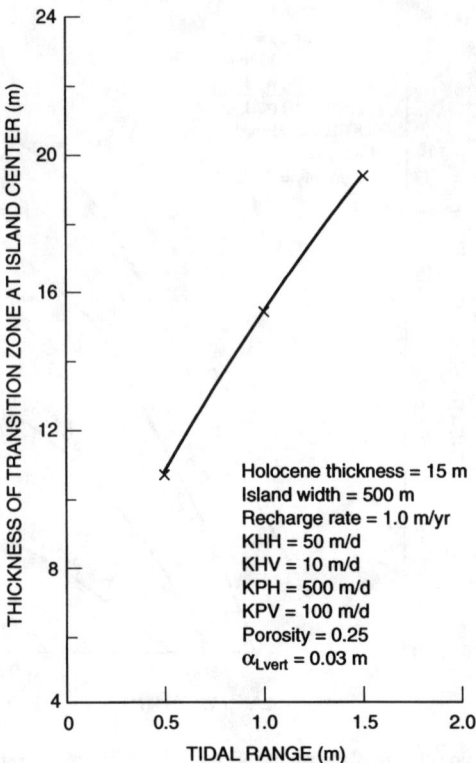

Figure 7. Relationship between tidal range and simulated steady-state transition zone thickness beneath island center. (Modified from Underwood, 1990.)

The same version of the SUTRA model described previously for the generic atoll island modeling was used for the Roi-Namur simulations; that is, variable-density saturated fluid flow and solute transport in a vertical cross-section through Roi-Namur Island were simulated. However, for this study aquifer and hydraulic parameters and responses specific to Roi-Namur were used. Thus the model simulated consisted of a four-layer system in which three different Holocene layers with a combined thickness of 20 m overlay approximately 900 m of highly permeable Pleistocene deposits (Figure 9).

Mesh Design and Boundary Conditions

The entire model mesh extends 8400 m laterally, from a point in the lagoon to the ocean side of the reef face, and 1000 m vertically, from sea level to the approximate volcanic basement. Figure 10 shows the portion of the mesh beneath and immediately adjacent to the island, which contains the freshwater lens. Boundary conditions for the Roi-Namur model (Figure 10) were similar to those for the generic atoll model. Specified pressure values equivalent to hydrostatic seawater and solute mass concentration equivalent to seawater were assigned to 74 nodes bounding the ocean and

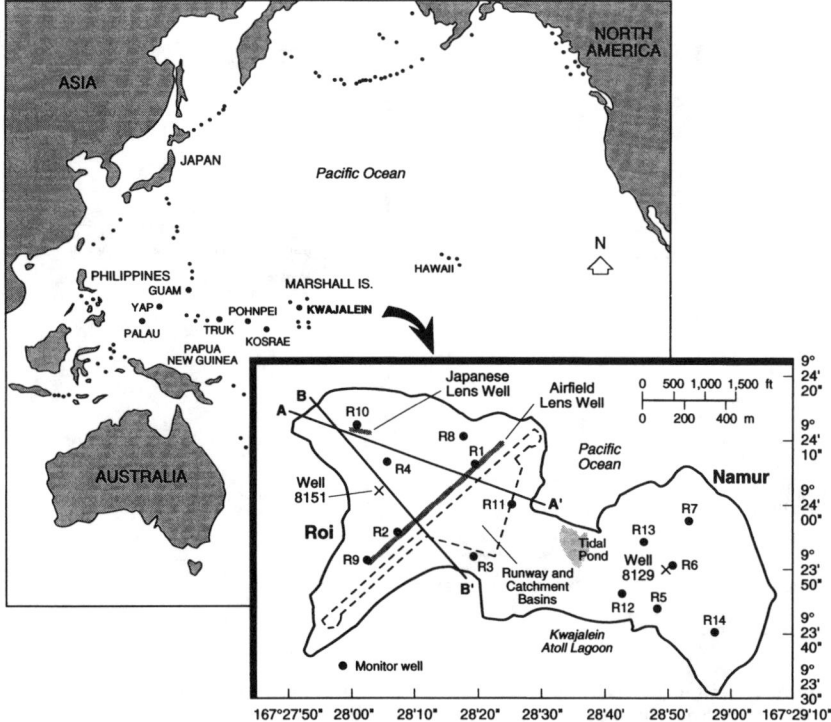

Figure 8. Location of Roi-Namur Island, Kwajalein Atoll, Republic of the Marshall Islands.

lagoon, and the ocean and lagoon were modeled as a fluctuating semidiurnal tide with a period of 12 h and a range of 1 m. Sources having a specified concentration equivalent to rainwater were assigned to 38 nodes at the water table where recharge enters the groundwater system (Figure 10). As for the generic model, source nodes were not placed in the 50-m buffer zone on either side of the island to avoid numerical instabilities. Seven nodes at the top of the mesh also did not receive recharge because they were overlain by the impermeable runway and rainwater catchments. Finally, one node near the top of the mesh was programmed as a sink node to simulate groundwater extraction (Figure 10). Extraction at this node was equivalent to pumping from an infinitely long horizontal gallery or skimming well oriented perpendicular to the mesh. The extraction volume was determined by dividing the total volume of water removed from the lens by the length of the gallery (Gingerich, 1992).

Input Data and Calibration

Input parameters to the model consisted of aquifer and fluid characteristics and the initial conditions of the freshwater lens. Physical properties of the aquifer matrix and fluid such as porous matrix compressibility (α), solid grain density (ρ_s), fluid viscosity (μ), fluid compressibility (β), fluid base density, and solute molecular diffusivity

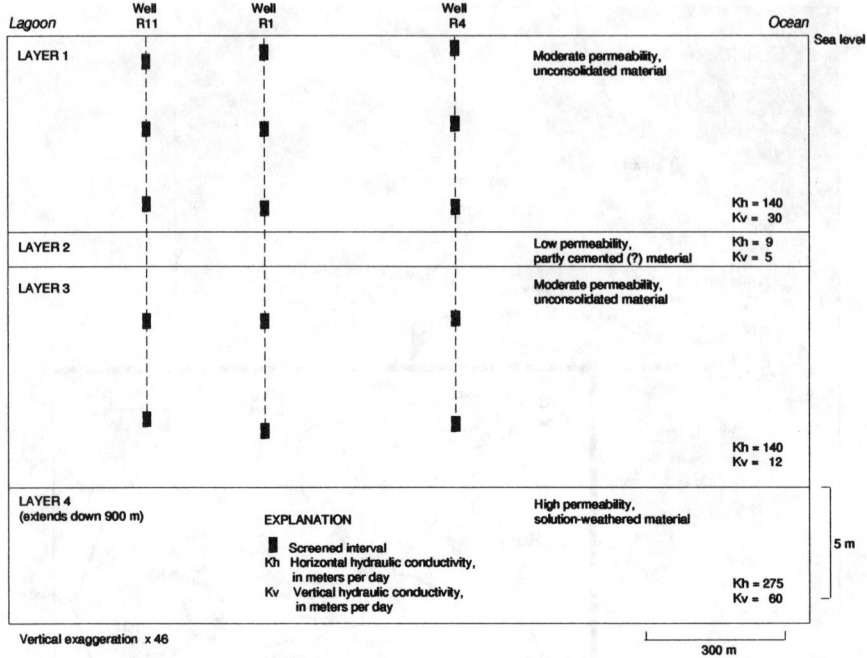

Figure 9. Hydrogeologic cross-section of Roi-Namur, Kwajalein Atoll, Republic of the Marshall Islands. (From Gingerich, S. B., M.S. Thesis, Dept. of Geology & Geophysics, Univ. of Hawaii, Honolulu, 1992.)

(σ_w) were assumed constant throughout space and time and were equal to common fluid and rock physical properties. Matrix porosity (ε) and specific storage were entered into the model on a node-wise basis, and vertical (k_v) and horizontal (k_h) permeability, longitudinal dispersivity (α_{Lhor} and α_{Lvert}), and transverse dispersivity (α_T) were entered on an element-wise basis. All the dispersivities were kept constant throughout the mesh.

The procedure described previously for calibrating the generic atoll model was also used for calibrating the Roi-Namur model. First, the position of the 50% seawater isopleth was calibrated by adjusting aquifer permeabilities. Next, tidal efficiencies and lags were calibrated with the tidal boundary condition. Finally, the salinity profile was calibrated by adjusting aquifer dispersivities. Gingerich (1992) described in detail the calibration procedure that resulted in the final parameter values for the calibrated Roi-Namur model (Table 3).

Model Results

Gingerich (1992) described all of the modeling simulations conducted for the Roi-Namur groundwater lens. In this paper we describe only the results of simulations that investigate the effects of extracting varying amounts of groundwater under several different pumping and recharge conditions.

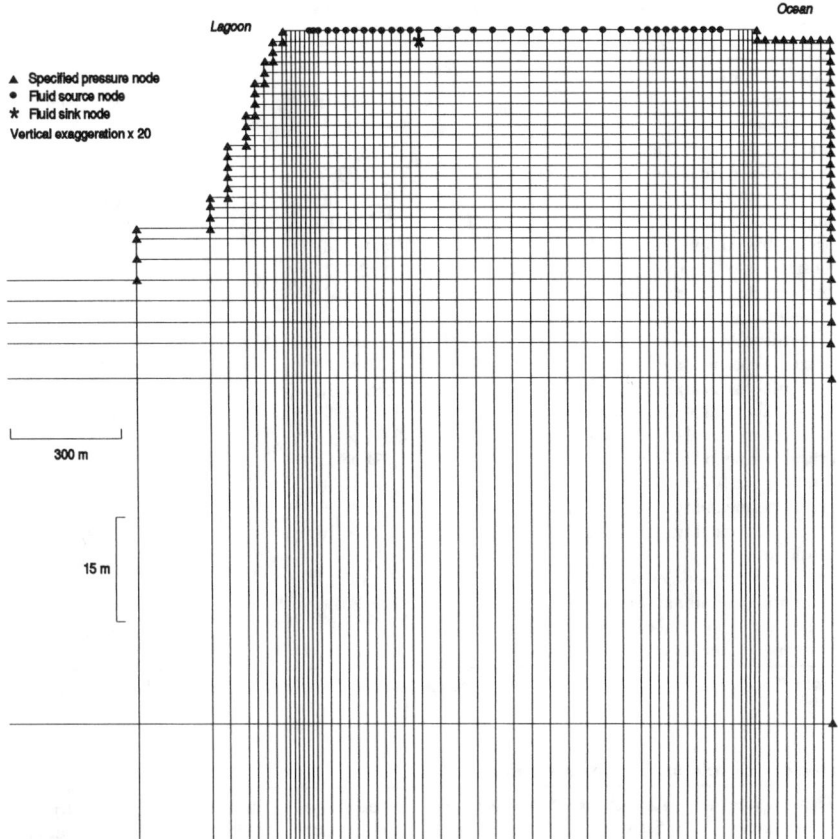

Figure 10. Finite-element mesh for Roi-Namur model. (Modified from Gingerich, 1992.)

Five different development scenarios (summarized in Table 4) involving three different recharge conditions and five different pumping conditions were investigated. Development scenario A assumes that the average annual recharge of 57.6 cm is distributed evenly throughout the year and the average annual pumpage of 8700 m^3, likewise, is extracted evenly throughout the year. Figure 11, which shows recharge, pumpage, and chloride concentration as a function of time for all development scenarios, illustrates that for scenario A the salinity of pumped groundwater increases only slightly throughout the year.

Development scenario B also uses the average annual recharge and pumping rates of 57.6 cm and 8700 m^3, respectively, but assumes a more realistic distribution of recharge and pumpage throughout the year. In this scenario the recharge is spread over a 9-month period, as shown in Figure 11, and the pumpage is evenly distributed over the 6-month dry period of December–May. As seen in Figure 11, under this development scenario the chloride concentration in the pumped groundwater rises and peaks in June at the end of the pumping season, but at a level well below the USEPA drinking water limit of 250 mg/L.

Table 3. Parameter Values for Calibrated Roi-Namur Model

Parameter	Value	Unit
Physical Constants		
Fluid compressibility (β)	4.47×10^{-10}	$[kg/(m\ s^2)]^{-1}$
Fluid density, seawater (ρ_s)	1,025	(kg/m^3)
Fluid density, freshwater (ρ_f)	1,000	(kg/m^3)
Concentration, seawater (C)	0.0357	(kg/kg)
Fluid diffusivity (σ_w)	1.0×10^{-9}	(m^2/s)
Fluid viscosity (μ)	8.3×10^{-4}	$[kg/(m\ s)]$
Solid matrix compressibility (α)	1.0×10^{-9}	$[kg/(m\ s^2)]^{-1}$
Density of a solid grain (ρ_{sg})	2,700	$(kg/(m^3)$
Component of gravity vector in y direction (g)	-9.81	(m/s^2)
Calibration Variables		
Horizontal permeability, layer 1 (k_{h1})	1.77×10^{-10}	(m^2)
Horizontal permeability, layer 2 (k_{h2})	1.18×10^{-11}	(m^2)
Horizontal permeability, layer 3 (k_{h3})	1.77×10^{-10}	(m^2)
Horizontal permeability, layer 4 (k_{h4})	3.54×10^{-10}	(m^2)
Vertical permeability, layer 1 (k_{v1})	3.54×10^{-11}	(m^2)
Vertical permeability, layer 2 (k_{v2})	5.90×10^{-12}	(m^2)
Vertical permeability, layer 3 (k_{v3})	1.42×10^{-11}	(m^2)
Vertical permeability, layer 4 (k_{v4})	8.26×10^{-11}	(m^2)
Porosity (ε)	0.3	(m^3/m^3)
Specific storage coefficient (ST)	0.33	(m^{-1})
Maximum longitudinal dispersivity (α_{Lmax})	3.0	(m)
Minimum longitudinal dispersivity (α_{Lmin})	0.02	(m)
Transverse dispersivity (α_T)	0.001	(m)

(From Gingerich, S. B., M.S. Thesis, Dept. of Geology & Geophysics, Univ. of Hawaii, Honolulu, 1992.)

Table 4. Development Scenario Information

Development Scenario	Length of Scenario	Months of Recharge	Recharge (cm/year)	Months of Withdrawal	Withdrawal (m³/year)
A	1 year	12	57.6	12	8,700
B	1 year	9	57.6	6	8,700
C	1 year	9	57.6	6	13,050
D	1 year	3	31.2	9	13,050
E	1 year	3	31.2	9	19,575

Thus it is concluded that during normal recharge years the current pumping rate of 8700 m³/year is well below the sustainable yield for the Roi-Namur groundwater system. Hence, to better evaluate the sustainable yield for the Roi-Namur groundwater system, three additional development scenarios were simulated. Development scenario C assumes the average annual recharge of 57.6 cm is distributed over the same 9-month period as for scenario B, but the annual pumpage distributed over the 6-month dry period of December–May is increased by 50% to 13,050 m³. As seen in Figure 11, this scenario causes the groundwater chloride concentration to peak at a level that is about double its original concentration but still slightly below the USEPA drinking water limit.

Development scenario D investigates the effects of a reduced recharge rate of 31.2 cm/year, which actually occurred during the drought year of 1984, applied over the 3-month period of September–November, with pumping at the increased average annual rate of 13,050 m³ spread over the nonrainy 9-month period of December–

MODELING ATOLL GROUNDWATER SYSTEMS

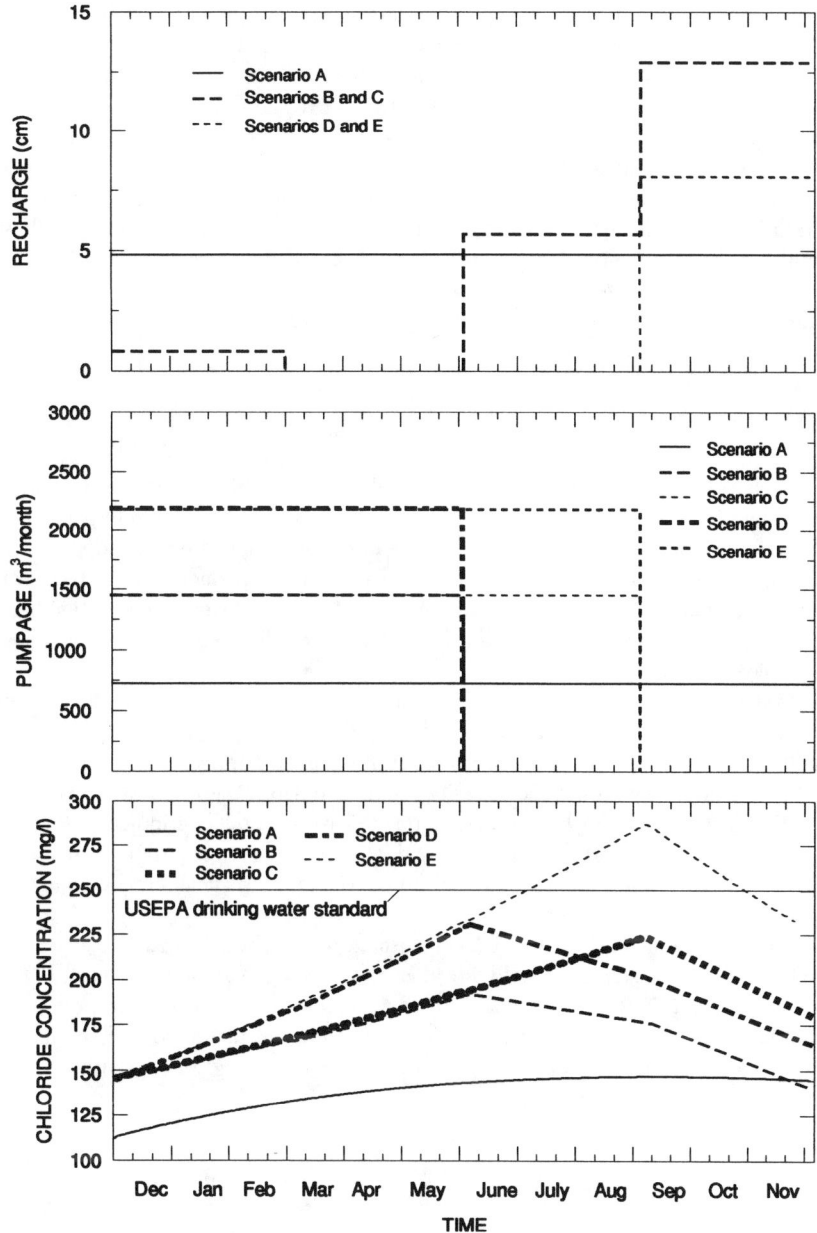

Figure 11. Development scenario information.

August. As shown in Figure 11, this scenario causes the groundwater chloride concentration to peak at a level that is approximately double its original concentration but still slightly below the USEPA drinking water limit.

Development scenario E assumes the drought recharge rate of 31.2 cm/year is applied over the 3-month period of September–November as in scenario D but with an increased pumping rate of 19,575 m^3/year distributed over the 9-month dry period of December–August. As seen in Figure 11, this scenario causes the chloride concentration to rise above the USEPA drinking water limit of 250 mg/L. Based on these simulated development scenarios, it is concluded that the sustainable yield for the Roi-Namur groundwater system is probably at least 50% greater that the current pumping rate of 8700 m^3/year, even during severe drought years.

CONCLUSIONS

Atoll island groundwater systems, although quite variable in detail, can most realistically be modeled with a multilayer system consisting of moderately permeable Holocene age aquifers overlying much more permeable Pleistocene age deposits. In this model the groundwater system has two basic flow components: short-term vertical fluctuations, on the scale of about 1 m or less, that are driven by ocean tides, and long-term average flow that is driven by recharge and discharge which occurs over the scale of hundreds of meters.

Simulation results using this model for a generic atoll island demonstrate that the thickness of the freshwater lens is controlled by the following aquifer parameters and boundary conditions: (1) Holocene aquifer horizontal permeability, (2) Holocene aquifer vertical longitudinal dispersivity, (3) tidal range, (4) groundwater recharge, and (5) island width. Furthermore, generic atoll island modeling indicates that the transition zone thickness is controlled by tidal fluctuations in a dispersive medium, which are a function of the longitudinal dispersivity in the vertical direction and tidal range.

Site-specific modeling for Roi-Namur island in the Kwajalein Atoll has demonstrated that the sustainable yield for this groundwater system is at least 13,000 m^3/year, which is 50% greater than current extraction rates. Furthermore, the modeling has demonstrated that groundwater pumping at this increased rate should not cause chloride concentrations to exceed the USEPA drinking water standards even during periods of substantial drought.

ACKNOWLEDGMENTS

The work upon which this paper is based was supported by federal and state grants for the following research projects: Modeling of Atoll Groundwater Systems (agreement no. 14–34–0001–0113, Office of Water Research and Technology, U.S. Department of the Interior), Atoll Groundwater Systems, Phase II (agreement no. 14–08–0001–G1221, U.S. Department of the Interior), and Groundwater Assessment for Bikini Atoll (grant from Bikini Atoll Rehabilitation Commission). The authors thank the Water Resources Research Center publications staff for their assistance in preparing the manuscript. This is contributed paper CP-94-04 of the Water Resources Research Center at the University of Hawaii at Manoa, Honolulu.

REFERENCES

Anderson, M. P., Unsteady groundwater flow beneath strip oceanic island, *Water Resour. Res.*, 12(4), 640, 1976.

Anthony, S. S., Peterson, F. L., MacKenzie, F. T., and Hamlin, S. N., Geohydrology of the Laura fresh-water lens, Majuro Atoll: A hydrogeochemical approach, *Geol. Soc. Am. Bull.*, 101, 1066, 1989.

Ayers, J. F. and Vacher, H. L., A numerical model describing unsteady flow in a fresh water lens, *Water Resour. Bull.*, 19(5), 785, 1983.

Ayers, J. F. and Vacher, H. L., Hydrogeology of an atoll island: A conceptual model from detailed study of a Micronesian example, *Ground Water*, 24, 185, 1986.

Buddemeier, R. W. and Holladay, G. L., Atoll hydrology, island groundwater characteristics and their relationship to diagenesis, in *Proc. Fourth Int. Coral Reef Symp.*, Rosenstiel Sch. of Mar. and Atmos. Sci., Univ. of Miami, Florida, 1977, 167.

Chidley, T. R. and Lloyd, J. L., A mathematical model study of freshwater lens, *Ground Water*, 15(3), 215, 1977.

Collins, M. A., The extended Boussinesq problem, *Water Resour. Res.*, 12(1), 54, 1976.

Falkland, A. C., Christmas Island water resources study, Vol. 1 of general report, Aust. Dept. Housing and Constr., Canberra, A.C.T., 1983.

Fetter, C. W., Position of the saline water interface beneath oceanic islands, *Water Resour. Res.*, 8(5), 1307, 1972.

Gingerich, S. B., Numerical simulation of the freshwater lens on Roi-Namur Island, Kwajalein Atoll, Republic of the Marshall Islands, M.S. thesis, Dept. of Geology & Geophysics, Univ. of Hawaii, Honolulu, 1992.

Griggs, J. E. and Peterson, F. L., Ground-water flow dynamics and development strategies at the atoll scale, *Ground Water*, 31(2), 209, 1993.

Hantush, M. S., Unsteady movement of fresh water in thick unconfined saline aquifers, *Bull. Int. Assoc. Sci. Hydrol.*, 13, 40, 1968.

Henry, H. R., Effects of dispersion on salt encroachment in coastal aquifers, Sea Water in Coastal Aquifers, U.S. Geol. Surv. Water Supp. Paper 1613-C, U.S. Gov. Print. Off., Washington, D.C., 1964, C71.

Herman, M. E. and Wheatcraft, S. W., Groundwater dynamics investigation of Enjebi Island, Enewetak Atoll: An interpretive computer model simulation, in *Finite Elements in Water Resources*, Springer-Verlag, New York, 1984, 133.

Hunt, C. D., Jr. and Peterson, F. L., Groundwater resources of Kwajalein Islands, Marshall Islands, Tech. Rep. No. 126, Water Resources Research Center, Univ. of Hawaii at Manoa, Honolulu, 1980.

Lam, R. K., Atoll permeability calculated from tidal diffusion, *J. Geophys. Res.*, 79(21), 3073, 1974.

Oberdorfer, J. A., Hogan, P. J., and Buddemeier, R. W., Atoll island hydrogeology: Flow and fresh water occurrence in a tidally dominated system, *J. Hydrol.*, 120, 327, 1990.

Peterson, F. L., Hydrogeology of the Marshall Islands, in *Geology and Hydrogeology of Carbonate Islands*, Vacher, H. L. and Quinn, T. M., Eds., Elsevier Press, in press.

Rumer, R. R. and Shiau, J. C., Salt water interface in a layered coastal aquifer, *Water Resour. Res.*, 4(6), 1235, 1968.

Underwood, M. R., Atoll island hydrogeology: Conceptual and numerical models, Ph.D. dissertation, Dept. of Geology & Geophysics, Univ. of Hawaii at Manoa, Honolulu, 1990.

Underwood, M. R., Peterson, F. L., and Voss, C. I., Groundwater lens dynamics of atoll islands, *Water Resour. Res.*, 28(11), 2889, 1992.

Vacher, H. L., Groundwater hydrology of Bermuda, Bermuda Public Works Dept., Hamilton, 1974.

Vacher, H. L., Dupuit-Ghyben-Herzberg analysis of strip-island, *Geol. Soc. Am. Bull.*, 100, 580, 1988.

Van der Veer, P., Analytical solution for steady interface flow in a coastal aquifer involving a phreatic surface with precipitation, *J. Hydrol.*, 34, 1, 1977.

Voss, C. I., SUTRA-Saturated-unsaturated transport-A finite element simulation model for saturated-unsaturated fluid density-dependent groundwater flow with energy transport or chemically-reactive single-species solute transport, U.S. Geol. Surv. Water Resour. Invest. Rept. 84–4369, 1984.

Wheatcraft, S. W. and Buddemeier, R. W., Atoll island hydrology, *Ground Water*, 19, 311, 1981.

SECTION 5

On Biodegradation/Virus Transport Modeling

CHAPTER 16

A Review of Biodegradation Models: Theories and Applications

H. S. Rifai and P. B. Bedient

INTRODUCTION

Biodegradation is an important attenuation mechanism of contaminant concentrations in groundwaters and soils. Biodegradation, when it occurs, limits pollutant migration and reduces contaminant mass in the subsurface. Biodegrading plumes are generally less widespread than their nonbiodegrading counterparts and are transported at a slower relative rate. In order to quantify the impact of biodegradation on contaminant transport and fate, it is necessary to develop a mathematical expression for biodegradation that can be incorporated into the advection-dispersion equation. The resulting model allows scientists and engineers to simulate natural or enhanced biodegradation and investigate their efficacy as remediation alternatives for contaminated aquifers.

This paper reviews the different mathematical expressions and models that have been proposed in the general literature for simulating biodegradation in groundwater. First, biodegradation processes are reviewed and the kinetic expressions that can be used to simulate biodegradation are presented. A discussion of the available models for biodegradation and their potential applications follows. The paper concludes with a number of case studies where models have been used to simulate biodegradation and bioremediation at the field scale.

OVERVIEW OF BIODEGRADATION PROCESSES

Biodegradation is a biochemical reaction which is mediated by microorganisms. In general, an organic compound is oxidized (loses electrons) by an electron acceptor which in itself is reduced (gains electrons). Under *aerobic* or *oxic* environmental conditions, oxygen acts as the electron acceptor. The oxidation of organic compounds coupled to the reduction of molecular oxygen is termed *aerobic respiration*. When oxygen is not present (*anoxic* conditions), microorganisms can use organic chemicals or inorganic anions as alternate electron acceptors under *anaerobic* conditions. Anaerobic biodegradation can occur under *fermentative, denitrifying, iron-reducing, sulfate-reducing,* or *methanogenic* conditions.

Microbial investigations of the subsurface have revealed that most aquifers support a microbial population. Typical microbial population numbers range from 1×10^5 to 1×10^7 cells per gram dry weight (cells/gdw) or cells/ml (Ghiorse and Balkwill, 1983; Wilson et al., 1983; Balkwill and Ghiorse, 1985; Beeman and Suflita, 1987). Relatively high numbers of microorganisms have been detected in both contaminated and pristine aquifers of varying depth and geological composition. The picture that emerges from microbiological studies is that subsurface microorganisms tend to be small, capable of response to the influx of nutrients, and primarily attached to solid surfaces (Suflita, 1989).

There are six basic requirements for biodegradation: (1) the presence of the appropriate organisms; (2) an energy source such as organic carbon that is used by the organisms for cell maintenance and growth; (3) a carbon source; (4) an electron acceptor such as O_2, NO_3^-, SO_4^{2-}, CO_2; (5) nutrients such as nitrogen, phosphorus, calcium, magnesium, iron, and trace elements; and (6) acceptable environmental conditions such as appropriate temperature, pH, and salinity levels.

The biodegradation of contaminants in groundwater is mainly controlled by the rate of the reaction and the availability of the electron acceptor. A mathematical expression that represents the chemical reaction can be written to account for the effect that the rate of the reaction has on biodegradation. This mathematical expression can then be combined with the transport equation to account for the electron acceptor limitation effect on the biodegradation process in the subsurface.

KINETICS OF BIODEGRADATION

There are several types of kinetic expressions which may be appropriate for predicting rates of biodegradation in groundwater aquifers. The most common expression being the hyperbolic saturation function presented by Monod (1942) and referred to as Monod or Michaelis-Menten kinetics:

$$\mu = \mu_{max} \frac{C}{K_c + C} \tag{1}$$

where μ is the growth rate (time^{-1}), μ_{max} is the maximum specific growth rate (time^{-1}), and C is the concentration of the growth-limiting substrate (mg/L). The term K_c is known as the half-saturation constant or the growth-limiting substrate concentration which allows the microorganism to grow at half the maximum specific growth rate.

The rate equation describing μ as a function of C contains first-order, mixed-order, and zero-order regions. When $C >> K_c$, $K_c + C$ is almost equal to C, and the reaction approaches zero-order with:

$$\mu = \mu_{max} \tag{2}$$

and μ_{max} becomes the limiting maximum reaction rate. When $C << K_c$, Equation (1) reduces to:

$$\mu = \frac{\mu_{max}}{K_c} \cdot C \tag{3}$$

and the reaction approaches first-order with $\frac{\mu_{max}}{K_c}$ equal to the first-order rate constant.

In groundwater, the Monod growth function is related to the rate of decrease of an organic compound. This is done by utilizing a yield coefficient, Y, where Y is a measure of the organisms formed per substrate utilized. The change in substrate concentration can then be expressed as follows:

$$\frac{dC}{dt} = \frac{\mu_{max} MC}{Y(K_c + C)} \tag{4}$$

where M is the microbial mass in mg/L. Because of the relationship between substrate utilization and the growth of microbial mass, Equation (4) is accompanied by an expression of the change in microbial mass as a function of time:

$$\frac{dM}{dt} = \mu_{max} M Y \frac{C}{(K_c + C)} - b \cdot M \tag{5}$$

where b is a first-order decay coefficient that accounts for cell death.

A simple alternative to using growth functions for determining the rate of degradation of a chemical involves the use of a first-order equation of the form:

$$C = C_o \cdot e^{-kt} \tag{6}$$

where C is the biodegraded concentration of the chemical, C_o is the starting concentration, and k is the rate of decrease of the chemical. First-order rate constants are particularly useful since they allow half-lives for chemicals to be estimated:

$$t_{1/2} = \frac{0.693}{k} \tag{7}$$

Although there are certain advantages to assuming first-order kinetics in laboratory systems, a more rigorous theoretical basis for extrapolating laboratory rate constants to the subsurface is necessary. The advantage of using Monod kinetics, for example, is that the constants K_c and μ_{max} uniquely define the rate equation for mineralization of a specific compound. The ratio $\frac{\mu_{max}}{K_c}$ also represents the first-order rate constant for degradation when $C \ll K_c$. This rate constant incorporates both the activity of the degrading population and the substrate dependency of the reaction. It therefore takes into account both population and substrate levels, and provides a theoretical basis for extrapolating laboratory rate data to the environment.

MODELING BIODEGRADATION

The problem of quantifying biodegradation in the subsurface can be addressed by using models which combine physical, chemical, and biological processes. Developing such models is not simple, however, due to the complex nature of microbial kinetics, the limitations of computer resources, the lack of field data on biodegradation, and the need for robust numerical schemes that can simulate the physical, chemical, and biological processes accurately.

The main expressions that have been utilized for modeling biodegradation kinetics include:

1. Monod kinetics
2. First-order decay kinetics
3. An instantaneous reaction assumption

Monod kinetics have been discussed in the previous section. The reduction of contaminant concentrations using Monod kinetics can be expressed as:

$$\Delta C = M_t \mu_{max} \left(\frac{C}{K_c + C} \right) \Delta t \tag{8}$$

where C is the contaminant concentration, M_t is the total microbial concentration, μ_{max} is the maximum contaminant utilization rate per unit mass microorganisms, K_c is the contaminant half-saturation constant, and Δt is the time interval being considered.

Incorporating Equation (8) into the one-dimensional transport equation, for example, results in:

$$\frac{\partial C}{\partial t} = D_x \frac{\partial^2 C}{\partial x^2} - v \frac{\partial C}{\partial x} - M_t \mu_{max} \left(\frac{C}{K_c + C} \right) \tag{9}$$

where v is the seepage velocity, and D_x is the dispersion coefficient.

First-order kinetics, as mentioned earlier, represent an exponential decay and are a simplification of Monod kinetics [see Equation (6)]. The instantaneous reaction expression, an expression first proposed by Borden and Bedient (1986), assumes that microbial biodegradation kinetics are fast in comparison with the transport of oxygen, and that the growth of microorganisms and utilization of oxygen and organics in the subsurface can be simulated as an instantaneous reaction between the organic contaminant and oxygen.

From a practical standpoint, the instantaneous reaction model assumes that the rate of utilization of the contaminant and oxygen by the microorganisms is very high, and that the time required to mineralize the contaminant is very small, or almost instantaneous. Biodegradation is calculated using the expression:

$$\Delta C_R = -\frac{O}{F} \tag{10}$$

A REVIEW OF BIODEGRADATION MODELS

Table 1. Biodegradation Models

Name	Description	Author(s)
—	1-D, aerobic, microcolony, Monod	Molz et al. (1986)
BIOPLUME	1-D, aerobic, Monod	Borden et al. (1986)
—	1-D, analytical first-order	Domenico (1987)
BIO1D	1-D, aerobic & anaerobic, Monod	Srinivasan and Mercer (1988)
—	1-D, cometabolic, Monod	Semprini and McCarty (1991)
—	1-D, aerobic anaerobic, nutrient limitations, microcolony, Monod	Widdowson et al. (1988)
—	1-D, aerobic, cometabolic, multiple substrates, fermentative, Monod	Celia et al. (1989)
BIOPLUME II	2-D, aerobic, instantaneous	Rifai et al. (1988)
—	2-D, Monod	MacQuarrie et al. (1990)
BIOPLUS	2-D, aerobic, Monod	Wheeler et al. (1987)
ULTRA	2-D, first order	Tucker et al. (1986)
—	2-D, denitrification	Kinzelbach et al. (1991)
—	2-D, Monod, Biofilm	Odencrantz et al. (1990)

(From Bedient et al., *Ground Water Contamination: Transport and Remediation*, Prentice Hall, NJ, 1994.)

where ΔC_R is the change in contaminant concentration due to biodegradation, O is the concentration of oxygen, and F is the ratio of oxygen to contaminant consumed.

In order to develop a biodegradation model for groundwater, one of the above expressions is combined with the advection-dispersion equation. A different transport equation is written for each of the following dissolved interacting components: the contaminants, the nutrients, the electron acceptor, and the microorganisms. The resulting system of equations is then solved through time. Most of the biodegradation models that have been developed to date use these basic procedures and they generally only vary in the biodegradation expression used or the methods used to solve the system of equations.

As will be seen in the following section, one notable difference between the different biodegradation models has to do with the method used to simulate the microbial population and its interactions with the dissolved components in the groundwater. The microscale approach simulates the microbial population as a biofilm or a microcolony which is limited in its growth by the rate of diffusion into the biofilm or the colony. The macroscale approach, on the other hand, assumes that the diffusion limitation is not significant and thus does not limit the access that the microorganisms have to the contaminants, nutrients, and electron acceptor. There is no consensus among researchers as to which of the two approaches is more appropriate for groundwater aquifers.

EXISTING BIODEGRADATION MODELS

Many biodegradation models have been developed in recent years, most of which utilize some form of the three expressions presented earlier (see Table 1). The models listed in Table 1 also simulate a number of aerobic and anaerobic biodegradation processes subject to specified conditions and assumptions. This section will

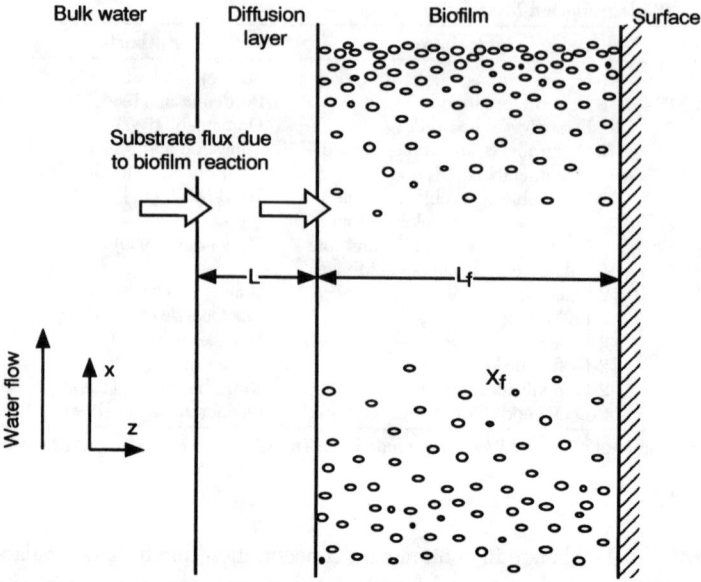

Figure 1. Idealized biofilm illustrating uniform cell density (X_f), thickness (L_f), water flow, and substrate flux into biofilm. (From McCarty, P. L., Rittman, B. E., and Bouwer, E. J., Microbiological processes affecting chemical transformations in groundwater, in *Groundwater Pollution Microbiology*, Bitton, G., and Gerba, C. P., Eds., Copyright © 1984 by John Wiley & Sons, pp. 89–115. Reprinted by permission of John Wiley & Sons, Inc.)

focus on the conceptual nature of the models rather than on the specific model implementations in order to illustrate the mathematical expressions used to simulate biodegradation.

The Biofilm Model

McCarty et al. (1984) believe that the nature of the groundwater environment (low substrate concentration and high specific surface area) dictates that the predominant type of bacterial activity will be bacteria attached to solid surfaces in the form of biofilm. The attached bacteria remain generally fixed in one place and obtain energy and nutrients from the groundwater that flows by.

Figure 1 is an illustration of an idealized biofilm which is a homogeneous matrix of bacteria and their extracellular polymers that bind them together and to the inert surface (McCarty et al., 1981). Groundwater flows past the biofilm in the x direction, while substrates are transported from the water to the biofilm in the z direction. The distance L represents the thickness of a mass-transport diffusion layer through which substrate must pass in order to go from the bulk liquid into the biofilm, where utilization occurs.

Within the biofilm, two processes occur simultaneously: namely, utilization of the substrate by the bacteria, assumed to follow a Monod-type relation, and diffusion of the substrate through the biofilm according to Fick's Law. Figure 2 shows the interaction of the three processes—substrate utilization, molecular diffusion within the

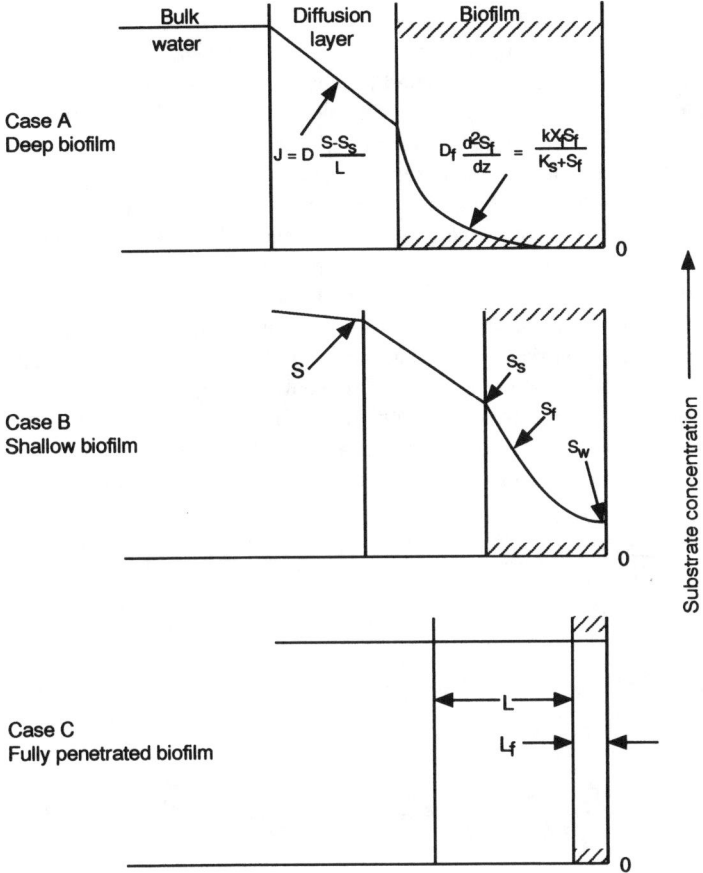

Figure 2. Interactions of substrate utilization, molecular diffusion within biofilm and mass transport across diffusion layer. (From McCarty, P. L., Rittman, B. E., and Bouwer, E. J., Microbiological processes affecting chemical transformations in groundwater, in *Groundwater Pollution Microbiology*, Bitton, G., and Gerba, C. P., Eds., Copyright © 1984 by John Wiley & Sons, pp. 89–115. Reprinted by permission of John Wiley & Sons, Inc.)

biofilm and mass transport across the diffusion layer. For a thick biofilm (Case A), the substrate concentration approaches zero and the biofilm is called deep. If the biofilm is very thin (Case C), almost no substrate utilization occurs and the biofilm is essentially fully penetrated at the surface concentration S_s. The remaining case is termed shallow (Case B). The biofilm concept has been adapted by Molz et al. (1986) and Widdowson et al. (1987) and used in their biodegradation model as will be seen in the following section.

Microcolony Models

Molz et al. (1986) and Widdowson et al. (1987) developed one-dimensional and two-dimensional models for aerobic biodegradation of organic contaminants in groundwater coupled with advective-dispersive transport. A microcolony approach was utilized in the modeling, microcolonies of bacteria were represented as disks of uniform radius and thickness attached to aquifer sediments. A boundary layer of a given thickness was associated with each colony across which substrate and oxygen are transported by diffusion to the colonies. Simulations of two-dimensional transport suggested that under aerobic conditions microbial degradation reduces the substrate concentration profile along longitudinal sections of the plume and retards the lateral spread of the plume. Anaerobic conditions developed in the plume center due to microbial consumption and limited oxygen diffusion into the plume interior.

Widdowson et al. (1988) extended their 1986 and 1987 studies to simulate oxygen- and/or nitrate-based respiration. Basic assumptions incorporated into the model included a simulated particle-bound microbial population comprised of heterotrophic, facultative bacteria in which metabolism is controlled by lack of either an organic carbon-electron donor source (substrate), electron acceptor (O_2 and/or NO_3), or mineral nutrient (NH_4^+), or all three simultaneously. Based on these assumptions, five coupled, nonlinear equations govern microbial growth dynamics in porous media:

$$\frac{\partial S}{\partial t}(1 + \alpha_s A_s/n) = -v_x \frac{\partial S}{\partial x} + D_s \frac{\partial^2 S}{\partial x^2} - \frac{D_{sb}}{n}\left(\frac{S-s}{\delta}\right) N_c \pi r_c^2 \quad (11)$$

$$\frac{\partial O}{\partial t}(1 + \alpha_o A_s/n) = -v_x \frac{\partial O}{\partial x} + D_o \frac{\partial^2 O}{\partial x^2} - \frac{D_{ob}}{n}\left(\frac{O-o}{\delta}\right) N_c \pi r_c^2 \quad (12)$$

$$D_{sb}\left(\frac{S-s}{\delta}\right)\pi r_c^2 = \frac{\mu_m m_c}{Y}\left(\frac{s}{K_s+s}\right)\left(\frac{o}{K_o+o}\right) \quad (13)$$

$$D_{ob}\left(\frac{O-o}{\delta}\right)\pi r_c^2 = \gamma\mu_m m_c\left(\frac{s}{K_s+s}\right)\left(\frac{o}{K_o+o}\right) + \alpha K_d m_c\left(\frac{o}{K_o'+o}\right) \quad (14)$$

$$\frac{1}{N_c}\frac{\partial N_c}{\partial t} = \mu_m\left(\frac{s}{K_s+s}\right)\left(\frac{o}{K_o+o}\right) - K_d \quad (15)$$

where S is the substrate concentration in the pore fluid, s is the substrate concentration within the colony, D_s is the substrate dispersion coefficient, n is the porosity, t is the time, x is length, O is the oxygen concentration in the pore fluid, o is the oxygen concentration within the colony, D_o is the oxygen dispersion coefficient, A_s is the effective specific surface of the aquifer matrix, N_c is the number of colonies per unit volume of aquifer, μ_m is the maximum specific growth rate of heterotrophic microorganisms, δ is the thickness of the boundary layer, γ is the oxygen use coefficient for synthesis of heterotrophic biomass, α_s and α_o are the ratios of the adsorbed concentration to that in solution for the substrate and oxygen, K_d is the distribution coefficient, K_s is the substrate half-saturation constant, K_o is the oxygen half-saturation constant, K_o' is the oxygen saturation constant for decay, D_{sb} and D_{ob} are diffusion coefficients in boundary layer for substrate and oxygen, and m_c is the cell mass per colony. Widdowson et al. (1988) used their model to simulate a number of laboratory column biodegradation scenarios.

Monod Kinetic Models

Monod kinetic expressions have been used in most of the developed biodegradation models to date as can be seen in Table 1. This section will focus on one of those models, namely the BIO1D model since it is a commercially available model. Other models which utilize similar concepts are those developed by MacQuarrie et al. (1990), Semprini and McCarty (1991), and Celia et al. (1989). The models developed by Semprini and McCarty (1991) and Celia et al. (1989) are of special interest because they simulate cometabolic processes.

BIO1D Model

Srinivasan and Mercer (1988) presented a one-dimensional, finite difference model for simulating biodegradation and sorption processes in saturated porous media. Aerobic biodegradation was modeled using a modified Monod function; anaerobic biodegradation was modeled using Michaelis-Menten kinetics. In addition, first-order degradation can be simulated for both substances. The Srinivasan and Mercer (1988) model is an extension of that presented by Borden and Bedient (1986). The governing equations are:

$$f = D\frac{\partial^2 S}{\partial x^2} - V\frac{\partial S}{\partial x} - B(S, O) - [1 + A(S)]\frac{\partial S}{\partial t} = 0 \tag{16}$$

$$g = D\frac{\partial^2 O}{\partial x^2} - V\frac{\partial O}{\partial x} - F \cdot B(S, O) - [1 + A(O)]\frac{\partial O}{\partial t} = 0 \tag{17}$$

For aerobic conditions:

$$B(S, O) = Mk\left(\frac{S}{k_s + S}\right)\left(\frac{O}{k_o + O}\right)\left(\frac{S - S_{min}}{S}\right) \tag{18}$$

for $S \geq S_{min}$ and $O \geq O_{min}$. Otherwise $B(S, O) = 0$.

For anaerobic conditions:
$B(S,O)$ reduces to $B(S)$ and only one equation is solved for S.

$$B(S) = M_n k_n \left(\frac{S}{k_{sn} + S}\right) \tag{19}$$

where S is the substrate concentration in the pore fluid (ML^{-3}), O is the oxygen concentration in the pore fluid (ML^{-3}), D is the longitudinal hydrodynamic dispersion coefficient (L^2T^{-1}), x is the distance, V is the interstitial fluid velocity (LT^{-1}), $B(S, O)$ is a biodegradation term expressed as a function of the dependent variables S and O ($ML^{-3}T^{-1}$), $A(S)$ is the adsorption term expressed as a function of S [the term $(1 + A(S))$ is the retardation factor], t is the time, M is the microbial mass, k is the maximum substrate utilization rate per unit mass of microorganisms, k_s is the substrate half-saturation constant, k_o is the oxygen half-saturation constant, S_{min} is the minimum substrate concentration that permits growth and decay, O_{min} is the minimum oxygen concentration that permits growth and decay, and F is the ratio of oxygen to substrate consumed. Note that M_n, k_n and k_{sn} are counterparts of M, k, and k_s under anaerobic conditions. The BIO1D model was used by Klecka et al. (1990) at the Dow-Cliffs Superfund site as will be seen in a subsequent section.

Instantaneous Reaction Models—The BIOPLUME II Model

The BIOPLUME II model was developed by modifying a two-dimensional transport model developed by the USGS and known as the method of characteristics (MOC) model (Konikow and Bredehoeft, 1978). The basic concept applied in the BIOPLUME II model includes the use of a dual-particle mover procedure to simulate the transport of oxygen and contaminants in the subsurface (Rifai et al., 1988). The transport equation is solved twice at every time step to calculate the oxygen and contaminant distributions:

$$\frac{\partial(Cb)}{\partial t} = \frac{1}{R_c}\left(\frac{\partial}{\partial x_i}\left(bD_{ij}\frac{\partial C}{\partial x_j}\right) - \frac{\partial}{\partial x_i}(bCV_i)\right) - \frac{C'W}{n} \quad (20)$$

$$\frac{\partial(Ob)}{\partial t} = \left(\frac{\partial}{\partial x_i}\left(bD_{ij}\frac{\partial O}{\partial x_j}\right) - \frac{\partial}{\partial x_i}(bOV_i)\right) - \frac{O'W}{n} \quad (21)$$

where C and O are concentrations of contaminant and oxygen, respectively, C' and O' are concentrations of contaminant and oxygen in a source or sink fluid, n is effective porosity, b is saturated thickness, t is time, x_i and x_j are cartesian coordinates, W is volume flux per unit area, V_i is seepage velocity in the direction of x_i, R_c is retardation factor for contaminant, and D_{ij} is coefficient of hydrodynamic dispersion.

The two plumes are combined using the principle of superposition to simulate the instantaneous reaction between oxygen and the contaminants, and the decrease in contaminant and oxygen concentrations is calculated from:

$$\Delta C_{RC} = O/F;\ O = 0 \text{ where } C > O/F \quad (22)$$

$$\Delta C_{RO} = C \cdot F;\ C = 0 \text{ where } O > C \cdot F \quad (23)$$

where ΔC_{RC}, ΔC_{RO} are the calculated changes in concentrations of contaminant and oxygen, respectively, due to biodegradation.

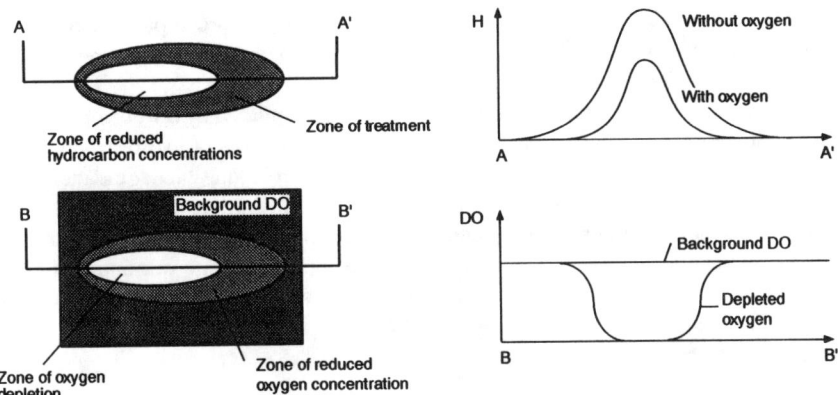

Figure 3. Principle of superposition for organics and oxygen in BIOPLUME II model. (From Rifai, H. S. et al., *J. Environ. Eng.*, 114(5), 1007, 1988. © 1988 by ASCE. Reproduced by permission of the publisher.)

Figure 3 is a conceptual schematic of the BIOPLUME II model. On the left side of the figure, a plan view of the contaminant and oxygen plumes with and without biodegradation are shown. After the two plumes are superimposed, the contaminant plume is reduced in size and concentration. The dissolved oxygen is depleted in zones of high contaminant concentrations and reduced in zones of relatively moderate contaminant concentrations. The schematics on the right in Figure 3 present transects down the plume centerline which help to illustrate the distributions of contaminant and oxygen concentration with and without biodegradation. The BIOPLUME II model has been applied to a number of sites: the Conroe Superfund site, the Traverse City-Michigan jet fuel spill site, and the Gas Plant Facility in Michigan.

MODELING OF BIODEGRADATION AT FIELD SITES

Conroe Superfund Site, Texas

Borden et al. (1986) applied the first version of the BIOPLUME model to simulate biodegradation at the Conroe Superfund site in Texas. The United Creosoting Company site was operated as a wood preserving facility from 1946 to 1972. In 1972, the facility was closed and the land sold off for use by a small commercial operation and later subdivided for a housing development. Monitoring of the site has shown elevated levels of organic contaminants in the soil and groundwater. Wastes generated by the wood creosoting operation were disposed in two unlined ponds. These wastes were composed of predominantly polycyclic aromatic hydrocarbons and pentachlorophenol (PCP). In addition to the organic wastes, elevated levels of chloride are present in the groundwater.

The geology at the UCC site is heterogeneous with numerous changes between relatively clean well-sorted sand and sandy clay. Two separate water bearing zones have been identified: (1) a lower permeability unconfined zone, and (2) a slightly higher permeability semiconfined zone. These two zones are separated by a thin semiconfining clay layer. The piezometric gradient in the area is low resulting in groundwater velocities in the range of 5 m/year.

Oxygen exchange with the unsaturated zone was simulated by Borden et al. (1986) as a first-order decay in hydrocarbon concentration. The loss of hydrocarbon due to horizontal mixing with oxygenated groundwater and resulting biodegradation were simulated by generating oxygen and hydrocarbon distributions independently and then combining by superposition. Simulated oxygen and hydrocarbon concentrations closely matched the observed values. The Conroe Superfund site was one of the first sites to be modeled using a biodegradation expression in a transport model.

Traverse City Site, Michigan

The Traverse City field site is a U.S. Coast Guard Air Station located in Grand Traverse County in the northwestern portion of the lower peninsula of Michigan (Figure 4). The groundwater at the site is contaminated with organic chemicals from a source near the Hangar/Administration building. The main contaminants at the site are benzene, toluene, and xylenes (BTX). The contaminant plume ranges from 150 to 400 ft wide and is about 4000 ft long. A pumping wellfield system was installed at the downgradient end of the dissolved plume to control off-site migration. A modeling effort of natural attenuation at the site was completed by Rifai et al. (1988) with the BIOPLUME II model.

The data in Figure 5 show the change in total BTX mass with time in the system. The data also show the mass removed with the pumping wellfield. The modeling at the site was performed for the period before the pumping wells were installed and also for the period after the wells were turned on. The data in Figure 6 show the results of the model simulation along the centerline of the plume for the period before the wellfield was turned on. The model predictions by Rifai et al. (1988) matched the observed concentrations at the monitoring wells reasonably well except in the vicinity of well M31. Rifai et al. (1988) indicated that this was because the simulation did not account for anaerobic biodegradation which was believed to be occurring in the interior of the plume.

Gas Plant Facility in Michigan

Soluble hydrocarbon and dissolved oxygen (DO) were characterized in a shallow aquifer beneath a gas plant facility in Michigan by Chiang et al. (1989). The distributions of benzene, toluene, and xylene (BTX) in the aquifer had been monitored in 42 wells for a period of 3 years. The site geology is characterized as a medium to coarse sand with interbeds of small gravel and cobbles. The general direction of groundwater flow is north to west. The depth to water ranges from 10 to 25 ft below land surface, and the slope of the water table was estimated as 0.006 ft/ft. Based on groundwater and soil sampling data, Chiang et al. (1989) concluded that the flare pit was the major source of the hydrocarbons found in the aquifer and the slop oil tank was the secondary source.

Results from the three-year sampling period showed a significant reduction in total benzene mass with time in groundwater. The plume sampled in 1984 contained

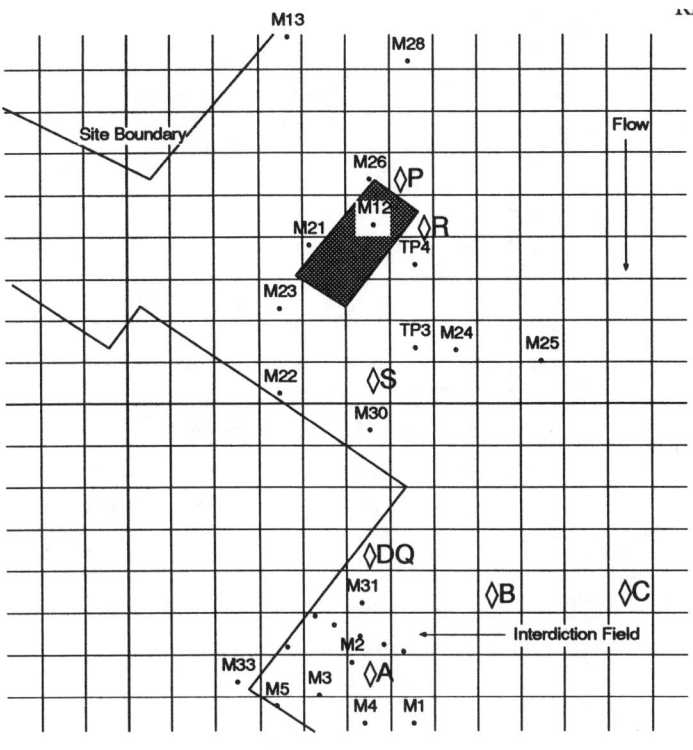

Figure 4. Traverse City field site map. (From Rifai, H. S. et al., *J. Environ. Eng.*, 114(5), 1007, 1988. © 1988 by ASCE. Reproduced by permission of the publisher.)

an approximate total mass of 9.83 kg, while the plume sampled in 1985 and 1986 contained 5.66 kg and 2.27 kg, respectively. Chiang et al. (1989) determined the attenuation rates of the soluble benzene, and determined the effects of DO on the biodegradation of BTX through a combination of material balance, statistical analyses, soluble transport modeling, and laboratory microcosm experiments.

Chiang et al. (1989) evaluated a first-order decay biodegradation approach and the BIOPLUME II model for simulating biodegration at the Gas Plant Facility. Using the model and assuming first-order decay, several simulations were made to match the observed benzene concentration distribution of 1/22/85 by setting the observed concentration distribution of 11/1/84 as the initial condition. The variables involved included the distribution of the leakage/spill rates between the flare pit and the slop oil tanks and macrodispersivities of the aquifer.

The BIOPLUME II model was used to simulate the July 1987 data by setting the observed concentration distribution of February 1987 as an initial condition.

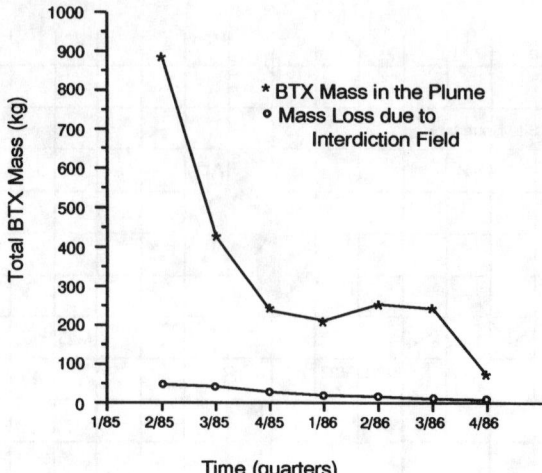

Figure 5. Variation in total BTX with time at Traverse City field site. (From Rifai, H. S. et al., *J. Environ. Eng.*, 114(5), 1007, 1988. © 1988 by ASCE. Reproduced by permission of the publisher.)

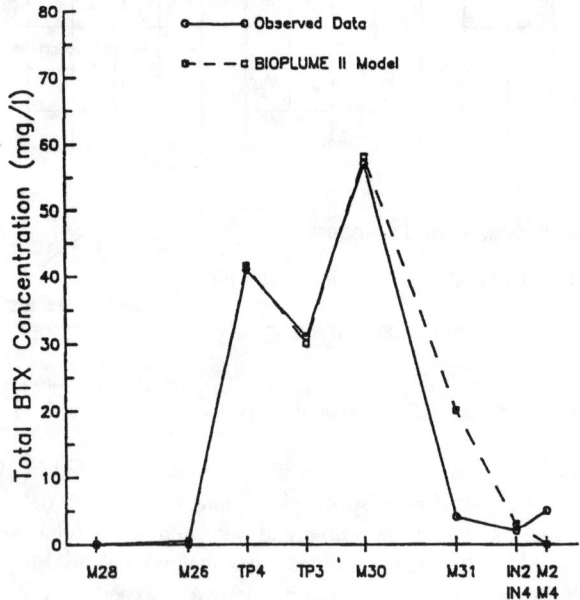

Figure 6. Calibrated versus observed BTX concentrations at Traverse City field site for period 1. (From Rifai, H. S. et al., *J. Environ. Eng.*, 114(5), 1007, 1988. © 1988 by ASCE. Reproduced by permission of the publisher.)

Figure 7 shows correlations between the measured and the simulated soluble BTX concentrations of July 1987. As can be seen from Figure 7, the correlations for BTX were reasonable. The correlations, however, for oxygen were not as similar. The authors attributed the differences to the fact that the BIOPLUME II model assumes a requirement of 3 ppm of oxygen for 1 ppm of benzene, whereas the actual requirement is in the range of 1–3 ppm.

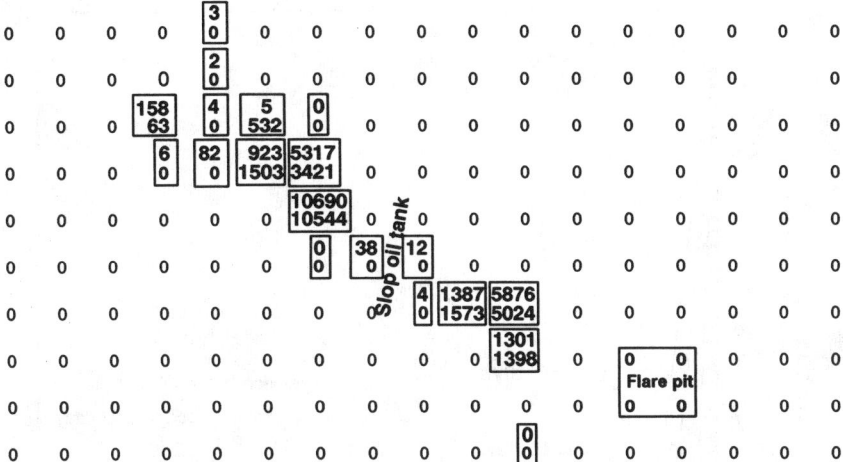

Figure 7. Predicted (bottom number) and observed (top number) BTX by BIOPLUME II (ppb). (From Chiang, C. Y., et al., *Ground Water*, 6, 823, 1989. With permission.)

Cliffs-Dow Superfund Site

Groundwater at the Cliffs-Dow site is contaminated with low levels of phenolic and polycyclic compounds. The aquifer sediments at the site consist of mostly coarse sands and gravels. The hydraulic conductivity ranges between 3.5×10^{-3} to 4.6×10^{-2} cm/s. The principal contaminants found at the site near the source area include phenol, several methyl-substituted phenols, and naphthalene at concentrations ranging from 220 to 860 µg/L.

Based on the analysis of samples obtained from monitoring wells, Klecka et al. (1990) found that the levels of organic contaminants are reduced to near or below the detection limit within a distance of 100 m downgradient from the source. Further analyses of the groundwater chemistry were used to verify that biodegradation was occurring at the site and causing the disappearance of the contaminants.

Klecka et al. (1990) attributed the reduction of dissolved oxygen from 1.2 mg/L in an upgradient well to less than 0.1 mg/L at downgradient wells to biodegradation. This conclusion, Klecka et al. (1990) argue, is supported by the increase in total inorganic carbon at well B-3A. Microcosm studies were used by Klecka et al. (1990) to simulate the aerobic biodegradation of phenols at the site.

The migration of organic constituents in the aquifer was simulated using the BIO1D model and assuming a first-order decay expression. Half-lives for the contaminants at the site were estimated from the results of soil microcosm experiments based on the time required for 50% disappearances of the parent compound. The velocity was varied over a range from 0.2 to 0.46 m/day, which is representative of the range of groundwater flow rates at the site.

Figure 8 illustrates the impact of biodegradation on contaminant concentrations at the site. Model simulations performed using a half-life of two days indicated that the levels of the phenolic components were reduced by more than 99% within a distance of 30 m downgradient of the source. When the half-life was increased by a factor

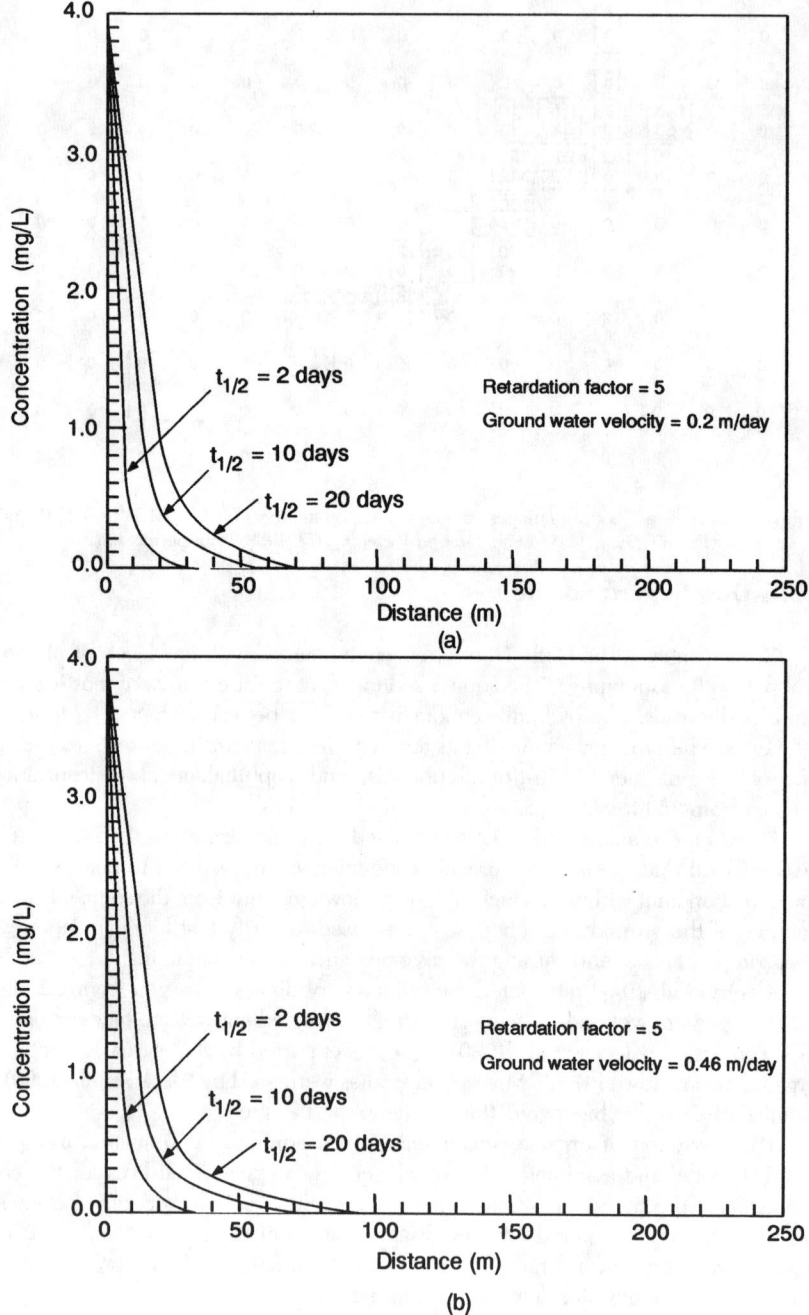

Figure 8. BIO1D simulations for the Cliff-Dow Superfund site. (From Klecka, G. M., et al., *Ground Water*, 4, 534, 1990. With permission.)

of 10, the concentrations were reduced to a similar extent within 75 m. Because of the dominance of biodegradation, increases in groundwater velocity from 0.2 to 0.46 m/day had minor effects on the level of attenuation predicted with the model.

CONCLUSIONS

A number of biodegradation models have been developed over the last ten years. These models are generally similar in that they simulate the transport and biodegradation of a number of components in the groundwater. The models differ in the mathematical biodegradation expressions that they use and in the numerical procedures used to solve the complicated system of equations. Applications of these models to field sites has proven to be complicated due to the lack of biodegradation parameters that can be measured in the field and input into the model. As a result, most modeling applications at the field scale have resorted to first-order decay or instantaneous representation of the biodegradation process.

REFERENCES

Balkwill, D. L. and Ghiorse, W. C., Characterization of subsurface bacteria associated with two shallow aquifers in Oklahoma, *Appl. Environ. Microb.*, 3(50), 580, 1985.

Bedient, P. B., Rifai, H. S., and Newell, C. J., *Ground Water Contamination: Transport and Remediation*, Prentice Hall, NJ, 1994.

Beeman, R. E. and Suflita, J. M., Microbial ecology of a shallow unconfined groundwater aquifer polluted by municipal landfill leachate, *Microbial Ecol.*, 14, 39, 1987.

Borden, R. C. and Bedient, P. B., Transport of dissolved hydrocarbons influenced by oxygen-limited biodegradation: 1. theoretical development, *Water Resour. Res.*, 13, 1973, 1986.

Borden, R. C., Bedient, P. B., Lee, M. D., Ward, C. H., and Wilson, J. T., Transport of dissolved hydrocarbons influenced by oxygen-limited biodegradation: 2. field application, *Water Resour. Res.*, 13, 1983, 1986.

Celia, M. A., Kindred, J. S., and Herrera, I., Contaminant transport and biodegradation: 1. a numerical model for reactive transport in porous media, *Water Resour. Res.*, 25(6), 1141, 1989.

Chiang, C. Y., Salanitro, J. P., Chai, E. Y., Colthart, J. D., and Klein, C. L., Aerobic biodegradation of benzene, toluene, and xylene in a sandy aquifer—data analysis and computer modeling, *Ground Water*, 6, 823, 1989.

Domenico, P. A., An analytical model for multidimensional transport of a decaying contaminant species, *J. Hydrol.*, 91, 49, 1987.

Ghiorse, W. C. and Balkwill, D. L., Enumeration and morphological characterization of bacteria indigenous to subsurface environments, *Dev. Ind. Microbiol.*, 24, 213, 1983.

Klecka, G. M., Davis, J. W., Gray, D. R., and Madsen, S. S., Natural bioremediation of organic contaminants in groundwater: Cliffs-Dow Superfund site, *Ground Water*, 4, 534, 1990.

Konikow, L. F. and Bredehoeft, J. D. Computer model of two-dimensional solute transport and dispersion in groundwater, automated data processing and computations, Techniques of Water Resources Investigations of the U.S.G.S., Washington, D.C., 1978, pp. 100.

MacQuarrie, K. T. B., Sudicky, E. A., and Frind, E. O., Simulation of biodegradable organic contaminants in groundwater: 1. numerical formulation in principal directions, *Water Resour. Res.*, 26(2), 207, 1990.

McCarty, P. L., Reinhard, M., and Rittman, B. E. Trace organics in groundwater, *Environ. Sci. and Technol.*, 15(1), 40, 1981.

McCarty, P. L., Rittman, B. E., and Bouwer, E. J., Microbiological processes affecting chemical transformations in groundwater, in *Groundwater Pollution Microbiology*, Bitton, G. and Gerba, C. P., Eds., John Wiley, New York, 1984, pp. 89–115.

Molz, F. J. , Widdowson, M. A., and Benefield, L. D., Simulation of microbial growth dynamics coupled to nutrient and oxygen transport in porous media, *Water Resour. Res.*, 22(8), 1207, 1986.

Monod, J., Recherches sur la croissance des cultures bacteriènnes, Herman & Cie, Paris, 1942.

Odencrantz, J. E., Valocchi, A. J., and Rittman, B. E., Modeling two-dimensional solute transport with different biodegradation kinetics, Proceedings of Petroleum Hydrocarbons and Organic Chemicals in Ground Water: Prevention, Detection and Restoration, October 31–November, 1990, NWWA, Houston, TX, 1990.

Rifai, H. S., Bedient, P. B., Wilson, J. T., Miller, K. M., and Armstrong, J. M., Biodegradation modeling at aviation fuel spill site, *J. Environ. Eng.*, 114(5), 1007, 1988.

Semprini, L. and McCarty, P. L., Comparison between model simulations and field results for in-situ biorestoration of chlorinated aliphatics: part 1. biostimulation of methanotrophic bacteria, *Ground Water*, 29(3), 365, 1991.

Srinivasan, P. and Mercer, J. W. Simulation of biodegradation and sorption processes in groundwater, *Ground Water*, 26(4), 475, 1988.

Suflita, J. M., Microbiological principles influencing the biorestoration of aquifers, in *Transport and Fate of Contaminants in the Subsurface*, EPA/625/4-89/019, Robert S. Kerr Environmental Research Laboratory, U.S. EPA, Ada, OK, 1989, pp. 85–99.

Wheeler, M. F., Dawson, C. N., Bedient, P. B., Chiang, C. Y., Borden, R. C., and Rifai, H. S., Numerical simulation of microbial biodegradation of hydrocarbons in groundwater, in Proceedings of the Solving Ground Water Problems with Models Conference, February 10–12, Denver, CO; National Water Well Association, Dublin, OH, 1987.

Widdowson, M. A., Molz, F. J., and Benfield, L. D., Development and application of a model for simulating microbial growth dynamics coupled to nutrient and oxygen transport in porous media, in Proceedings of the Solving Ground Water Problems with Models Conference, February 10–12, Denver, CO; National Water Well Association, Dublin, OH, 1987, pp. 28–51.

Widdowson, M. A., Molz, F. J., and Benefield, L. D., A numerical transport model for oxygen- and nitrate-based respiration linked to substrate and nutrient availability in porous media, *Water Resour. Res.*, 24(9), 1553, 1988.

Wilson, J. T., McNabb, J. F., Balkwill, D. L., and Ghiorse, W. C., Enumeration and characterization of bacteria indigenous to a shallow water-table aquifer, *Ground Water*, 2,134, 1983.

CHAPTER 17

Evaluation of the Groundwater Disinfection Rule "Natural Disinfection" Criteria Using Field Data

Marylynn V. Yates

INTRODUCTION

In 1985, the United States Environmental Protection Agency (USEPA) proposed maximum contaminant level goals (MCLGs) for viruses and *Giardia*, a protozoan parasite (USEPA, 1985). These standards are in addition to the standard for the indicator microorganism, total coliform. Rather than require public water systems to monitor the water for the presence of these pathogenic microorganisms, the EPA proposed treatment requirements for groundwater with the goal that the level of pathogenic viruses and *Giardia* in the treated water would result in a risk of less than one infection per 10,000 persons per year (USEPA, 1992). Water utilities may avoid chemical disinfection of the source water if they meet one of the EPA's "natural disinfection" criteria. The criteria include setback distance, depth to well screen or thickness of unsaturated zone, groundwater travel time, and virus travel time. The numerical values for the criteria are based on an acceptable virus concentration of 2 viruses/10^7 liters at the wellhead. This concentration was calculated using a risk of less than one infection per 10,000 people per year (Regli et al., 1991).

There are approximately 180,000 community and noncommunity public water supply systems with wells or well fields that are not groundwater under the influence (GWUI) wells. The draft proposed groundwater disinfection rule (GWDR) will allow such systems to avoid disinfecting provided they can demonstrate that adequate "natural disinfection" occurs between potential sources of fecal contamination (e.g., septic tanks, sewer lines) and wells. One method for determining if adequate "natural disinfection" occurs is to use a virus transport model such as VIRALT, which was developed for the EPA's office of groundwater and drinking water (Park et al., 1991). However, the model has never been tested using field data to determine whether it can accurately predict virus transport. The objective of this project was to determine whether VIRALT can accurately predict virus contamination of groundwater, and thus accurately determine whether a utility can avoid disinfection of their water supply wells.

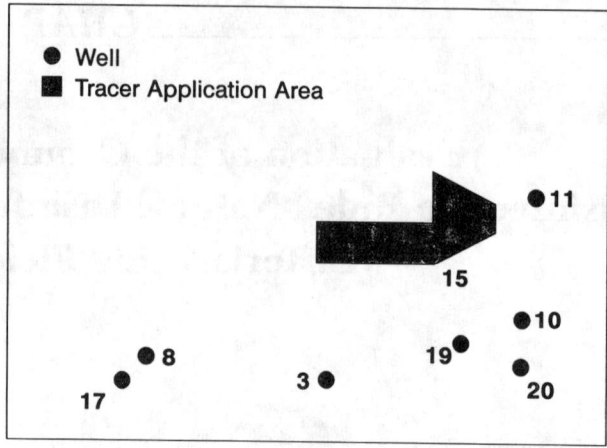

Figure 1. Diagram of virus tracer study site, Fort Devins, MA.

METHODS

In order to examine the "natural disinfection" criteria, several sites where virus contamination of groundwater has been documented were found. Because VIRALT requires values for several hydrogeologic properties, it was necessary to locate sites for which that information was known. As a result, the sites were research sites where known quantities of viruses were added to the soil and their movement to groundwater followed. The necessary hydrogeologic and microbiological information from these sites were obtained and used as input values for VIRALT. The model's predictions were then compared with the measured virus concentrations in wells at the sites.

Fort Devins, Massachusetts

The first location chosen for testing was a site in Fort Devins, Massachusetts, where an experiment to monitor the movement of coliphages during the rapid infiltration of wastewater was conducted (Schaub et al., 1975). High concentrations of f2 coliphage [approximately 10^8 plaque-forming units (pfu) per liter] were added to the effluent as it was applied to the soil. The concentrations of the phage in several wells on and near the site were monitored for 21 days. A schematic of the site showing the location of the recharge site and the monitoring wells is given in Figure 1.

In order to test the ability of VIRALT to predict the measured values, several pieces of information characterizing the site had to be input into the model. These input data are shown in Table 1. In several instances, the required information was not reported by the investigators; in those cases, the default values provided by VIRALT were used.

Vosen, Wisconsin

The virus transport data from Vosen was obtained in a study of virus transport from a septic tank (Stramer, 1984). Poliovirus-containing stools from infants who had

Table 1. Input Parameters for Fort Devins, MA Site

Parameter	Value	Source of Value
Saturated Zone		
Transmissivity	104 m² day⁻¹	measured
Aquifer thickness	12.2 m	measured
Porosity	0.46	measured
Hydraulic gradient	0.0148	measured
Angle of flow	315°	measured
Unsaturated Zone		
No. of soil layers	1	known
Soil type	sand	known
Thickness	19.5 m	measured
Saturated hydraulic conductivity	8.5 m day⁻¹	measured
Saturated moisture content	0.46	measured
Residual moisture content	0.045	default
Source		
Time period	7 days	known
Leakage rate	1.13 m day⁻¹	known
Virus concentration	2.5×10^8 pfu l⁻¹	known
Source area	5018 m²	known
Contaminant Transport		
Soil bulk density	1.43	measured
Molecular diffusion	0	default
Saturated dispersivity	1.2 m	default
Unsaturated dispersivity	0.12 m	assumed
Temperature	15 C	default
Virus adsorption	0 cm³ g⁻¹	measured
Virus inactivation rate	0.29 day⁻¹	default
Retardation coefficient	1	default
Well pumping rate	0.012 m³ day⁻¹	known

recently been vaccinated with the Sabin vaccine were flushed down the toilet in a single-family dwelling. The movement of the viruses was followed to the septic tank, into the leach field, and into several monitoring wells that had been installed at the site downgradient from the leach field. The soil at the site is composed of a sandy loam topsoil in the upper 71 cm, which is underlain by a medium to fine sandy soil. The water table at this site had a mean depth of 1.6 m below the soil surface. Figure 2 shows a schematic of the site.

The values used as input parameters for VIRALT are given in Table 2. In this case, values for many of the required input parameters were not measured by the investigator, and the program's default values were used. In the case of the virus adsorption coefficient and retardation factor, the use of the program default values (i.e., no adsorption) was justified in that little adsorption of poliovirus would be expected to occur in a sandy soil.

Lake Redstone, Wisconsin

This study of virus transport was conducted in a manner similar to that described for Vosen (Stramer, 1984). Poliovirus-containing stools from infants who had been vaccinated with the oral poliovirus vaccine (strain LSc2ab) were flushed down the toilet at a single-family dwelling. Again, virus concentrations were measured in the septic tank, leach field, monitoring wells, and a nearby lake. A schematic of the site

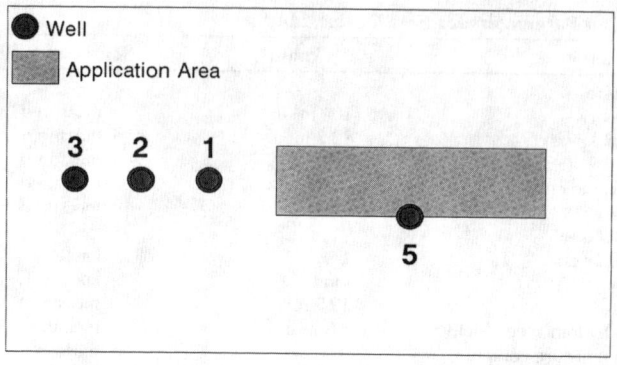

Figure 2. Virus transport study site, Vosen, WI.

Table 2. Input Parameters for Vosen, WI Site

Parameter	Value	Source of Value
Saturated Zone		
Transmissivity	7.128 m² day⁻¹	calculated
Aquifer thickness	1 m	measured
Porosity	0.45	default
Hydraulic gradient	0.03	measured
Angle of flow	175°	measured
Unsaturated Zone		
No. of soil layers	1	known
Soil type	sand	known
Thickness	1.6 m	measured
Saturated hydraulic conductivity	7.128 m day⁻¹	measured
Saturated moisture content	0.43	default
Residual moisture content	0.045	default
Source		
Time period	7.2 days	known
Leakage rate	0.0147 m day⁻¹	known
Virus concentration	8.2 × 10⁴ pfu l⁻¹	known
Source area	3.34 m²	known
Contaminant Transport		
Soil bulk density	1.51	default
Molecular diffusion	0	default
Saturated dispersivity	0.1 m	default
Unsaturated dispersivity	0.1 m	assumed
Temperature	15 C	known
Virus adsorption	0 cm³ g⁻¹	default
Virus inactivation rate	0.29 day⁻¹	default
Retardation coefficient	1	default
Well pumping rate	0.1 m³ day⁻¹	known

is shown in Figure 3, and values for the required model input parameters are shown in Table 3.

The soil at this site is classified as a poorly structured Jackson series silt loam with pockets of a well-structured sandy loam. The soil was reported to contain 33.5% sand, 51.5% silt, and 15% clay. Because of the high clay content of this soil, it was felt that

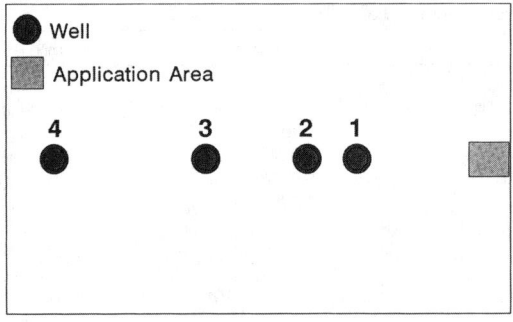

Figure 3. Virus transport study site, Lake Redstone, WI.

VIRALT's default value of zero for virus adsorption was not appropriate. The value used for the adsorption coefficient was obtained from an experiment conducted by Goyal and Gerba (1979) in which the adsorption of poliovirus 1 LSc to a soil of similar texture was measured.

At the time the viruses were added to the septic tank system, the water table level had risen to a point above the bottom of the leach field. Thus, it was assumed that there was no virus transport through the unsaturated zone, and only the saturated zone module of VIRALT was run.

RESULTS

Fort Devins, Massachusetts

VIRALT's prediction of f2 coliphage transport to Well 15 as compared to the measured values is shown in Figure 4. It can be seen that, although the model predicted that it would take longer for the virus to arrive at the well than it actually did, the predicted values agree very well with the measured values after approximately Day 10. The model predictions for Well 19, located approximately 30 m from the infiltration site, are shown in Figure 5. In this case, the model predicted that the viruses would not reach the well in 21 days, although they were detected in that well soon after application. The model predictions for Wells 10 and 20 (not shown), were essentially the same as for Well 19: The model predicted that the viruses would not reach the wells in 21 days, although they were detected by the investigators.

Vosen, Wisconsin

The predicted concentrations of viruses in Wells 1, 2, 3, and 5 at the Vosen site are compared to the measured concentrations in Figures 6 through 9. At Well 1, VIRALT predicted that the maximum virus concentration would be approximately

Table 3. Input Parameters for Lake Redstone, WI Site

Parameter	Value	Source of Value
Saturated Zone		
Transmissivity	0.24 m² day⁻¹	calculated
Aquifer thickness	2 m	measured
Porosity	0.4	estimated
Hydraulic gradient	0.0583	measured
Angle of flow	175°	measured
Source		
Time period	8.33 days	known
Leakage rate	0.12 m day⁻¹	known
Virus concentration	2.58.2 × 10⁴ pfu l⁻¹	known
Source area	3 m²	known
Contaminant Transport		
Soil bulk density	1.5	default
Molecular diffusion	0	default
Saturated dispersivity	0.6 m	default
Unsaturated dispersivity	0.1 m	assumed
Temperature	21 C	known
Virus adsorption	4.5 cm³ g⁻¹	assumed
Virus inactivation rate	0.538 day⁻¹	default
Retardation coefficient	17.87	default
Well pumping rate	0.1 m³ day⁻¹	known

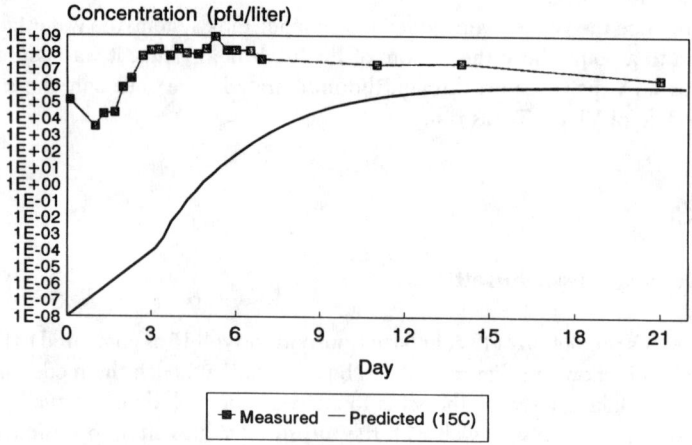

Figure 4. Virus movement during tracer study, Well 15, Fort Devins, MA.

300 viruses per liter on Day 18. The measured concentration on Day 20 was approximately 30 per liter; the maximum concentration (3000 viruses per liter) was measured on Day 80. For Wells 2, 3, and 5, VIRALT underpredicted the maximum virus concentrations at the wells by 1 to 2 orders of magnitude.

Lake Redstone, Wisconsin

Virus concentrations at Wells 1, 2, and 3 predicted by VIRALT are compared with those measured by Stramer (1984) in Figures 10 through 12. In all cases, VIRALT underpredicted the measured concentrations by at least 10 orders of magnitude.

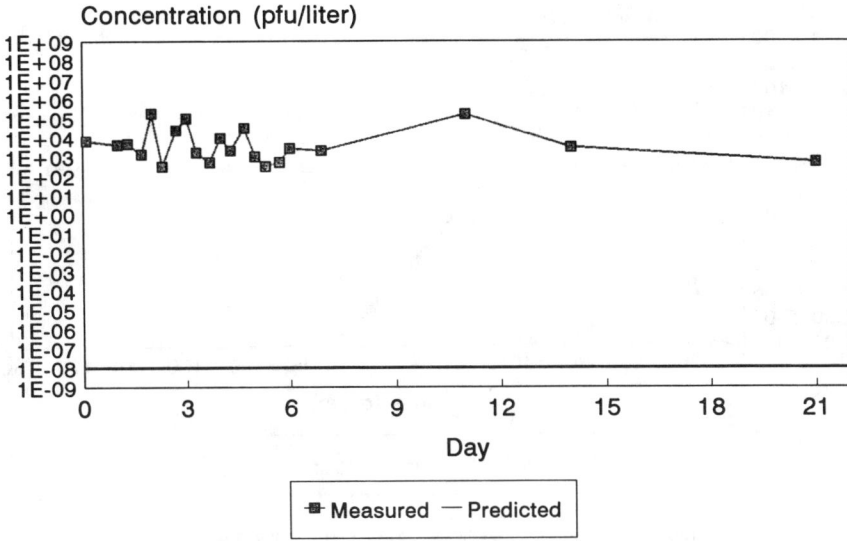

Figure 5. Virus movement during tracer study, Well 19, Fort Devins, MA.

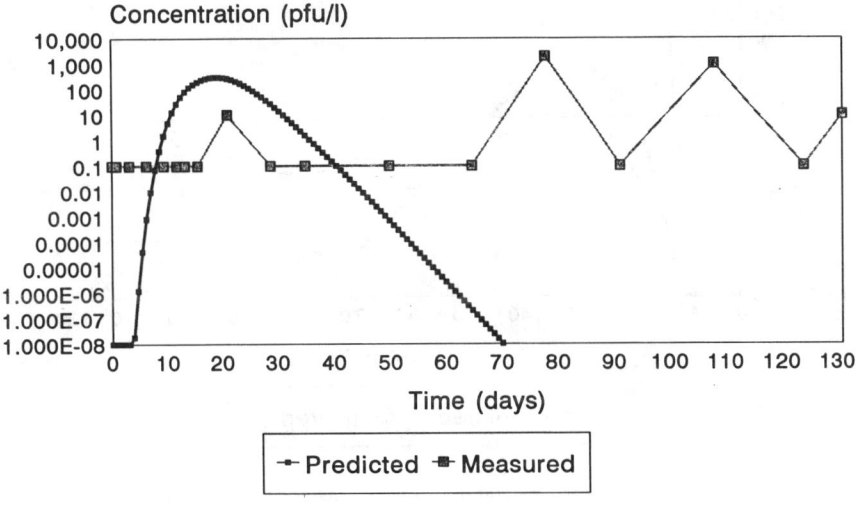

Figure 6. Virus transport from a septic tank, Well 1, Vosen, WI.

DISCUSSION

The results of this project indicate that VIRALT, the model being considered by the USEPA to allow water utilities to determine whether their wells would qualify as meeting natural disinfection (and thus not require chemical or physical disinfection), does not always match field measurements of virus transport in the subsurface. It must be emphasized that the model has only been tested at five sites, three of which are

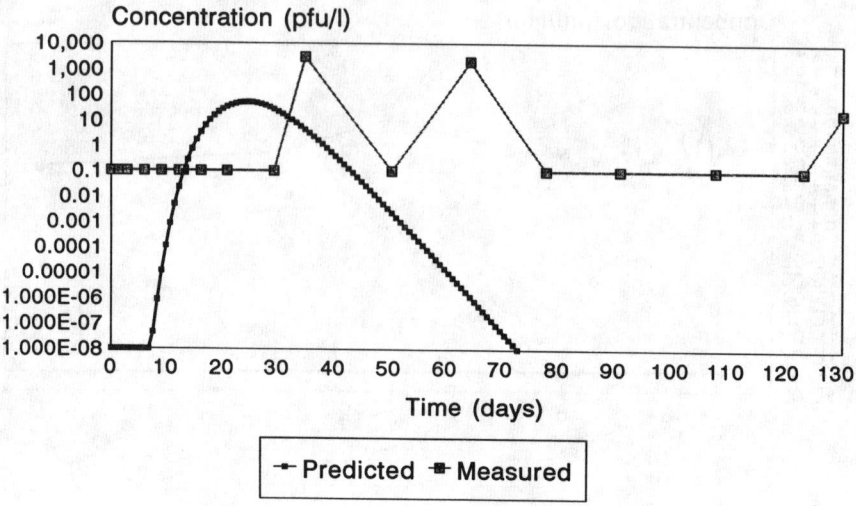

Figure 7. Virus transport from a septic tank, Well 2, Vosen, WI.

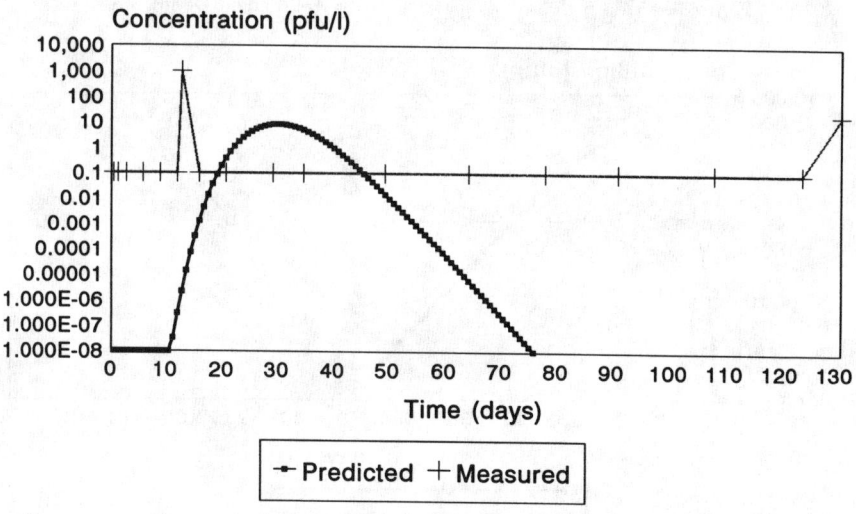

Figure 8. Virus transport from a septic tank, Well 3, Vosen, WI.

described herein. None of these sites have continuously pumping wells, which would be the case in a public water utility. The model must be tested at several sites where continuously pumping production wells are operating prior to complete an evaluation of its predictive capabilities. However, at this time, there are no public water supply wells for which virus occurrence data as well as the necessary hydrogeologic data are available.

In at least one situation, namely the Vosen experiment, the model's ability to accurately predict virus transport was hampered by the user's inability to simulate

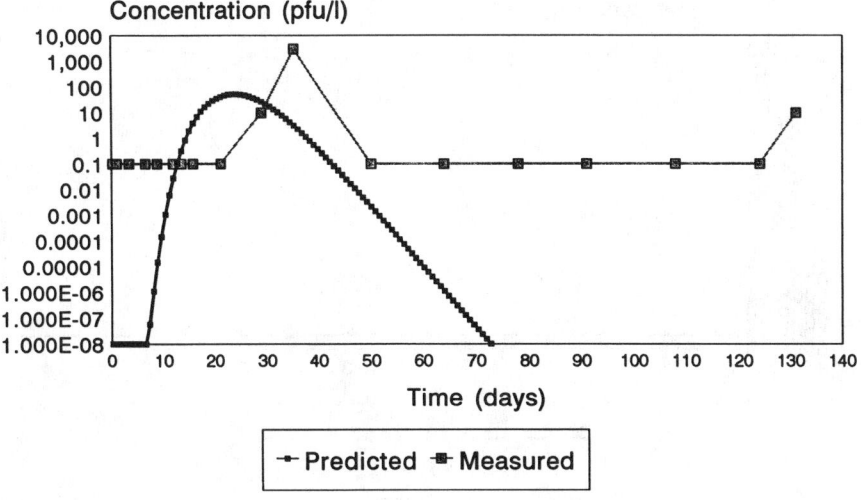

Figure 9. Virus transport from a septic tank, Well 5, Vosen, WI.

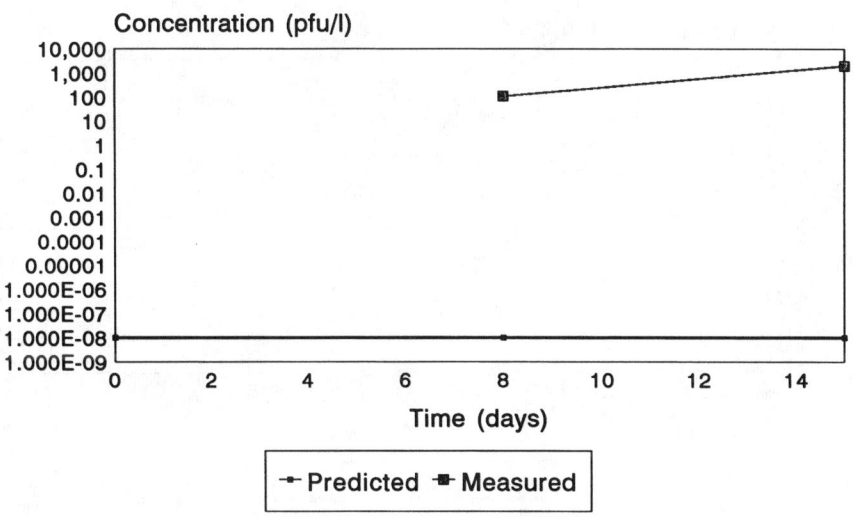

Figure 10. Virus transport from a septic tank, Well 1, Lake Redstone, WI.

rainfall events. The spikes in virus concentration in the wells at the site occurred after rainfall events, which presumably caused virus desorption from the soil and allowed transport to the water table.

Another important facet of this project is to determine the sensitivity of the model to each of the input parameters. In other words, the impact of each of the input parameters on the model's predictions must be determined. This will allow the water industry to know how important it is to know the value of each of the input parameters

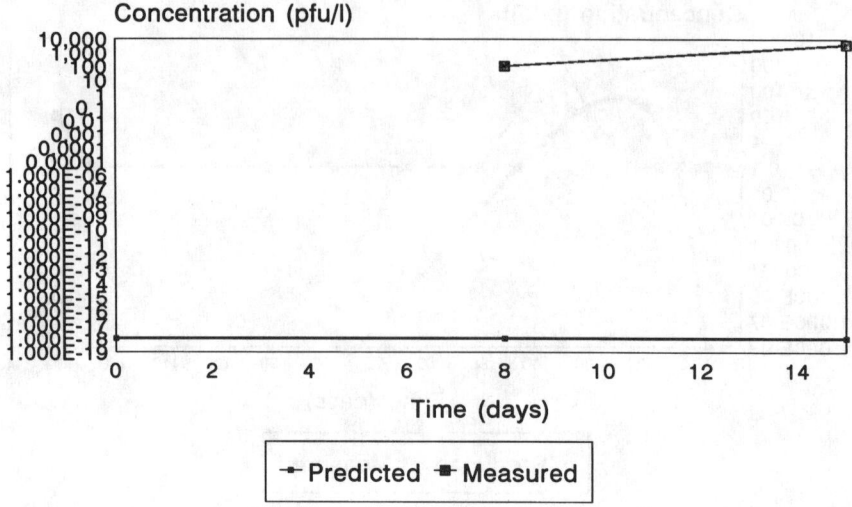

Figure 11. Virus transport from a septic tank, Well 2, Lake Redstone, WI.

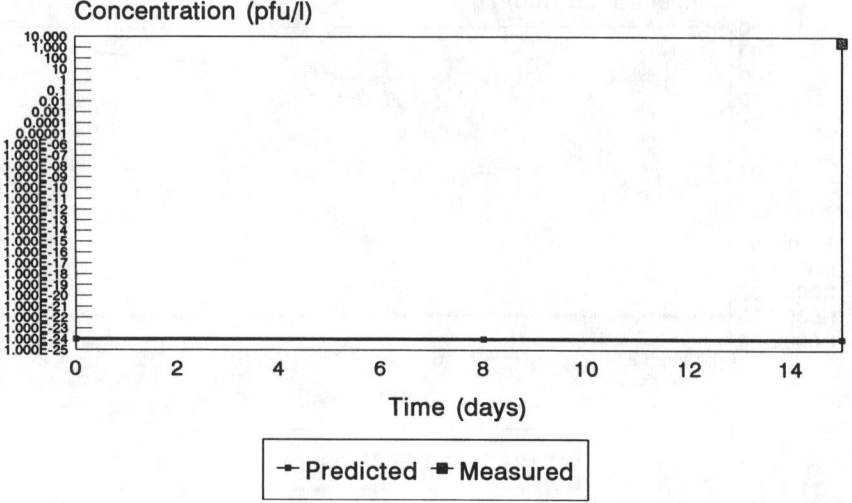

Figure 12. Virus transport from a septic tank, Well 3, Lake Redstone, WI.

to a high degree of accuracy. For example, if changing the value of the hydraulic conductivity by 50% changes the model's predictions by 50%, it will be important to have a very accurate value of the hydraulic conductivity when using the model. On the other hand, if changing the value of the hydraulic conductivity by 50% changes the model prediction by 1%, it might not be as important to know the hydraulic conductivity with a high degree of accuracy. A sensitivity analysis of the model is currently underway.

REFERENCES

Goyal, S. M. and Gerba, C. P., Comparative adsorption of human enteroviruses, simian rotavirus, and selected bacteriophages to soils, *Appl. Environ. Microbiol.*, 38, 241, 1979.

Park, N.-S., Blanford, T. N., and Huyakorn, P. S., VIRALT—a model for simulating viral transport in ground water, documentation and user's guide, Version 2.0. HydroGeoLogic, Inc., Herndon, VA, 1991.

Regli, S., Rose, J. B., Haas, C. N., and Gerba, C. P., Modeling the risk from *Giardia* and viruses in drinking water, *J. Amer. Water Works Assoc.*, 83, 76, 1991.

Schaub, S. A., Meier, E. P., Kolmer, J. R., and Sorber, C. A., Land Application of wastewater: the fate of viruses, bacteria, and heavy metals at a rapid infiltration site, Report TR 7504, U.S. Army Medical Research and Development Command, Washington, D.C., 1975.

Stramer, S. L., Fates of poliovirus and enteric indicator bacteria during treatment in a septic tank system including septage disinfection, Ph.D. dissertation, Department of Bacteriology, University of Wisconsin-Madison, WI, 1984.

United States Environmental Protection Agency, National primary drinking water regulations; synthetic organic chemicals, inorganic chemicals, and microorganisms. *Fed. Regist.*, 50, 46936, 1985.

United States Environmental Protection Agency, Draft ground-water disinfection rule. Office of Ground Water and Drinking Water, Washington, D.C., 1992.

SECTION 6

On Fracture Flow Modeling

CHAPTER 18

A New Modeling Approach for Predicting Flow in Fractured Formations

T. J. Cheema and M. R. Islam

ABSTRACT

Prediction of aquifer performance in complex hydrogeologic settings often requires numerical simulation. To date, no comprehensive simulation package has been developed for modeling groundwater. Most available simulators are limited and are not applicable to fractured formations. The presence of fracture is associated with high water productivity as well as high contaminant vulnerability. Consequently, it is important to develop a technique for properly modeling fractured formations.

This paper presents a method which uses the MODFLOW simulator with data provided by experimental models. In most cases parallel-plate model data on anisotropy are adequate. In all cases, however, dual-porosity data should be used to determine the maximum hydraulic conductivity. This method allows one to avoid conducting expensive field experiments and provides one with a relatively easy-to-use tool for predicting fluid flow in a fractured formation. The method is applied in the Black Hills region which shows intense fracturing. The method is proven to be effective in several sites. Data obtained through laboratory experiments are used in MODFLOW simulator to predict fluid flow performance in a formation characterized by the presence of fractures and caves. The potentiometric surface computed by local models compared very well with observed field results.

INTRODUCTION

Huntoon (1974) simply described a model as a statement of a worker's conception of a phenomenon. Simulation models are miniature conceptualizations of real physical systems. They provide a mathematical description form to model the behavior of the system under various conditions. In general, the final results of a model are primarily determined by the familiar groundwater flow and transport equations. These equations are often simplified when site-specific conditions are applied. During the last 15 years, numerical models have been accepted as intelligently applied and useful tools (Lehr, 1979). Numerical modeling provides the most general tool for the quantitative analysis of various groundwater applications (Faust and Mercer, 1980). In particular, digital computer models and simulators are useful in the evaluation of aquifer parameters and characteristics, in the estimation of aquifer stresses, in resolving data inconsistencies, and in determining the optimal allocation of exploration funds (Remson et al., 1980).

The most commonly used groundwater numerical simulators are MODFLOW (USGS modular three-dimensional finite difference program, by McDonald and Harbaugh, 1984) and method of characteristics (MOC) (Konikow and Bredehoeft, 1978). The first simulator is capable of modeling only single-phase flow. Local anomalies can be modeled with this simulator, but only by adding a general anisotropy parameter and vertical leakage parameter. This simulator solves the following single-phase partial differential equation:

$$\frac{\partial}{\partial x}\left(K_{xx}\frac{\partial h}{\partial x}\right) + \frac{\partial}{\partial y}\left(K_{yy}\frac{\partial h}{\partial y}\right) + \frac{\partial}{\partial z}\left(K_{zz}\frac{\partial h}{\partial z}\right) - W = S_s\frac{\partial h}{\partial t} \qquad (1)$$

where K_{xx}, K_{yy}, and K_{zz} are values of hydraulic conductivity along the x, y, and z coordinate axes, which are assumed to be parallel to the major axes of hydraulic conductivity, h is the potentiometric head, W is the volumetric flux, S is the specific storage of the porous material; and t is the time. A modified version of the same (MODPATH) attempts to model contaminant transport but the application is very limited. It is important to note that these MODFLOW models are not meant for modeling fracture flow. However, there is no commercially available groundwater model that incorporates fracture/matrix flow.

This paper is aimed at using the MODFLOW simulator for modeling fractured formations where anisotropy data, as well as hydraulic conductivity, are provided through laboratory experimentation.

THEORY

In two-dimensional flow, Darcy's law can be written as:

$$q_i = K_{ij}J_i \qquad (2)$$

where q is the specific flux vector with components q_x and q_y in x and y directions, respectively, K_{ij} the hydraulic conductivity (scalar in an isotropic medium and a tensor in an anisotropic medium), and J_i is the hydraulic gradient with components, $J_x = \partial h/\partial h$ and $J_y = \partial h/\partial y$ in x and y directions, respectively. In the above equation, the vectors q and J are nonlinear except when they are in the direction of one of the principal axes. The directional hydraulic conductivity, J_q, is defined by the directional hydraulic conductivity along the direction of the flow and is defined by the ratio between the specific discharge at a point and the component of the gradient in the direction of the flow. On the other hand, the directional hydraulic gradient, K_j, along the direction of the gradient, is defined by the ratio between the component of the specific discharge q along the direction of the gradient and the gradient itself. For the case studied in this paper, we deal with a system for which the hydraulic gradient is known

PREDICTING FLOW IN FRACTURED FORMATIONS

and the fluid is free to choose the path of least resistance (along the direction of highest conductivity). This leads to the measurement of the directional hydraulic conductivity in the direction of the gradient. If θ is the angle between the vectors q and J, the following relationships hold:

$$K_j = \frac{q \cos \theta}{J} \tag{3}$$

$$\cos \theta = \frac{q \cdot J}{|q| \cdot |J|} \tag{4}$$

For a two-dimensional system, the components of J are given by

$$J_x = J \cdot \cos \alpha \tag{5}$$

$$J_y = J \cdot \cos \beta \tag{6}$$

where α and β are the angles between the directions of J and x, y, respectively. Equation (3) can be rewritten as:

$$\cos \theta = (K_x J_x^2 + K_y J_y^2) \cdot \frac{1}{|q| \cdot |J|} \tag{7}$$

Therefore, K_j becomes:

$$K_j = \frac{K_x \cdot J_x^2 + K_y \cdot J_y^2}{J^2} \tag{8}$$

Substituting these into Equations (5) and (6) results in:

$$K_j = \frac{K_x \cdot J^2 \cdot \cos^2 \alpha + K_y \cdot J^2 \cdot \cos^2 \beta}{J^2} \tag{9}$$

For a polar coordinate system, the above equation becomes:

$$K_j = \left(K_x \cdot \frac{x^2}{r^2} + K_y \cdot \frac{y^2}{r^2}\right) \tag{10}$$

Therefore, by drawing $r = K_j^{1/2}$ in the direction of J, one obtains the following equation:

$$\frac{x^2}{1/K_x} + \frac{y^2}{1/K_y} = 1 \tag{11}$$

This is an equation of an ellipse. Consequently, if the fractured medium is considered to be a homogeneous and anisotropic medium, the plot of $r = 1/K_j^{1/2}$ versus α in a polar coordinate system will take the shape of an ellipse.

EXPERIMENTAL PROCEDURE

Several sites of the Madison and Minnelusa formations, located in the Black Hills region of South Dakota, were selected as the prototype. Geology of these formations has been described in detail elsewhere (Cheema et al., 1994; Cheema and Islam, 1994). Initially fracture trace maps were constructed for the areas of study. The presence of the fractures in the formation was verified using ground probing radar survey and resistivity survey (Cheema et al., 1994). Major trends of fracture traces were also compared with actual fractures by statistical analysis (Cheema and Islam, 1994a).

Two-dimensional models were used for the laboratory tests conducted in this study. A square flow cell with 15 in each side was constructed. This flow cell was equipped with 16 outlets and inlets, four on each side of the square, which are connected to the groove through 1/8-in diameter holes within the body of the cell. The experiment was conducted by taking square regions from the fracture trace map. The square region was then rotated clockwise and at every 22.5° rotation, the fracture trace pattern duly drawn. The fracture trace configuration was then transferred to the impervious vinyl sheet that was cut along the fracture traces. Each piece was then glued firmly to the flow cell. For the dual porosity case, the same procedure was followed except that the impervious vinyl was replaced by a porous polyethylene sheet. The driving force behind the water flow was the hydraulic gradient which was simulated in the laboratory using different measuring flasks at different levels. A constant head was maintained at the top of each flask. Once the outflow, Q_{out}, was measured, the model was then rotated counterclockwise at an angle of 90°, 180°, and 270° measuring Q_{out} each time. In total, 16 outflow measurements were taken for each study area. The hydraulic conductivity in the direction of the gradient (K_g) was calculated using the following equation:

$$K_g = \frac{Q_{out}}{A\frac{\Delta h}{L}} \quad (12)$$

The resulting $K_g^{-1/2}$ values were plotted against the angle of orientation for each fractured formations modeled. These data were curve fitted using nonlinear regression so that the best-fit ellipse emerged.

RESULTS AND DISCUSSION

Figures 1 through 4 show different cases of parallel-plate model results. Results with the dual-porosity model are shown in Figures 5 through 7. The major and minor

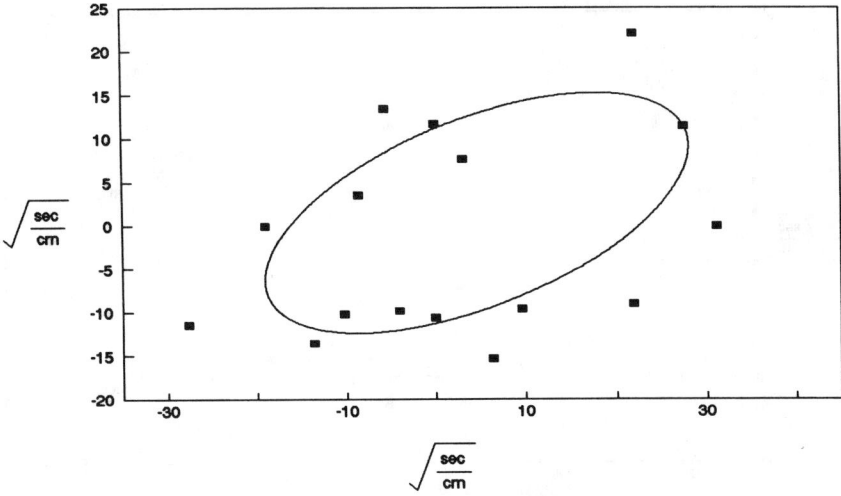

Figure 1. Best-fit ellipse for the Minnelusa formation (parallel plate).

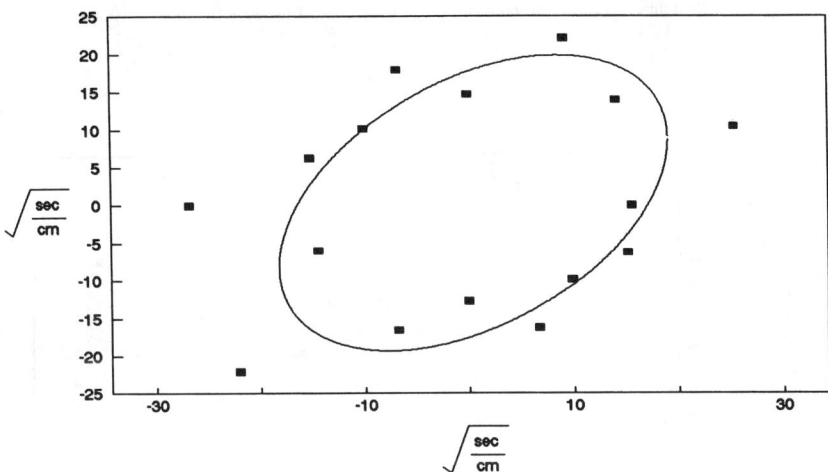

Figure 2. Best-fit ellipse for the Madison formation (parallel plate).

principal axes of the ellipses indicate the maximum and minimum hydraulic conductivities. Table 1 lists the conductivity values as well as anisotropy values for different cases studied. Numerical modeling using the MODFLOW simulator was conducted using the anisotropy value listed in Table 1. These results are discussed in the following section for various areas studied.

Determination of Hydraulic Conductivity

Maximum and minimum hydraulic conductivity values are listed in Table 1. Note that the values obtained with the dual-porosity model are substantially higher than

332 GROUNDWATER MODELS FOR RESOURCES ANALYSIS AND MANAGEMENT

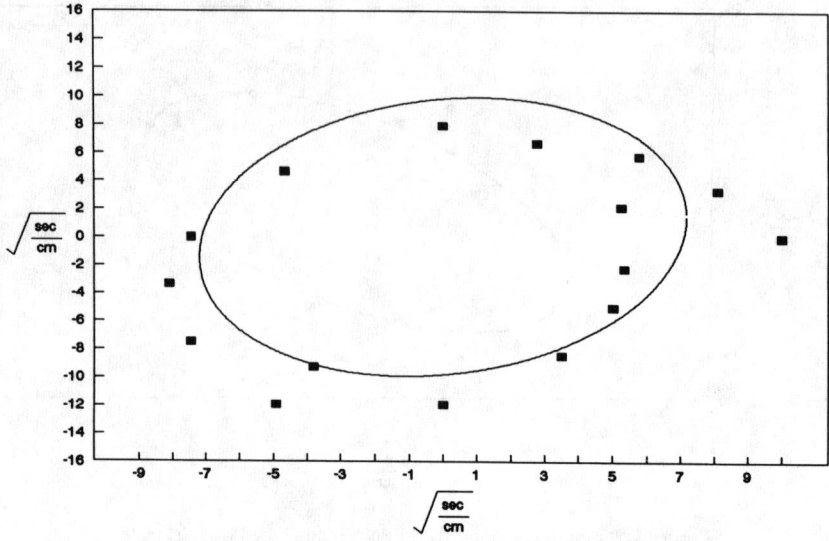

Figure 3. Best-fit ellipse for the combined Madison and Minnelusa formations (parallel plate).

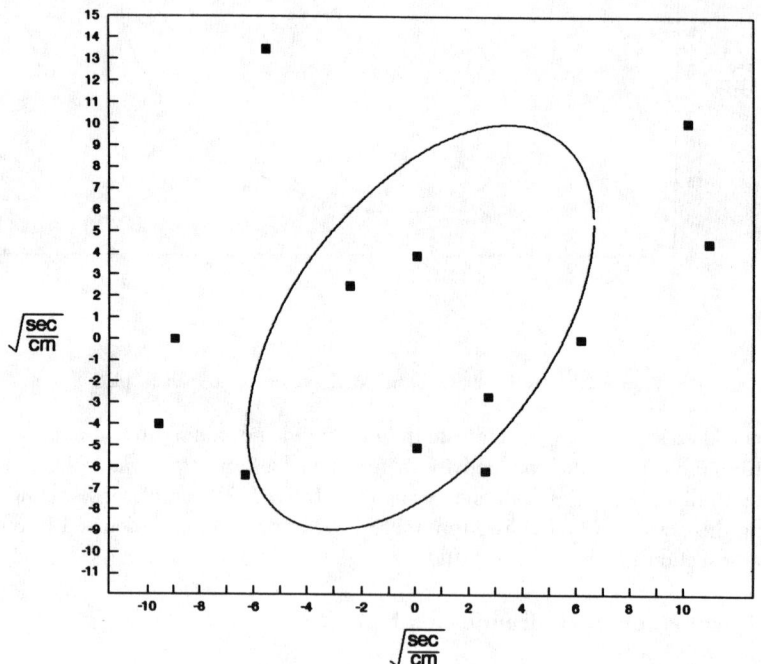

Figure 4. Best-fit ellipse for the Rapid City area (parallel plate).

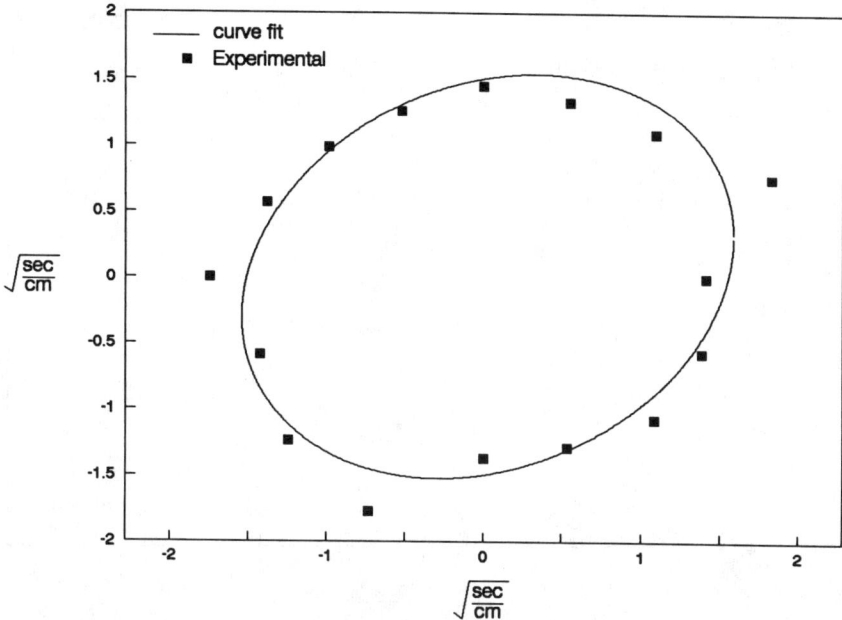

Figure 5. Best-fit ellipse for the Minnelusa formation (dual porosity).

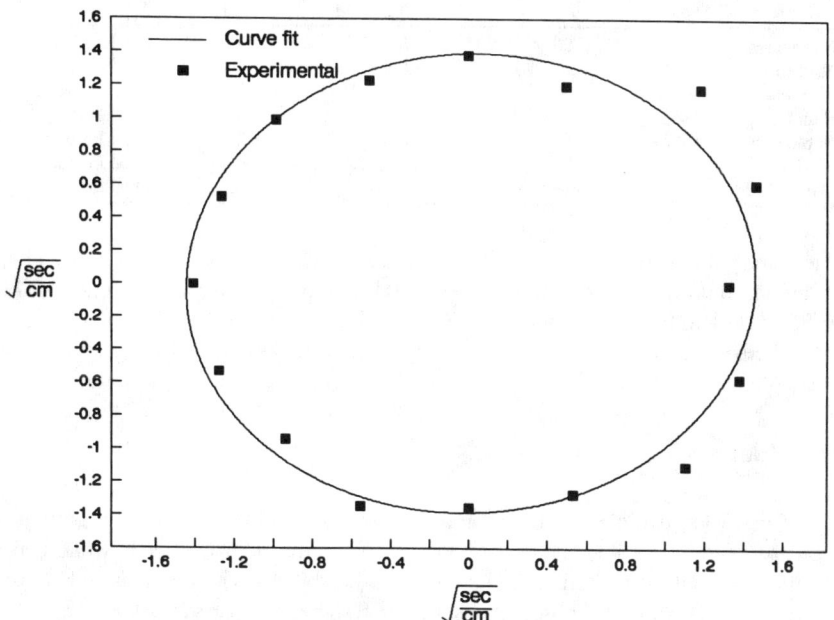

Figure 6. Best-fit ellipse for the Madison formation (dual porosity).

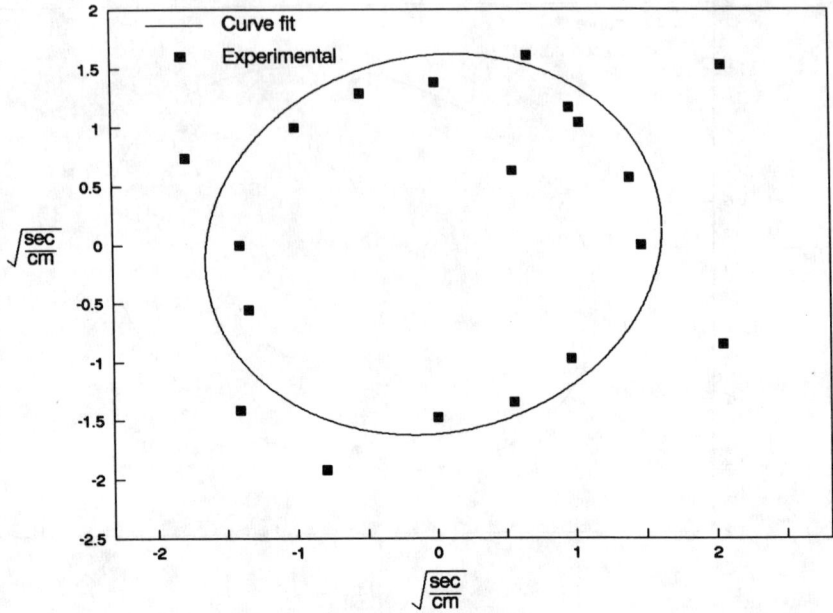

Figure 7. Best-fit ellipse for the Wind Cave area (dual porosity).

Table 1. Experimental Results of Parallel-Plate and Dual-Porosity Models

Simulated Area	K_{max} cm/s	K_{min} cm/s	Ratio	Type of Model
Minnelusa	7.93×10^{-3}	1.6×10^{-4}	4.94	Parallel Plate
Madison	3.27×10^{-3}	2.63×10^{-3}	1.24	Parallel Plate
Minnelusa + Madison	1.89×10^{-2}	1.05×10^{-2}	1.8	Parallel Plate
Rapid City (Madison)	6.4×10^{-2}	3.2×10^{-2}	2.0	Parallel Plate
Rapid City (Madison)	0.465	0.373	1.25	Dual Porosity
Madison	0.510	0.474	1.08	Dual Porosity
Wind Cave	0.387	0.380	1.02	Dual Porosity

the values obtained with the parallel-plate model. The only field hydraulic conductivity (maximum) value reported in the area is 0.46 cm/s for the Madison aquifer. This value is close to the value determined with the dual-porosity model. This finding gave confidence in the results from the dual-porosity model. Consequently, in all numerical simulations, dual-porosity results were used for hydraulic conductivity values.

Boxelder Area

This region was discretized using a grid distribution of 45 × 30. This particular zone was modeled using a no-flow boundary along the right-hand side while maintaining constant heads and recharge cells in other blocks. The boundary conditions were used following field observations. Figure 8 shows comparison between field observations and numerical results. In this particular case, note that the anisotropy value

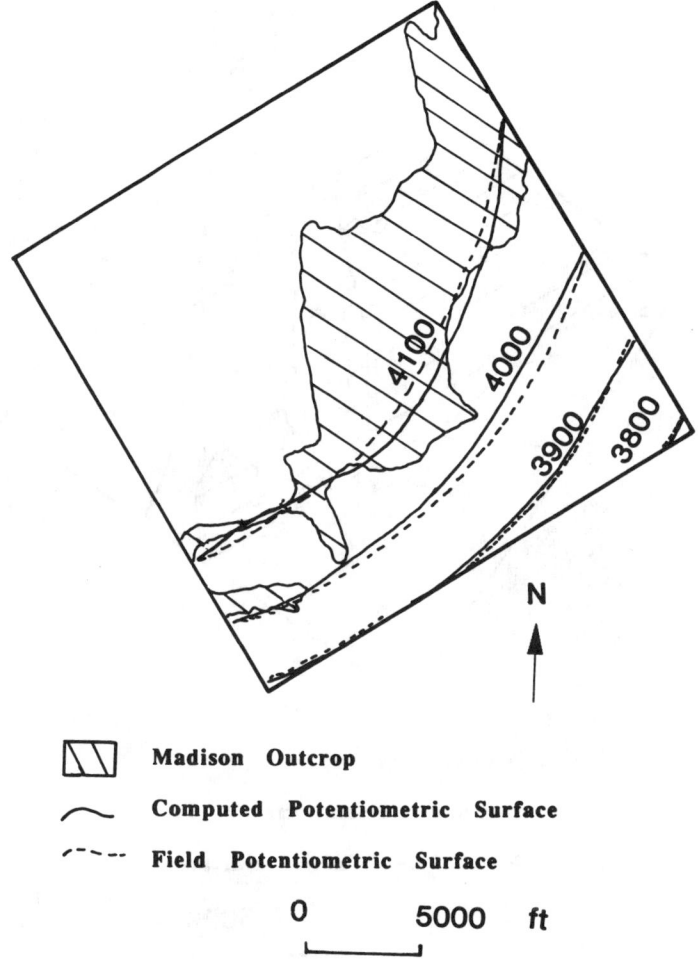

Figure 8. Comparison of numerical and field observations for the Boxelder area. (After Ghannam, 1992.)

determined from the parallel plate model appears to give adequate result. This area is known to have high degree of fracturization with low matrix permeability.

Rapid City Area

A 30 × 15 grid distribution was also used for this area. However, constant head cells were used along the right-hand side while maintaining no-flow boundary conditions and recharge cells were used for other sides. Figure 9 compares numerical and field results for this area. Note that for this particular region, the best suited anisotropy value was obtained through the parallel-plate modeling. However, the hydraulic conductivity value, used for this case, was provided by the dual-porosity model.

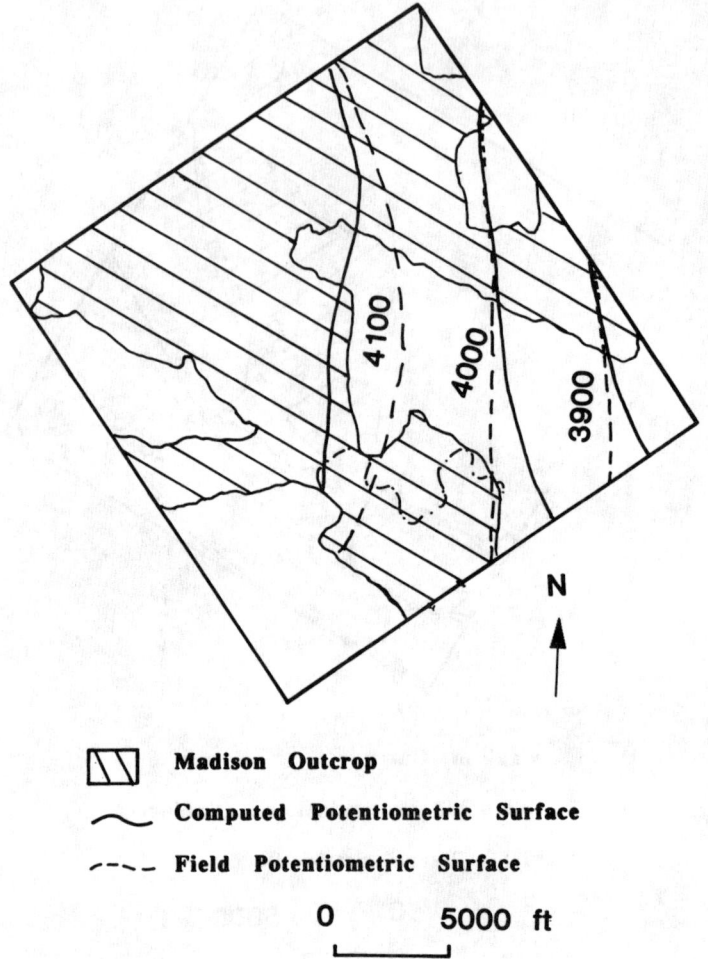

Figure 9. Comparison of numerical and field observations for the Rapid City area. (After Ghannam, 1992.)

Wind Cave Area

Boundary conditions for this case were denoted by no-flow cells as well as constant head cells. Field results are compared with numerical results in Figure 10. Note that the good agreement was achieved using an optimum anisotropy value (3.8) quite different from that of both parallel-plate and dual-porosity models. However, the anisotropy was close to the value obtained if the Minnelusa formation, underlying the Madison formation, was used to determine the anisotropy. If both Madison and Minnelusa formations were combined, the anisotropy value was 1.77, much lower than the optimum value of 3.8. This Wind Cave area is well known for its extreme variations in anisotropy values, due to the presence of discontinues, e.g. caves, which are not accounted for in the fracture trace analysis.

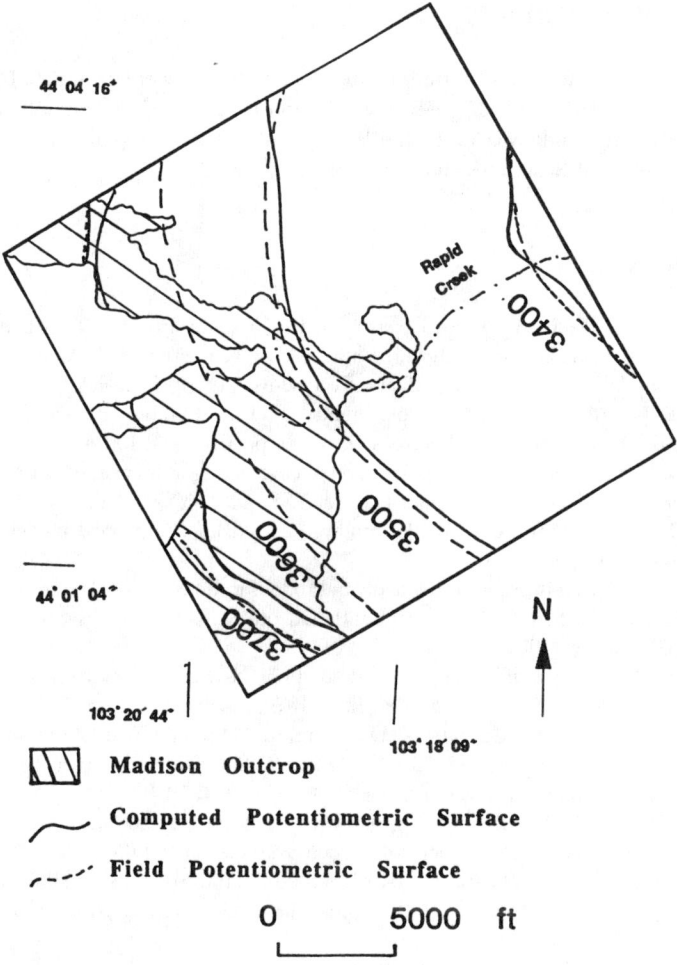

Figure 10. Comparison of numerical (using optimum anisotropy) and field observations for the Wind Cave area. (After Ghannam, 1992.)

CONCLUSIONS

A new modeling approach allows one to use the USGS MODFLOW simulator for predicting flow in fractured formations. The method, which uses laboratory-determined anisotropy values in the MODFLOW simulator, provides good agreement between numerical and field observations. The method was applied in different locations of three different areas of the Black Hills region of South Dakota. Only one area (Wind Cave) showed signs of inconsistencies. This area, however, has shown evidence of extremely variable anisotropy values in different locations, even when field values are determined.

ACKNOWLEDGMENTS

Initial experiments and numerical modeling studies were conducted by Jihad Ghannam and Ahmet Onak. Their contribution to this work is gratefully acknowledged. Also, this study was jointly funded by South Dakota Department of Environment and Natural Resources and the Rapid City USGS.

REFERENCES

Cheema, T. J., Davis, A. D., and Islam, M. R., Fracture, Fracture Traces, and Directional Permeability—An Interdisciplinary Approach, Proc. 14th Annual American Geophysical Union Hydrology Days, Ft. Collins, CO, April 1994, pp. 31–42.

Cheema, T. and Islam, M. R., Experimental determination of hydraulic anisotropy in fractured formation, *AEG Bull.,* XXXI, No. 3, pp. 329–341, 1994a.

Cheema, T. and Islam, M. R., Comparison of cave passageways with fracture traces and joints, Black Hills Region, South Dakota, *NSS Bull.,* 56(2), 1994b.

Faust C. R. and Mercer, J. W., Groundwater modeling: numerical models, *Ground Water,* 18(4), 395, 1980.

Ghannam, J. N., Anisotropic Transmissivity Model for the Madison Aquifer in the Black Hills, unpublished Ph.D. thesis, South Dakota School of Mines and Technology, Rapid City, SD, 1992, 252 pp.

Huntoon, P. W., Finite difference methods as applied to the solution of groundwater flow problems, *Wyo. Water Resour. Res. Inst.,* 1974, 107 pp.

Konikow, L. F. and Bredehoeft, J. D., Computer Model of Two-Dimensional Solute Transport and Dispersion in Ground Water: U.S. Geological Survey Techniques of Water Resources Investigations, Book 7, Chap. C2, 1978, 90 pp.

Lehr, J. H., Editorial mathematical groundwater models may be intellectual toys today, but they should be useful tools tomorrow, *Ground Water,* 17(5), 418, 1979.

McDonald, M. G. and Harbaugh, A. W., A modular three dimensional finite difference groundwater flow model, U. S. Geological Survey Open File Report 83875, 1984, 528 pp.

Remson, I., Gorelick, M., and Fliegner, J. F., Computer models in ground-water explorations, *Ground Water,* 18(5), 1980.

CHAPTER 19

Is It Appropriate to Apply Porous Media Groundwater Circulation Models to Karstic Aquifers?

Peter W. Huntoon

INTRODUCTION

This chapter contends that (1) it is essential to recognize that there is an hierarchical organization within the dissolution conduits which dominates the permeability structure of karst aquifers and (2) there is a consequent difficulty in modeling karstic hydraulic behavior. This discussion will reveal that the all-important dissolution voids found in karst aquifers were actually created by the flow system and that the resulting permeability structure is inherently anisotropic.

An assumption underpinning this discussion is that karst aquifers comprise a very distinct class of aquifers, be they confined or unconfined. Consequently, it will be helpful to formally define a karst aquifer. This definition will emphasize that the distinguishing characteristic of a karst aquifer is that its permeability structure was created by the flow system rather than being an inherited, static attribute of the host rock. A discussion of karst genesis leads to the conclusion that dissolution permeability is inherently anisotropic in all karst aquifers. The reason that typical digital modeling approaches fail to predict flow rates or flow directions is the extreme anisotropy of karst aquifers and our inability to characterize the complex permeability structure present through point sampling.

The purpose of this paper is not only to challenge several conventional wisdoms, but also to provide insights into why they are wrong. Documentation of specific examples that illustrate points, concepts, or issues is provided by the authors of the many articles cited.

Dye traces provide direct information on velocities and migration pathways in karst aquifers. Quantitative analyses of dye concentrations can reveal information on the storage characteristics of many karst systems. Such information is exactly what a digital model is being asked to provide.

Groundwater modelers commonly carry a bias that devolves from sound training in physics and mathematics: a desire to characterize the media that they are treating as continuous. This bias leads to two serious flaws in modeling flow through karst aquifers: (1) the notion that aquifer parameters for karst aquifers can be characterized by point sampling techniques such as pump and slug tests, and (2) the belief that those systems can be adequately described using sophisticated two-dimensional partial differential equations such as:

$$\frac{\partial}{\partial x}\left(T_x\frac{\partial h}{\partial x}\right) + \frac{\partial}{\partial y}\left(T_y\frac{\partial h}{\partial y}\right) = S\frac{\partial h}{\partial t} + W \qquad (1)$$

where

b = saturated thickness
K = hydraulic conductivity
S = storage coefficient
t = time
T = Kb
T_x, T_y = anisotropic transmissivities at x,y
x, y = spatial coordinates
W = source-sink term

or comparable three-dimensional representations (Scarrow, 1989).

In general, such approaches fail when applied to karst aquifers, especially unconfined systems. To appreciate the limitations of conventional modeling approaches, one must acknowledge the peculiarities of a karst aquifer. Next one must develop hydrogeologic definitions for both the terms karst and karst aquifer. Significant insights can be gained by considering how permeability evolves in a karst aquifer.

PERMEABILITY STRUCTURE IN SOLUBLE ROCKS

From a hydrologic perspective, the most distinctive characteristic of a carbonate aquifer is that the medium is readily soluble (Palmer, 1991). Consequently the permeability structure in a carbonate aquifer is highly dynamic over geologic time.

The single most distinguishing feature of karst aquifers when contrasted with other aquifer types is that the voids in a soluble aquifer are predominantly created by the flow regime. Thus, the permeability structure in a soluble aquifer is: (1) a consequence of the circulation system imposed on the aquifer and (2) organized to facilitate the circulation of fluids in the downgradient direction (Figure 1). See Bakalowicz and Mangin (1980) for further discussion. The permeability structure in a karst aquifer is not an independent, inherited static attribute of the rock. The rule of thumb for determining if you are dealing with a karst aquifer is: If carbonate rocks are present, assume you are dealing with karst even if no karst morphologic features are obvious.

Permeabilities of all rocks change through diagenesis. Dissolution and cementation processes can respectively enhance or destroy porosity and thereby affect permeability. However, it is the magnitude of such changes that makes characterizing soluble aquifers especially challenging. A 100% increase in permeability resulting

Figure 1. Dissolved rills in marble organized to move and concentrate water in the downgradient direction, Marble Mountains, CA. Flow is toward the top. Similar subsurface permeability organization at all scales is a primary characteristic of karsts.

from dissolution is considered significant in an indurated sandstone. In contrast, dissolution of carbonate rocks commonly produces permeability increases that are many orders of magnitude greater than initial values (Figure 2).

An equally important corollary is that the permeability structure can adjust as the flow regime responds to changing boundary conditions. Thus, the ultimate hierarchy of permeability pathways and their organization within the aquifer are dictated by the hydrodynamic characteristics of the current flow system and, to a variable degree, the characteristics of past flow systems.

Karst

The word "karst" is the German form of the word "kras", which has been variously translated as "bare, stony ground" (Sweeting, 1981) or "a bleak, waterless place" (Monroe, 1970). The term originated from the geographic name for the landscape in

Figure 2. Caves in dense limestone totally dominate the permeability structure within a karstic aquifer because cave permeabilities are many orders of magnitude greater than those in the large volumes of surrounding host rocks. Thunder cave, Grand Canyon, AZ.

the vicinity of Trieste, Italy, and adjacent Slovenia, a limestone terrane where subsurface drainage occurs through caves. Karst subsequently evolved as a descriptive term for areas throughout the world with geomorphologic and hydrologic characteristics similar to the Trieste karst (Gams, 1993).

Although the origin of the word is clear, there is no universally accepted geologic definition of karst, and certainly no hydrogeologic definition. Karst is most often used in a geomorphologic sense to describe landscapes that result from dissolution and subsurface drainage of carbonate terranes. Some authors require that particular topographic features be present, such as closed depressions, disrupted surface drainage, caves, and subsurface drainage (Dilamarter and Csallany, 1976; Jennings, 1985; White, 1988; Ford and Williams, 1989). The term karst is also employed in a broader sense to describe any of the characteristic landforms and subsurface features produced by processes that include the dissolution, reprecipitation, and associated collapse of soluble rocks (Quinlan, 1978; Milanovic, 1981; Bonacci, 1987; James and Choquette, 1988). Karst is thus used to describe the terrain, the rock, and the solution cavities.

Most authors of karst literature recognize that dissolution is the fundamental karstification process, but the tendency of many has been to focus on morphology rather than process. For hydrogeologists, the association between the term karst and dissolution landform morphology obfuscates and inhibits ready extension of the term to aquifer systems. The concept of a confined (artesian) karst aquifer sharply illustrates the dilemma. By using karst as an adjective and extending from its morphologic

Figure 3. Exceptionally well-developed karst in the metaquartzites of the Precambrian Roraima Formation, Auyan Tepui, Guayana shield, Venezuela. The depth of karstic circulation below the plateau surface exceeds 1500 ft based on the positions of springs discharging from the flanks of the plateau.

roots to aquifers, there exists the implication that the circulation systems under consideration are directly connected to, or a subsurface extension of, an overlying karstified land surface. What is the meaning of a "confined karst aquifer" where the dissolution permeability structure is isolated from the land surface by confining strata and is also spatially separated by great lateral distance from outcrops of the soluble host rock which may or may not exhibit karst features?

It is appropriate to adopt a process-oriented definition for karst that emphasizes its hydrologic function and geohydrologic uniqueness, rather than its ambiguous morphologic character.

> Karst is a geologic environment containing soluble rocks with a permeability structure dominated by interconnected conduits dissolved from the host rock which are organized to facilitate the circulation of fluid in the downgradient direction wherein the permeability structure evolved as a consequence of dissolution by the fluid.

This definition does not require that the rocks have a specific lithology (Figure 3), nor does it limit the circulating fluid to water. It is, however, sufficiently broad to encompass circulation systems in the unsaturated zone, as well as in unconfined and confined aquifer systems.

Karst Aquifer

The hydrogeologic definition for karst can be extended readily to the concept of a karst aquifer by linking it with the widely adopted definition of an aquifer proposed by Lohman et al. (1972): An aquifer is a formation, a part of a formation, or group of formations that contains sufficient saturated permeable material to yield significant quantities of water to wells and springs. The key word in this definition is saturated.

> A karst aquifer is an aquifer containing soluble rocks with a permeability structure dominated by interconnected conduits dissolved from the host rock which are organized to facilitate the circulation of fluid in the downgradient direction wherein the permeability structure evolved as a consequence of dissolution by the fluid.

This definition allows the karst aquifer to be confined or unconfined. The difference between a karst landscape and a karst aquifer is that the aquifer is saturated, whereas the karst landscape can be partially or entirely unsaturated. The karst aquifer need not have any surface morphologic manifestation whatever, a point that is important when treating most confined karst aquifers.

Epikarst

Epikarst aquifer was first defined by Mangin (1974–1975) to describe saturated zones within the intensely dissolved veneers found on the tops of carbonate sections beneath soils and on exposed carbonate outcrops. His usage of epikarst aquifer, as well as by most subsequent workers, is restricted to the saturated parts of dissolution veneers occurring on elevated outcrops which are separated from underlying aerially extensive aquifers by a vadose zone. Although there are important lateral circulation components within the elevated epikarst aquifers, they were initially conceptualized as compartmentalized collector systems which ultimately funnel water to infiltration conduits through the vadose zone (Williams, 1985). Consequently, vertical flow was emphasized over lateral circulation.

Since introduction of the concept of an epikarst aquifer, epikarst has been widely adopted as a noun to denote the morphology of the highly dissolved veneer itself (Figure 4). This usage is adopted here.

> Epikarst is an intensely dissolved zone consisting of an intricate network of intersecting roofless dissolution-widened fissures, cavities, and tubes dissolved into the uppermost part of the carbonate bedrock. The dissolution features in the epikarst zone are organized to move infiltrating water laterally to downgradient seeps and springs or to collector structures such as shafts that conduct the water farther into the subsurface.

Lateral groundwater movement predominates over vertical movement in most epikarst zones. In many low-lying environments, the epikarst serves as a host for shallow, thin, unconfined aquifers characterized by extreme lateral permeabilities and poor storage attributes (Figure 5) (Huntoon, 1992). Epikarst zones are commonly ephemerally or partially saturated (Mills, 1989).

APPLYING GROUNDWATER MODELS TO KARSTIC AQUIFERS 345

Figure 4. Alpine epikarst, Perue Peak, Prince of Whales Island, southeast AK.

Figure 5. Spectacular epikarst in the Kunming stone forest, Yunnan Province, south China. Where buried by clastic sediments, the stone pillars comprise the bedrock framework for shallow unconfined karst aquifers common throughout the region. Lateral circulation dominates in such aquifers.

Figure 6. Paleokarst cavity infilled by collapse breccia near the top of the Redwall Limestone, western Grand Canyon, AZ. Paleokarst is distinguished from karst by destruction of the organized karstic permeability structure. Notice the syndepositional thickening of the prominent ledge in the overlying Supai Group revealing that some of the collapse was occurring as that unit was being deposited.

Paleokarst

The term paleokarst has entered the literature in so many contexts that its meaning has become ambiguous. The primary theme linking the various usages of the term involves the hydrologic inactivation of a karst through destruction of the organized permeability structure that had been imposed by some former circulation system (Figure 6). A secondary theme involves inactivation of a karst as a consequence of dewatering. One should use a slightly restrictive definition that emphasizes the destruction of the permeability structure.

> Paleokarst is a karst in which the organized permeability structure has been destroyed through processes such as burial, infilling, collapse, compaction, brecciation, cementation, or structural fragmentation.

The term unsaturated karst can be used to identify karsts that have been deactivated through dewatering.

HYDRAULIC GRADIENT AND KARSTIC PERMEABILITY

One conventional wisdom in hydrogeology is that in order to locate cavernous permeability in dense limestones the job can be accomplished by exploring for zones of increased fracture density. Implicit is the assumption that the dissolution cavities and caves are localized along the fractures because fracture permeability allowed for circulation which in turn localized dissolution (Eraso, 1985). Another conventional wisdom is that exploration efforts should focus on limestones rather than dolomites in heterogeneous carbonate sections because the limestones have greater solubilities than the dolomites. The fact is that exploration strategies based on these and similar wisdoms, at best, deliver mixed results.

Examination of caves and dye tracing tests reveals that the passages commonly crosscut both strata with different solubilities and tectonic structures with little regard for their presence. Obviously, the presence of structures and different rock types were not the key independent variables that led to localization of karst conduits. In one well-documented example from France, Blavoux et al. (1992) demonstrated through dye tracing that karst circulation is perpendicular to densely spaced compressional and extensional faults and related folds.

Spelunkers produce cave maps in great numbers. Those for caves in near-horizontal strata commonly reveal an ordered network of passages that obviously reflects the local joint pattern. The common interpretation in such simple settings is that the passages are joint controlled. This conclusion misses the point and trivializes the issue. The pertinent question is: Why is the cave localized on these particular joints when vast volumes of the surrounding rock are ubiquitously jointed but largely undissolved? Obviously, in these cases, the passages are localized on some joints but the joints do not control the location of the passages in the larger rock mass.

The ultimate hierarchy of permeability pathways and their organization within a karst aquifer are dictated by the hydrodynamic characteristics of the flow system, and not some inherited geologic fabric (Ewers, 1982; Huntoon, 1985). This idea can be cast in a formulation that allows the permeability within a soluble aquifer to adjust with time as a function of circulation through the aquifer. The following simplistic presentation, without getting unduly mathematical, illustrates how the permeability structure within a soluble aquifer evolves.

Linkage between permeability and circulation in a soluble aquifer takes a form such as:

$$q = -K\frac{\partial h}{\partial s} \qquad (2)$$

where

h = hydraulic head
K = hydraulic conductivity
q = specific discharge
s = general spatial coordinate
$\partial h/\partial s$ = hydraulic gradient

but wherein

$$K_t = f(q)K_{(t-1)} \qquad (3)$$

where $t-1$ and t indicate time progression.

Notice that Equations (2) and (3) comprise a feedback loop where the hydraulic conductivity is adjusted between the time steps as some function of the latest available specific discharge through the rock. This natural feedback phenomenon will inevitably produce a highly anisotropic permeability structure regardless of the permeability configuration in the initial state.

A model that describes the evolution and localization of conduits in a soluble aquifer can be constructed utilizing a reasonably general groundwater equation such as Equation (1). Using appropriate boundary conditions, the solution to Equation (1) yields a map of the hydraulic head for a given time step. Those heads are used in Equation (2) to calculate the current specific discharges for the x and y directions, and the specific discharges are used in Equation (3) to adjust T_x and T_y throughout the flow field. Successive views of the permeability structure in the aquifer are simulated by cycling through a loop comprised of Equations (2), (3), and (1).

Adjustments in T_x and T_y result in progressive increases in transmissivity and reorientations of the pathways. This process is accomplished in nature by successive imprinting of new, ever more favorably oriented dissolution voids through the rock mass, as well as by enlargement of those tubes that have the most favorable orientations. As the favorably oriented conduits become established, their size will grow, allowing them to accommodate increasingly greater percentages of the total circulation through the aquifer. Ultimately, the progressive capture of flow by the most favorably oriented tubes speeds the rate of dissolution enlargement within them. As a result, their locations become increasingly fixed within the rock mass. The earliest tubes can be abandoned or hydraulically deactivated entirely.

The term $f(q)$ in Equation (3) undoubtedly has a linear form early in the development of a karst system. It takes on nonlinear characteristics as karst conduits enlarge within the aquifer, particularly along the uppermost parts of unconfined aquifers (Worthington, 1991). Nonlinear characteristics also prevail in the vadose zones above aquifers. Please see Dreybrodt (1988) for a considerably more detailed discussion of the carbonate dissolution kinetics involved.

Conduit localization in soluble aquifers is a function of specific discharge. Consequently the primary variables governing the location of the conduits are permeability and gradient as shown in Equation (2). However, hydraulic conductivity in a soluble aquifer depends on the hydraulic gradient. Accordingly, the hydraulic gradient is the independent variable that dictates ultimate conduit localization.

Important in the early stages of karstification is the fact that both permeability and hydraulic gradient compete for dominance in determining where the conduits will localize in the aquifer. As time goes on, the hydraulic gradient increasingly wins dominance, a condition forced by Equation (3).

Notice, of course, that hydraulic conductivity and transmissivity are tensor quantities so the imposed permeability structure can be extremely anisotropic during the initial stages in the life of the circulation system or in the initial stages following a reorganization of hydraulic boundary conditions. As the dissolution voids reorganize during subsequent karstification, the permeability structure will change radically, yet inevitably retain strong anisotropy.

From this cursory analysis, it is apparent that conduits should develop in areas where gradients are steepest because those regions are subjected to the greatest flow. Greater flow, in turn, leads to the enhanced rates of dissolution of the host rock. Conduits will tend to adjust to orientations that parallel flowlines, given a reasonable amount of geologic time. Therefore the search for conduits in old karst systems will resolve itself to a search for locations in the aquifer where flow has been concentrated.

The permeability structure of karst aquifers will respond dynamically as the boundary conditions on the aquifer change through such external influences as tectonism, geomorphologic adjustments in base levels, imposition of changing source-sink configurations, etc. Existing conduit networks will be abandoned and supplanted as new networks develop and adjust to the new flow regime. In the extreme, new conduits will crosscut older networks.

Karst genesis and conduit distribution is, in fact, a complex hydrodynamic boundary value problem in space and time. Hydraulic gradient is the independent variable that controls the permeability structure within a geologically mature, soluble aquifer. Heterogeneous geologic fabrics within aquifers, including tectonic structures and rocks with differing solubilities, ultimately become of secondary concern in karstification processes.

In contrast, hydraulic gradient is of only incidental concern when predicting the permeability structure within nonsoluble aquifers. Geologic variables such as lithology and tectonic structure are of supreme importance in the exploration for permeability in well-indurated nonsoluble rocks. Therefore, exploration procedures developed for clastic rocks are often, if not usually, inappropriate for aquifers comprised of soluble rocks, and vice versa.

KARST PERMEABILITY

The transmissive characteristics of a soluble aquifer are overwhelmed by the presence of integrated networks of dissolved conduits. There is an organizational hierarchy of the dissolution tubes within the networks, with a tendency for those in the downstream direction to exhibit progressively greater degrees of organization. The resulting organized structure functions to move water in the downgradient direction.

The consequence of imposing such tube networks on soluble rocks is to create an extremely anisotropic permeability architecture wherein the tubes commonly exhibit hydraulic conductivities that are six or many more orders of magnitude greater

Figure 7. Parallel dissolved slots in an exhumed confined karst aquifer that are localized along extensional fractures, Redwall Limestone, Marble Canyon, AZ. The permeability structure in this aquifer is highly anisotropic with the maximum transmissivity tensor oriented parallel to the slots. Top to bottom is about 200 ft.

than the undissolved host rock (Figure 7). Obviously, the bulk of the circulation and transport of contaminants through such aquifers occurs in the tube networks.

Commonly, there are organizational distinctions between conduit networks found in unconfined and confined karst aquifers. Unconfined karst systems usually have a permeability structure that has some similarities to a surface water system. Specifically, unconfined systems contain collector structures at or just below the surface such as sinkholes and epikarst zones, many of which occur in the unsaturated zone, that feed low order tubes. These tubes coalesce downgradient into successively better organized and higher order caves quite similar in form and function to tributary streams found in surface water basins (Quinlan and Ray, 1989).

Karst networks differ from surface stream networks in one important way. In karst systems there are multiple interconnections between tubes of similar rank. Consequently, there are numerous opportunities for water to shunt between tubes in the network. This gives the appearance of cross-divide circulation, which is a particularly characteristic hallmark of unconfined karst circulation systems. The type of shunt that actually functions is commonly stage dependent. The result is that a tracer introduced at a single point can appear in many springs, often at widely divergent compass directions from the point of injection. See Aley (1988) for a dramatic example from Arkansas. Furthermore, there can be substantial variability in which springs discharge the tracer as water levels in the unconfined aquifer vary. This situation has dire consequences. At a low stage, pollutants may not reach a given spring which will then be certified as safe, but at flood stages pollutants will readily find their way to the site.

From the perspective of modeling, unconfined karst networks produce the most complex, highly anisotropic permeability regimes found in nature. Making matters worse is the fact that there are on-again, off-again permeability linkages of considerable hydraulic importance that are water level dependent. Consequently, the permeability anisotropy is a variable function of stage instead of being a fixed attribute.

Linkages between the dissolved cavities found in confined karst aquifers are commonly more pervasive than those in unconfined aquifers. The reason for this is that hydraulic gradients in confined systems tend to be substantially less than those found in unconfined aquifers. Therefore, dissolutional enhancement of permeabilities is more uniform in the two spatial dimensions parallel to the confining strata, or even three dimensions in thick carbonate sequences. Consequently, fully integrated maze-type tube networks tend to develop (Scheltens, 1984). The interconnections between voids in conduit mazes can closely approximate the linkages found between pores in intergranular media; they differ primarily in scale. Commonly, there is a moderate to strong degree of anisotropy with the maximum principal permeability tensor aligned roughly parallel to the modern gradient, some paleo-hydraulic gradient, or a geologic fabric within the host rock that caused localization of dissolution.

ANISOTROPY VERSUS HETEROGENEITY

Modelers favor approaches that treat the radical spatial permeability variations found in karst aquifers as extremely heterogeneous but isotropic. This approach is, of course, driven by the wish that once head maps are computed, (1) flow vectors can be superimposed perpendicular to the potential lines, (2) velocities can be computed, and (3) migration routes can be mapped out. Flow is perpendicular to potential lines only when the aquifer is isotropic with respect to permeability.

The concept of homogeneity refers to the sameness of a physical property between points in space. On the other hand, isotropy refers to the similarity of a physical property with respect to direction about a single point in space. The two concepts have nothing to do with each other. In karst aquifers, permeabilities are extremely anisotropic with those parallel to preferred tube orientations commonly being many orders of magnitude greater than those perpendicular to the tubes. Anisotropy is especially acute in unconfined aquifers. This situation has two consequences. First, it is inappropriate to model karst aquifers by casting permeabilities as extremely heterogeneous as an approximation for what is truly anisotropic. Second, flow vectors deduced from heterogeneous models are generally fiction. Flow in highly anisotropic karst aquifers can be subparallel to the potential lines, thus it is virtually impossible to use such models to predict migration directions. See Arnow (1963) for validation of this situation from the Edwards aquifer of Texas.

PARAMETER SAMPLING

The hydrologist who is responsible for characterizing the flow of groundwater through karst aquifers has two demanding jobs. First, this professional must devise an aquifer sampling strategy that captures the essence of the hydraulic conductivities of the dominant tube networks. Second, and even more fundamental, the hydrologist is obligated to identify and characterize the hierarchical organization of the conduit system because it defines the anisotropic permeability architecture that often overwhelms the mechanical responses of the aquifer.

Point sampling that is predicated on the invalid assumption that the medium is continuous is certain to fail. The tubes in most karsts comprise a small to tiny fraction of the total volume so any point sampling technique is guaranteed to produce values for transmissivities that are heavily weighted toward the minimum values associated with the undissolved host rock. Our concern is transmissivity underestimates of many orders of magnitude. Inspection of the Darcy equation [Equation (2)] reveals that flow rates, be they cast as specific discharges or velocities, in a digital model will be underestimated in exact proportion—by several orders of magnitude. Flow velocities through unconfined karst systems are commonly equivalent to rates found in surface streams. Consequently it is clear why errors in estimating travel times in karst aquifers are so severe; that is, why modelers predict travel times in years to centuries only to discover contaminant arrival in hours to days.

Point sampling techniques do not possess the ability to discriminate the organized permeability structure embedded within carbonate aquifers. Furthermore, point sampling cannot provide information on the organizational hierarchy of the various types of tubes present. What is required is an holistic strategy wherein the integrated karstic permeability structure within the entire aquifer is characterized. The focus of such an inquiry is to: (1) identify the types of, hierarchical rank of, and nature of interconnections between conduits present, (2) discern conduit organization within the karst aquifer, and (3) determine flow velocities. Central to such an investigation is careful delineation of hydraulic boundaries, inventory of point recharge and discharge features, and documentation of the effectiveness of confining strata within and bounding the aquifer.

This information will provide insights on which hydraulic, geologic, and geomorphologic features and factors led to the localization of conduits within the aquifer. Clearly the emphasis in a properly formulated characterization program is on identifying the attributes of the interconnected highly transmissive features within the aquifer. The remaining permeability characteristics of the host rock are largely immaterial except for possible concerns about their contributions to storage.

A popular concept with modelers is that of a minimum representative volume that will allow the essence of the permeability structure within an aquifer to be captured. Bulk variations within the aquifer parameters are assumed to be insignificant in such a volume. Once ascertained, the minimum volume allows for the development of a sampling strategy that is appropriately scaled to quantify the physical parameters of the aquifer. In reality, it is not an overstatement to say that the minimum representative volume for many karst aquifers is at least the entire ground water basin under consideration (Ford and Williams, 1989) and, for some problems, the entire aquifer. The point being that the integrated conduit structure in a karst aquifer must be characterized in its entirety, similar to the way one would view a surface drainage system.

SCALE CONSIDERATIONS

Even neglecting the supremely important matter of anisotropy, the issue of characterizing permeabilities in karst aquifer models can be addressed by considering the problem of collecting permeability data required to construct the model relative to the scale of the model. Using prodigious numbers of hydraulic conductivities reported in the literature, Sauter (1993) and Quinlan et al. (1992) have found that reported karst hydraulic conductivities are a function of the volume of the aquifer material being sampled. Permeability data can be collected through a variety of means including core tests (laboratory), slug and packer well tests (sublocal), pump tests (local), and dye tracing (regional). As illustrated on Figure 8, representative hydraulic conductivities for these methods from karstified aquifers are respectively 10^{-8}, $10^{-5.5}$, 10^{-4}, and 10^{-2} m/s.

Most digital models address problems on the regional or subregional scale. Only the dye tests are capable of providing hydraulic conductivities that adequately characterize conditions at such model scales. At best, on average, pump tests provide hydraulic conductivities that are two orders of magnitude too small. Slug, packer, and core tests are worthless sources of permeability data for modeling purposes.

Regional models using dye-trace determined permeabilities have a better chance of simulating regional responses than do small-scale models built to simulate local responses utilizing point-sampled permeabilities. As the scale of the model decreases, the need to characterize the fine structure of the permeability architecture increases. However, the adequacy of methods used to sample permeabilities deteriorates as scale decreases. Thus, as model scales decrease, we are simultaneously stretching our ability to adequately sample and characterize the aquifer parameters required in those models. There is an obvious conclusion. For karst aquifers, the ability of digital models to validly predict flow directions and velocities deteriorates as the scales of the models decrease. Karst aquifer models of subregional and smaller scale have thus far proven to be highly unreliable.

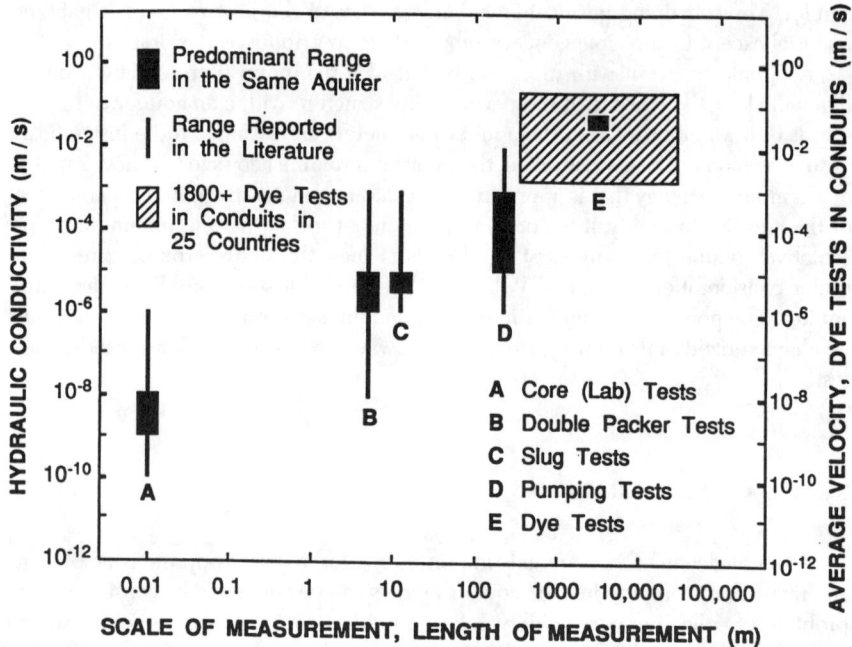

Figure 8. Ranges of average velocities and hydraulic conductivities in numerous karst aquifers as a function of the scale of measurement. (Figure authored by and reproduced here with the permission of James Quinlan as modified from Quinlan et al. (1992), with additional data from Sauter (1993).)

HYDRAULIC GRADIENTS

Circulation through karst aquifers is dominated by the extremely anisotropic transmissivities of the tube systems imbedded in the aquifers. Point sampling of heads predominantly measures the heads in the host rocks or fractures, but rarely in the conduits. Such sampling is used to construct potentiometric maps which too often are used in turn to predict flow directions by assuming flow is perpendicular to the potential lines. However, flow commonly diverges from perpendicular owing to the extreme anisotropy of the conduit transmissivities.

The problem of predicting flow directions is compounded by crossover and distributary linkages between tubes in the conduit network. It is common to observe the movement of dyes or contaminants in several compass directions from an injection point despite the presence of a uniform head gradient deduced from point sampling of wells drilled into the host rock.

In unconfined aquifers, there is the added problem that flow directions are stage dependent. As water levels in the conduit systems rise, higher level tubes partially or fully saturate which can dramatically alter the permeability architecture. The water thus flows in unexpected directions as contrasted to documented flow directions at lower stages. In typical cases, flows take place in more than one direction at once.

State-of-the-art digital models have been constructed by Sauter (1993) and Teutsch and Sauter (1992) utilizing a double continuum permeability approach to characterize separately the conduits and fractured host rocks in a karst aquifer. They linked two layers, one representing highly transmissive conduits and the other poorly transmissive fractures. Each layer could be highly anisotropic with respect to transmissivity. Total system response was then assessed as these layers responded to imposed scenarios. Such tentative steps laudably address the complex realities of karst aquifer permeability, but even these models appear simplistic and the data requirements to construct them are necessarily excessive.

MODELING OBJECTIVES

At the outset, it is crucial to define the objectives of a groundwater assessment or monitoring program before committing to the construction of a digital model. Two paramount questions emerge in karst aquifer problems: (1) What are the rates and directions of contaminant transport through the aquifer? (2) What are the hydraulic response characteristics of the aquifer (Dreiss, 1989a,b)? Both demand predictions of system responses cast in terms of fluid travel times and directions. Is a digital model capable of yielding the desired results? Is the typical input data set adequate to construct the model, particularly if transmissivities were developed using point-sampling techniques? The answers to both are no, even as a first-order qualitative approximation.

The data requirements to adequately model a karst aquifer, particularly unconfined systems, are so demanding that the questions posed necessarily will be answered by an adequate field data collection program before the model can be built. There are two reasons for this startling conclusion. First, to correctly model the aquifer, the transmissivity structure of the aquifer must be viewed holistically as a system which will require unusual care in delineating boundary conditions in the field. Second, the hierarchy, organization, and hydraulic responses of the all important dissolved permeability pathways will have to be deduced. By the time data documenting these system attributes are collected in the field, system response already will be thoroughly documented and modeling will prove to be a redundant exercise. These conclusions were reached by Palmer (1992) and Teutsch and Sauter (1992).

DYE TRACING

Sophisticated dye-tracing technologies are now routinely applied to karst aquifers in order to document system response characteristics, flow directions, and storage characteristics (Quinlan, 1989). These technologies are particularly useful and necessary in characterizing system responses for unconfined karst aquifers. The gradients in such aquifers tend to be steep, and dissolved pathways tend to be partially or ephemerally saturated. Circulation is usually dominated by hydraulic flow-through responses rather than pulse-through responses to recharge events. Consequently, travel times are rapid.

Dye tracing is valuable in assessing interconnections between points in confined karst aquifers as well. However, owing to low gradients and large storage, appropriate detection strategies must be developed. The sampling duration can span considerable lengths of time.

DISCUSSION

In cases involving karst aquifers, the ability of digital models to predict flow directions and velocities deteriorates as the scale of the models decrease. Subregional and smaller scale models have therefore proven to be highly unreliable. This unfavorable situation has developed because small-scale karst aquifer models require excessive data on permeability architecture from exactly the scale domain where permeability sampling methods are least reliable. The problem is further exacerbated by the reticence of most modelers to address the inherent highly anisotropic character of karst permeability, and then to perform the necessary field studies required to characterize the anisotropic structure present.

In assessing the migration of contaminants in karst systems, Quinlan (1989) correctly observed:

> . . . one well-designed tracer test, properly done and correctly interpreted, is worth 1000 expert opinions, or 100 computer simulations of ground water flow. The only disagreement that colleagues have expressed with this statement is to jocularly suggest that the two numbers should be reversed.

ACKNOWLEDGMENTS

James Quinlan and Gareth Davies conducted extensive reviews of this article and their constructive and insightful comments led to substantial conceptual improvements and clarifications in the presentation. Karst dye-tracing specialists Tom Aley, Ralph Ewers, and James Quinlan generously provided numerous case histories from their own work and the literature that support the concepts summarized here. Discussions with Michel Bakalowicz helped me to articulate the overriding importance of delineating the organized permeability structure within karst aquifers when assessing hydraulic responses.

REFERENCES

Aley, T., Complex radial flow of ground water in flat-lying, residuum-mantled limestone in the Arkansas Ozarks, Proc. 2nd Conf. on Environmental Problems in Karst Terranes and Their Solutions, National Water Well Association, Dublin, OH, 1988, 159.

Arnow, T., Ground water geology of Bexar County, Texas: U.S. Geological Survey, Water Supply Paper 1588, 1963, 36.

Bakalowicz, M. and Mangin, A., L'aquifere karstique, sa definition, ses characteristiques et son identification: Memoires hors Serie Société Geologique de France, no. 11, 1980, 71.

Blavoux, B., Mudry, J., and Puig, J., The karst system of the Fontaine de Vaucluse, southeastern France, *Environ. Geol. Water Sci.*, 19, 215, 1992.

Bonacci, O., *Karst Hydrology, With Special Reference to the Dinaric Karst:* Springer-Verlag, New York, 1987, 184.

Dilamarter, R. C. and Csallany, S. C., Eds., Hydrologic problems in karst regions: International Symposium on Hydrologic Problems in Karst Regions, Western Kentucky University, Bowling Green, 1976, 481.

Dreiss, S. J., Regional scale transport in a karst aquifer, 1. Component separation of spring flow hydrographs; *Water Resour. Res.*, 25, 117, 1989a.

Dreiss, S. J., Regional scale transport in a karst aquifer, 2. Linear systems and time moment analysis, *Water Resour. Res.*, 25, 126, 1989b.

Dreybrodt, W., *Processes in Karst Systems, Physics, Chemistry, and Geology*, Springer-Verlag, Berlin, 1988, 288.

Eraso, A., Metodo de prediccion de las direcciones principales de drenaje en el karst: Kobie, Serie Ciencias Naturales, Bilbao, Spain, no. 15, 1985, 15.

Ewers, R. O., An Analysis of Solution Cavern Development in the Dimensions of Length and Breadth, Ph.D. thesis, McMaster University, Hamilton, Ontario, 1982, 398.

Ford, D. L. and Williams, P. W., *Karst Geomorphology and Hydrology*, Unwin Hyman, London, 1989, 601.

Gams, I., Origin of the term "karst" and the transformation of the classical karst (kras), *Environ. Geol.*, 21, 110, 1993.

Huntoon, P. W., Gradient controlled caves, Trapper-Medicine Lodge area, Bighorn Basin, Wyoming, *Ground Water*, 23, 443, 1985.

Huntoon, P. W., Hydrogeologic characteristics and deforestation of the stone forest karst aquifers of south China, *Ground Water*, 30, 167, 1992.

James, N. P. and Choquette, P. W., Eds., *Paleokarst*, Springer-Verlag, New York, 1988, 416.

Jennings, J. N., *Karst Geomorphology*, 2nd ed., Basil Blackwell, New York, 1985, 293.

Lohman, S. W., et al., Definitions of groundwater terms, revisions and conceptual refinements, U.S. Geological Survey, Water Supply Paper 1988, 1972, 21.

Mangin, A., Contribution à l'étude hydrodynamique des aquiferes karstiques, *Annules de Speleologie*, 29 (3), 283, 1974; 29 (4) 495, 1974; 30 (1), 21, 1975.

Milanovic, P. T., *Karst hydrology*, Water Resources Publications, Littleton, CO, 1981, 434.

Mills, J. P., Foreland structure and karstic ground water circulation in the eastern Gros Ventre Range, Wyoming, M.S. thesis, University of Wyoming at Laramie, 1989, 101.

Monroe, W. H., A glossary of karst terminology, U.S. Geological Survey, Water Supply Paper 1899K, 1970, 26.

Palmer, A. N., Origin and morphology of limestone caves, *Geol. Soc. Am. Bull.*, 103, 1, 1991.

Palmer, A. N., Opportunities and pitfalls in the computer modeling of karst aquifers, Ogden, A. E., Ed., Abstracts of the 1992 Friends of Karst, Tennessee Technological University, Cookville, TN, 1992, 20.

Quinlan, J. F., Types of karst, with emphasis on cover beds in their classification and development, Ph.D. dissertation, University of Texas, Austin, 1978, 325.

Quinlan, J. F., Ground-water monitoring in karst terranes, recommended protocols and implicit assumptions: U.S. Environmental Protection Agency, Environmental Monitoring Systems Laboratory, Las Vegas, NV, EPA 600/X–89/050, 1989, 79.

Quinlan, J. F., Davis, G. J., and Worthington, S. R. H., Rationale for the design of cost-effective monitoring systems in limestone and dolomite terranes, Proc. 8th Waste Testing and Quality Assurances Symposium, U.S. Environmental Protection Agency, Office of Solid Waste and Emergency Response, and Office of Research and Development, Washington, D.C., 1992, 552.

Quinlan, J. F. and Ray, J. A., Groundwater basins in the Mammoth Cave region, Kentucky showing springs, major caves, flow routes, and potentiometric surface, Friends of the Karst, Mammoth Cave, KY, Occasional Publication 2, map, 1989.

Sauter, M., Double porosity models in karstified limestone aquifers, field validation and data provision, Hydrologic Processes in Karst Terranes, International Association of Hydrologic Sciences Publication 207, 1993, 261.

Scarrow, J. W., The use of models in groundwater contamination cases; *Virginia Environ. L. J.*, 9, 185, 1989.

Scheltens, J. P., Wind Cave, Wind Cave National Park, Hot Springs, South Dakota, Wind Cave/Jewel Cave Natural History Association, map, 1984.

Sweeting, M. M., *Karst Geomorphology*, Academic Press, New York, 1981, 427.

Teutsch, G. and Sauter, M., Groundwater modeling in karst terranes, scale effects, data acquisition and field validation, Proc. 3rd Conf. on Hydrology, Ecology, Monitoring, and Management of Ground Water in Karst Terranes, Nashville, TN, December 4–6, 1991, U.S. Environmental Protection Agency and National Ground Water Association, 1992, 17.

White, W. B., *Geomorphology and Hydrology of Karst Terrains*, Oxford University Press, New York, 1988, 464.

Williams, P. W., Subcutaneous hydrology and the development of doline and cockpit karst, *Z. Geomorphol., N.F.*, 29, 463, 1985.

Worthington, S. R. H., Karst Hydrology of the Canadian Rocky Mountains, Ph.D. dissertation, McMaster University, Hamilton, Ontario, Canada, 1991, 227.

Index

Activated carbon treatment, 267
ADE, see Advection-dispersion equation
Advection
  dispersion with, 264
  simulation of, 264
Advection-dispersion equation (ADE), 25
  model error in use of, 25–38
    limiting conditions for ADE, 29–30
    rudimentary examination of errors, 30–32
    simulation exercises and results, 32–36
    statistical origins of ADE, 27–29
    true vision of mass transport, 26–27
  statistical origins of, 27
  viability of, 29
Aerobic respiration, 295
Agricultural production, 109
Alluvium
  intermediate, 259
  older, 258
  recent, 259
Anaerobic biodegradation, 295
Analytical solutions, 47
Andesitic lavas, 256
Animators, 140
Anisotropy(ic), 353, 356
  medium, 330
  permeability regimes, 351
Applicability assessment, 42
Aquifer(s)
  atoll, 278
  basaltic, 260
  carbonate, 340
  code, 260
  epikarst, 344
  high-level, 259
  identification, 260
  multilayer system of, 290
  multiple, 277
  permeability structure in, 340
  sector, 260
  simulation model, 77
  sustainable yield, 264, 270
  systems, 260
  test, 256
  unconfined, 351, 354
Aquitards, 255, 257
Arkansas River valley, Colorado, 72
Artesian
  groundwater, 253, 261
  karst aquifer, 343
  water, 262
Ash, 256, 260
Asymptotic solution, 35
Atoll groundwater systems, modeling, 275–292
  atoll modeling, 276–277
  development and sustainable yield modeling for Roi-Namur Island, 283–290
    input data and calibration, 285–286
    mesh design and boundary conditions, 284–285
    model results, 286–290
  generic modeling of groundwater lens dynamics, 277–282
    input data and calibration, 279–280
    mesh design and boundary conditions, 279
    model results, 280–282
  hydrogeology of atoll islands, 276
Atrazine, 266
Autocorrelation, 29
Autocovariance function, 209

Background organics, 267
Bacterial indicators, 268
Basalt(s), 258
  aquifers, 259
  groundwater, 259
  lens, 259–261, 265
  olivine, 256
  primitive, 254, 256
  water, nomenclature of, 262
Benchmarks, 44
Benzene, toluene and xylenes (BTX), 306–308
Beretania aquifer, 264
Best linear unbiased estimator (BLUE), 212
Best management practice, 269
Bio-attenuation, natural, 130
Biodegradation models, review of, 295–312
  existing biodegradation models, 299–305
    biofilm model, 300–301
    instantaneous reactions models, 304–305
    microcolony models, 301–303
    Monod kinetic models, 303–304
  kinetics of biodegradation, 296–297
  modeling biodegradation, 298–299
  modeling of biodegradation at field sites, 305–311
    Cliffs-Dow Superfund site, 309–311
    Conroe Superfund site, Texas, 305–306

gas plant facility in Michigan, 306–308
  Traverse City site, Michigan, 306
 overview of biodegradation processes,
   295–296
BIO1D model, 299, 303, 309
Biofilm model, 300
BIOPLUME model, 299
BIOPLUME II model, 299, 304, 305
BIOPLUS model, 299
BLUE, see Best linear unbiased estimator
BOD, 268
Bottom storage of thick lenses, 263
Boundary condition, 162
 initial conditions and, 151, 152
 treatment of, 157
Breccia, 258
BTX, see Benzene, toluene and xylenes
Buoyancy, 264

Caldera, 254, 255, 258
Calibration, 61, 69, 86
Caprock, 257, 258
Carbonate aquifer, 340
Case preparation, 139
Caves, joint controlled passages, 347
Central Aquifer System, basal water in, 266
Central limit theorem, 35
Chemical leaching models, consequences of
   scale-dependency on application of,
   169–184
 modeling approaches, 169–171
 representation of basic processes,
   172–175
 spatial and temporal scale, 176–180
 types of users, 171–172
Chemical reactions, 130
Cholesky decomposition, 31
Cinder, 256
Cliffs-Dow Superfund site, 309–311
Code
 intercomparison, 47
 testing, 40
   limitation of, 43
   strategy, 41, 44
   verification, 42, 52
Cokriging, theory of, 213, 214
Common Sense Pathway, 138
Comparative verification, 49
Computer code inspection, 42
Computers, 81
Conceptual model, 138
 of flow dynamics, 262
 identification, 102
Conceptual tests, 44

Conditional simulation, 83, 85
Conduit(s)
 localization of, 352
 networks, 350
 system, hierarchical organization of, 352
Confined karst aquifer, 343
Conroe Superfund Site, 305, 306
Contaminant
 chemical, physical, and biological
   properties of, 123
 migration, 60
 plumes, 201
 transport, 25, 123, 355
Contamination
 protection from, 261
 risk, 123
Convection-dispersion equation, 187
Co number, see Courant number
Cook, Captain James, 253
Core tests, 353
Correlation, 29, 128
Coupled model, 268
Courant (Co) number, 161, 162
Court of law, testimony given in, 137
Courtroom, 139
Crank-Nicolson scheme, 153
Cross-bedding, 196
Cross-examination, 142, 143

Darcy's law, 149, 185, 262, 328
Database, to enhance geohydrological
   investigations, 270
DBCP, see 1,2-Dibromo-3-chloropropane
Degree of correlation, 51
Del Monte Kunia Well, 269
Depositions, 137, 140
Design point, 126
Development scenarios, different, 287
1,2-Dibromo-3-chloropropane (DBCP), 265,
   266
Digital models, 353, 355
Dike(s), 258
 complex, 260
 zone, marginal, 260
Direct testimony, 142
Discharge predictions, results of, 9
Dispersion, 80
Dispersion coefficient tensor, 187
Dispersivity, 70, 87, 267
Dissolution voids, 349
Diuron, 266
Dolomites, 347
Dual-aquifer system, 277
Dual-porosity model, 331, 334

INDEX

Dupuit method, 7, 265
Dye tracing, 347, 353, 355

EDB, see Ethylene dibromide
Einstein summation convention, 197
Epikarst, 344, 350
Equivalent homogeneous porous medium, 83
Ergodicity, 195
Ergodic random space function, 26
Erosional unconformities, 260
Error detection, 42
Ethylbenzene, 123, 133
Ethylene dibromide (EDB), 265
Eulerian velocity field, 26
Expert witness, 139, 141
Explicit linearization scheme, 155
Extrusive rocks, 255, 257

Facies models, 84
Failure
   domain, 124
   probability of, 124
   state, 124
Faults, compressional and extensional, 347
Fenamiphos, 266
Fick's law, 190
Finite difference techniques, 153
Finite-element technique, 153, 155, 160
First-order
   kinetics, 130, 298
   rate constants, 297
   reliability method (FORM), 123, 125, 126, 130
Flow
   density-dependent, 277
   dynamics, in thick lenses, 262
   model(s), 262
      available unsaturated, 164
      calibration of, 82
      robust analytical, 264
      prediction of, 220
      preferential, 267
   -through responses, 355
Flux, dispersive, 190
Fokker-Plank equation, 28
FORM, see First-order reliability method
Fossil coral reefs, 259
Four-layer system, 284
Fractured formations, new modeling approach for predicting flow in, 327–338
   experimental procedure, 330
   results and discussion, 330–337
      Boxelder area, 334–335
      determination of hydraulic conductivity, 331–334
      Rapid City area, 335–336
      Wind Cave area, 336–337
   theory, 328–330
Fracture trace maps, 330
Free product, 240
   recovery, 245
   thickness, variation of, 247
Free-rider problem, 113
Freshwater
   buoyancy of, 262
   -saltwater interface, 263
   thickness of, 290
Functionality
   analysis, 42
   of code, 44
   matrix, 44, 46

Galerkin method, 155
Galleries, 256
Gasoline
   components, 238
   spill, multicomponent approach to, 235
Gas plant facility, Michigan, 306, 308
Geographic information system (GIS), 180
Geological simulation model, 84
Geophysics, 263
Geostatistics, 83, 269
Ghyben-Herzberg lens, 256
GIS, see Geographic information system
Goodness-of-fit, 54
Graphical representation, visual inspection of, 53
Groundwater
   basal, 260
   contamination, 261
   extractions, 270
   flow
      investigations in, 261
      mathematical models simulating, 3
      models, 123
   high-level, 260
   investigations, history of, 260
   management, optimal, 101
   mining, 104
   modeling, 39, 123
   monitoring of potable, 268
   pesticide contamination of, 179
   quantity and quality management, 101
   recharge, 179, 269, 280
   sources, monitoring in potable, 267
   system, saltwater contamination of, 81
Groundwater aquifers, modeling multiphase contaminant flow in, 231–250
   multiphase multicomponent models, 232–240

sharp interface models, 240–247
    formulation of, 241–244
    recovery of established oil mound on water table, 244–247
Groundwater modeling, in Hawaii, 253–274
    current and future role of models in Hawaii, 269–270
    geological and hydrological environment, 254–260
        aquifer identification and classification, 260
        groundwater occurrence, 259–260
        hydrological character of Hawaiian rocks, 254–259
    history of groundwater investigations and models in Hawaii, 260–269
        early flow models, 262
        early recognition of groundwater occurrence, 261–262
        flow dynamics in thick lenses, 262–265
        organic contamination of basal lenses, 265–267
        potpourri, 267–269
Groundwater modeling, in 21st century, 79–93
    directions, 81–87
        geological heterogeneity, 83–85
        parameter estimation and model reliability, 81–83
        plume characterization, 85–87
    milestones of 20th century, 79–81
Groundwater modeling, litigation and, 137–145
    case background, 137
    litigative approach, 139–144
    modeling approach, 138

Half-saturation constant, 296
Hawaii(an), 253, 256
    history of models in, 260
    islands, 254, 261
Hawaiite, 255
Head variance, 199, 200
Hearing
    common occurrences prior to, 141
    experiences, 137
    public, 137
Helium detector, 267
Henry's law, 236
Heterogeneity
    of aquifer formation, 123
    geological, 83
    scales of, 186
    multiscale, 188
    statistical representation of, 193
    structural, 25

Holocene age
    aquifers, 277, 290
    layers, different, 284
    sediments, 276
Homogeneity, 351
Honolulu aquifers, multiple correlation for, 269
Horizontal permeability, upper aquifer, 280
HPS model, 127
Human enteric viruses, 268
Hydraulic conductivity, 30, 150, 163, 256, 331
    data, 12
    of rock mass, 258
Hydraulic gradient, 349
Hydrodynamic dispersion, 25
Hydrogeology, stochastic methods in, 83
Hypersurface, 124

ICPP, see Idaho Chemical Processing Plant
Idaho Chemical Processing Plant (ICPP), 68
Idaho National Engineering Laboratory (INEL), 67, 68
Implicit scheme, 153, 154
Implicit-time iteration, 159
INEL, see Idaho National Engineering Laboratory
Infiltration, 149
Interblock parameter estimation, 159
Intermediate extrusive rocks, 257
Intermediate series, 255
Interview, 139
Intrusive rocks, 255, 257
Inverse modeling, 81
Irrigation
    drainage, restriction in, 114
    wastewater reuse for, 268
    water, 268
Island width, 280
Isotropy, 351
Iterative conjugate gradient algorithm, 157

Joint probability density function, 124
Judge, trial, 140
Jury, 140

Kahana tunnel, 268
Kalauao Springs, 264
Karst, 341, 343
    aquifer, 340, 344, 351, 353, 355
        anisotropic, 352
        integrated conduit structure in, 353
        permeability structure of, 349
    networks, 351
    unconfined, 351
    unsaturated, 347

INDEX 363

Karstic aquifers, application of porous media groundwater circulation models to, 339–358
  anisotropy versus heterogeneity, 351–352
  dye tracing, 355–356
  hydraulic gradient and karstic permeability, 347–349
  hydraulic gradients, 354–355
  karst permeability, 349–351
  modeling objectives, 355
  parameter sampling, 352–353
  permeability structure in soluble rocks, 340–347
    epikarst, 344–345
    karst, 341–343
    karst aquifer, 344
    paleokarst, 346–347
  scale considerations, 353–354
Kauai, 255
Kinetic expressions, 296
KISS rule, 139
Kona coast, 261
Koolau basalt, 256, 257
Koolau Range, 258
Kriging, 211
Kurtosis, 33

Laboratory tests, 192, 330
Lag time, 263
Laminar flow, 262
Land subsidence, 64
Land use patterns, 269
Laplace transformation technique, 245
Lava tubes, 256
Lawyers, 139
Leaching
  concentration, dynamic model for predicting, 266
  index, 266
Leibnitz' rules, 242
Limestones, cavernous permeability in, 347
Limit-state surface, 124
Litigative approach, 139
Lysimeter, 268

Macrodispersivity
  equation, 203
  estimation of, 202
MAE, see Mean absolute error
Marine sediments, 258, 259
Mathematical model, 40
Maui, 256
MCS, see Monte Carlo simulation
ME, see Mean error
Mean absolute error (MAE), 55

Mean error (ME), 55
Mean error ratio (MER), 55
Mean square error (MSE), 34, 210
Mechanical mixing, 26
Mechanical testing program, 262
MER, see Mean error ratio
Metamorphic rocks, 258
Metamorphic sequences, 255
Method of characteristics (MOC), 268, 328
Microcolony models, 301
Minimum representative volume, 353
Mixing cell model, 263
MOC, see Method of characteristics
Model(s), 270
  analysis, deterministic, distributed-parameter, 62
  analytical, 124, 264
  approach, 138
  calibration, 138, 201
  complexity, 4
  deterministic, 60
  development, 40
  documentation, evaluation of, 42
  dual-porosity, 331
  examination, 41
  instantaneous reaction model, 298, 304
  mass transfer function, 191
  numerical, 264, 270, 327
  parallel-plate, 330
  perfect, 143
  pesticide leaching, 173
  physical, 264
  planning, 268
  PRZM, 268
  testing, 39–58
    assessment criteria, 52–55
    code testing strategy, 44–52
    functionality analysis, performance evaluation, and applicability assessment protocol, 42–44
    model evaluation and testing procedure, 40–42
  two-dimensional, 205
  types of, 79
  validation, 82, 143
Models, needs for next generation of, 95–98
  applications, 97–98
    modeling, 97–98
    model validation, 98
  research, 96–97
    computation, 97
    conceptual models, 96–97
    field techniques, 96
MODFLOW, 328, 337
MODPATH, 328

Moisture movement, 149
Monod kinetics, 296, 298, 303
Monte Carlo simulation (MCS), 31, 83, 131, 215
MSE, see Mean square error
Mugearite, 255
Multi-aquifer system, 276
Multiobjective programming models, 269
Multiphase flow, 87
Multiphase multicomponent models, 232
Multiple mixing cells, 266

NAPL, see Nonaqueous phase liquids
National Priorities List (superfund), 265
Natural disinfection criteria, evaluation of groundwater, 313–323
   methods, 314–317
      Fort Devins, Massachusetts, 314
      Lake Redstone, Wisconsin, 315–317
      Vosen, Wisconsin, 314–315
   results, 317–319
      Fort Devins, Massachusetts, 317
      Lake Redstone, Wisconsin, 318–319
      Vosen, Wisconsin, 317–318
Navier-Stokes equations, 189
Negative mean error (NME), 55
Negotiations, 141
Nematocides, 265
Newton-Raphson method, 157, 158
Nitrogen, 268
NME, see Negative mean error
Nonaqueous phase liquids (NAPL), 231, 240
Non-Gaussian
   correlated random walk, 31, 32
   model, 30
   non-white process, 27
Nonlinear equations, solving system of, 157
Nonlinear terms, 126
Nonpoint source pollution, 266, 267
North Aquifer System, basal water in, 266
Numerical models, 124, 164, 327
Nuuanu aquifer system, 264

Oahu, 253, 256, 257
   southern, 254
   urban development in central, 269
Oil
   mound, 244
   recovery of, 241
Olivine basalts, 256
Ordinary Kriging, 211
Organic chemicals, 123, 261, 265
Organic pesticides, 113

Packer well tests, 353
Pahoehoe, 255, 256
Paired-data performance, 54
Palagonite, 256
Paleokarst, 346
Palolo aquifer system, 264
Parallel-plate model, 330
Parameter(s)
   estimation, model reliability and, 81
   identification, 102
   reliance on effective, 83
   uncertainty, 4, 123
PCE, see Tetrachloroethylene
PCP, see Pentachlorophenol
Pearl Harbor aquifer, 257, 261, 267
   basal water in, 266
   organic chemical contamination in, 265
Peclet number, 161–163
Pentachlorophenol (PCP), 305
Performance
   criteria, 52
   evaluation, 40, 42, 44, 46
Permeability, double continuum, 355
Permeability structure
   anisotropic, 348, 349, 352
   destruction of organized, 346
Perturbation analysis, 204
Pesticide(s)
   ecosystem response to, 179
   leaching, 177
      estimating, 169, 178
   index for, 266
Petroleum hydrocarbon compounds, 265
Petroleum products, 231
Phosphorus, 268
Picard method, 157
Pineapple fields, 265
Pioneer Well, 269
Pit craters, 255
Planning model, 267
Pleistocene age
   aquifer, 277
   deposits, 276, 284
Plume characterization, 85
PME, see Positive mean error
Point sampling, 352, 354
Ponding, infiltration under, 151
Porous media, 132, 269
Positive mean error (PME), 55
Postaudits, value of in groundwater model applications, 59–78
   diffuse source of contamination, 72–75
   point source of contamination, 67–71
   regional water-level changes, 61–67

INDEX

Prediction
 accuracy, 185
 errors, 3
Preferential flow, 84
Pressure head, 150, 159
Primitive basalts, 256
PROBAN, 128
PRZM, 266
Pulse-through responses, 355
Pump tests, 256, 353
Pyroclastic rocks, 256
Pyroclastics, 255, 258

QA, see Quality assurance
Quality assurance (QA), 40, 41
Quantity management, maximizing net benefits of, 105

Radioactive decay, 130
Random
 advection, 27
 function, 208
 space function, 27
 variables, 124, 125
 walk, 27
Rao's attenuation factor, 266
Raoult's law, 235
Reactive solutes, uncertainty analysis of subsurface transport of, 123–135
 FORM, 125–126
 procedures, 127–132
  description of models used, 127–128
  input parameters, 129–132
  problem formulation, 128–129
 SORM, 126–127
  determination of design point, 126
  sensitivity measures, 126–127
 theoretical background, 124–125
 uncertainty analysis method, 123–124
Rebuttal testimony, 142
Recovery
 efficiency, 247
 well, 244
Regional groundwater management, benefits from, 101–121
 groundwater management objectives, 104–112
  maximizing net benefits, 105–109
  minimizing costs, 109–110
  summary of regional quantity problems, 110–112
 model classification, 102–103
 potential problems associated with large-scale groundwater development, 103–104

regional quality problems, 112–115
 groundwater versus surface water quality, 113
 regional quality models, 114–115
Regional models, 353
Regional-quality problems, 112
Regulatory records, 268
Reliability
 analysis, 130
 index, 125
Representative elementary volume (REV), 188, 190, 191
REV, see Representative elementary volume
Richards equation, 149
 mixed form of, 150
 numerical solutions of, 153
 pressure-based form of, 150
 transformed form of, 160
 water-content based form of, 150
Rift zones, 254, 255, 257, 258
RMSE, see Root-mean-squared error
Roi-Namur model, 284, 286, 288
Root-mean-squared error (RMSE), 70

Safe domain, 124
Salinity, 72
Salt River Valley, Arizona, 61, 65
Saltwater flow, 265
Sandbox experiments, 264
Sandstone, vertical bore hole in, 194
Saprolite, 257
Schofield high-level water body, 268
Scientist, naive, 143
Seawater encroachment, 261
Second-order reliability method (SORM), 123, 125, 126, 130
Sedimentary
 rocks, 258
 sequences, 255
Sensitivity measures, 124, 126
Sequential quadratic programming (SQP) method, 126
Sewage effluent, 268
Sharp interface, 265
 approach, 241
 models, 240
Sills, 258
Simulation code, 39
Sinkholes, 350
Skewness, 33
Skimming, 285
Slug tests, 353
Snake River Plain, Idaho, 67

Soil(s)
  hydraulic property of surface, 269, see also Soil properties
  properties, 149, 152
  solute transport in upper, 266
Soluble aquifer, 340, 347, 349
Solute transport
  analysis, 14
  mathematical models simulating, 3
  model, 75
  prediction of, 220
  in vadose zone, 187
SORM, see Second-order reliability method
Source term, 123
Southeast Oahu, 254
Specific yield, 257
SQP, see Sequential quadratic programming method
Standard normal
  density, 125
  space, 125
State Water Code, 254
Stationarity, 195
Statistical diffusion, Markov process, 28
Statistical information, 124
Stochastic subsurface hydrology, 269
Stocks, 258
Stratification, 196
Stream-aquifer systems, 72
Stream flow, minimum, 261
Subsidence, 64, 254
Sugarcane, 253, 261, 263
Superposition, principle of, 304
Survival state, 124
Suspended solids, 268
Sustainable water supply, 270
SUTRA, 264, 265, 275, 284
Synthetic soil columns, 269
System theory, 266

Taro culture, 261
TCE, see Trichloroethylene
TCP, see 1,2,3-Trichloropropane
Team theory, 140
Tempo model, 264
Test objectives, 47
Test reactor area (TRA), 68
Tetrachloroethylene (PCE), 231
Thick lens, 264
Thomas algorithm, 155
Tidal
  boundaries, fluctuating, 277
  range, 280
TRA, see Test reactor area

Trachytic lavas, 256
Transformed equation, 150
Transition
  probability density function, 28
  zone, 262, 263, 265, 290
Transmissivity, 218, 257
Transport
  code, 133
  dynamic model, 266
  models, 264, 266
  phenomena, 266
  predictions, 17, 18
Transverse dispersion, 264
Travel time predictions, 11
Traverse City site, Michigan, 306
Trial, 139
  characteristics, 137
  common occurrences prior to, 141
Trichloroethylene (TCE), 231, 265
1,2,3-Trichloropropane (TCP), 265
Tube networks, 349
Tuff, 256, 260
Turbulent flow, 262
Two-pump recovery operation, 240, 245

ULTRA model, 299
Uncertainties, comparison of model and parameter, 3–24
  in groundwater flow predictions, 4–13
    description of base case and parameter variations for flow, 6
    description of discharge models, 6–7
    discussion of hydraulic conductivity range used, 12–13
    levels of complexity, 6
    methods for evaluating results, 8
    results of discharge predictions, 8–11
  in groundwater transport predictions, 13–21
    analysis of complexity level, 6, 19–20
    description of base case and parameter variations for transport, 13–15
    description of transport models, 15–16
    discussion of transport results, 17–19
    methods for evaluating transport results, 16–17
Uncertainty importance factors, 126
Unconfined aquifers, 351
Universal Kriging, 211
University of Hawaii, 266
Unsaturated water flow, 149
Unsaturated zone, 149
  contaminant transport in, 270
  investigations and models of, 266

INDEX

Unsaturated zone, numerical solutions of one-dimensional flow in, 149–167
  available numerical models, 164
  conventional numerical solutions, 153–157
    finite-difference technique, 153–155
    finite-element technique, 155–156
    treatment of boundary conditions, 157
  mathematical formulation, 150–153
    boundary and initial conditions, 151–152
    governing equation, 150–151
    soil hydraulic properties, 152–153
  procedures for solving system of nonlinear equations, 157–159
  recent advances in numerical solutions, 159–164
U.S. Soil Survey, 266

Vadose zone, stochastic modeling of water flow and solute transport in, 185–230
  approaches for investigation of spatial variability, 191–221
    deterministic approach, 192–193
    statistical representation of heterogeneity, 193–196
    stochastic approach, 193
    stochastic modeling of flow and solute transport, 196–221
  concepts of water flow and solute transport modeling, 186–191
Vadose zones, 348
Validation, 51
van Genuchten equations, 152
Variables, mutually dependent, 127
Verification, 43

Vertical longitudinal dispersivity, upper aquifer, 280
Virus
  assessment, 268
  monitoring, 268
Volcanic shields, 254, 255

Waialae aquifer system, 265
Waiawa
  infiltration gallery, 268
  investigation at, 267
Waihee tunnel, 268
Waipahu, basaltic aquifer in, 267
Wastewater, 267
Wastewater injection, 264
Water
  environment, sustenance of, 261
  high-level, 261, 267
  resources, managing, 101
  supply, 261
Waterworks, operations experience of major, 270
Weathered zones, 257
Wellhead protection area, 267
Wells, 256
Wentworth, 257
  aquifer test conducted by, 256
  parameters of alluvium measured by, 259
  range of porosity found by, 256
  reanalysis of Hoyt's data by, 262
  rinsing hypothesis by, 263
Wildlife preserve, 114

Yield, sustainable, 283